MW01278585

ALGEBRAIC LOGIC

PAUL R. HALMOS

DOVER PUBLICATIONS, INC.
MINEOLA, NEW YORK

Bibliographical Note

This Dover edition, first published in 2016, is an unabridged republication of the work originally published by Chelsea Publishing Company, New York, in 1962.

Library of Congress Cataloging-in-Publication Data

Halmos, Paul R. (Paul Richard), 1916–2006.
Algebraic logic / Paul R. Halmos.—Dover edition.
p. cm.
Originally published: New York : Chelsea Publishing Company, 1962.
Includes bibliographical references and index.
ISBN-13: 978-0-486-80145-2
ISBN-10: 0-486-80145-4
1. Algebraic logic. I. Title.

QA266.H3 2016
511.3'24—dc23

2015025415

Manufactured in the United States by RR Donnelley
80145401 2016
www.doverpublications.com

PREFACE

Partly because of its perpetual intrinsic interest and partly because of its surprising new applications, mathematical logic is a much discussed subject nowadays. Algebraic logic is a modern approach to some of the problems of mathematical logic, and the theory of polyadic Boolean algebras, with which this volume is mostly concerned, is intended to be an efficient way of treating algebraic logic in a unified manner.

A study of the table of contents, or even just a quick riffle through the pages, will show the reader that this is a somewhat unusual sort of book. It is, in fact, a collection of papers, namely all my papers on algebraic logic. The reasoning that brought it into existence goes as follows. The literature of polyadic Boolean algebras is, so far, not very much more extensive than this set of papers. (See the Additional Bibliography on page 265.) Any book I could write on the subject now would therefore consist, essentially, of just these papers, with, naturally, some connective tissue, and, possibly, an occasional change of terminology and notation. To print such a small modification of something already in print would cost a prohibitive amount in comparison with the novelty of the contribution it could make. On the other hand, to learn the subject from the papers in their original sources would put the learner to an unreasonable inconvenience. There are ten papers, ranging in length from one page to 70 pages; they appeared between 1954 and 1959, in eight journals, in four countries. This diversity of sources makes access to the work as a whole inconvenient and sometimes impossible. Conclusion: a good purpose might be served by publishing the papers as they stand, but collected together.

The order in which the papers appear here is one in which a reader unfamiliar with the subject might want to read them; it is not the same as their chronological order. The introductory paper, for instance, was written after several of the others were in print already, and the summary at the end consists of the abstracts and announcements that were the first to appear.

The collection in its present form and order is accessible to a general mathematical audience; no vast knowledge of algebra or logic is required. The main prerequisite is familiarity with the elementary (algebraic) properties of Boolean algebras; a few of the results are proved by reference to the (topological) duality theory of Boolean spaces. Except for this slight Boolean foundation, the volume is essentially self-contained.

I am grateful to the American Mathematical Society, the Mathematical Association of America, the National Academy of Sciences, and to the editorial committees of Compositio Mathematica, the Duke Mathematical Journal, and Fundamenta Mathematicae for permission to reproduce in this form the papers that they first published.

P. R. H.

CONTENTS

I. INTRODUCTION 9
The basic concepts of algebraic logic. *American Mathematical Monthly,* vol. 53 (1956), pp. 363-387.

PART ONE

MONADIC ALGEBRAS

II. GENERAL THEORY 37
Algebraic logic, I. Monadic Boolean algebras. *Compositio Mathematica,* vol. 12 (1955), pp. 217-249.

III. REPRESENTATION 75
The representation of monadic Boolean algebras. *Duke Mathematical Journal,* vol. 26 (1959), pp. 447-454.

IV. FREEDOM 85
Free monadic algebras. *Proceedings of the American Mathematical Society,* vol. 10 (1959), pp. 219-227.

PART TWO

POLYADIC ALGEBRAS

V. GENERAL THEORY 97
Algebraic logic, II. Homogeneous locally finite polyadic Boolean algebras of infinite degree. *Fundamenta Mathematicae,* vol. 43 (1956), pp. 255-325.

VI. TERMS 169
Algebraic logic, III. Predicates, terms, and operations in polyadic algebras. *Transactions of the American Mathematical Society,* vol. 83 (1956), pp. 430-470.

VII. EQUALITY 213
Algebraic logic, IV. Equality in polyadic algebras. *Transactions of the American Mathematical Society,* vol. 86 (1957), pp. 1-27.

PART THREE

SUMMARY

VIII. General Theory 243

Polyadic Boolean algebras. *Proceedings of the National Academy of Sciences,* vol. 40 (1954), pp. 296-301.

IX. Terms and Equality 251

Predicates, terms, operations, and equality in polyadic Boolean algebras. *Proceedings of the National Academy of Sciences,* vol. 42 (1956), pp. 130-136.

X. Brief Summary 261

Polyadic Boolean algebras. *Proceedings of the International Congress of Mathematicians, Amsterdam,* vol. 2 (1954), pp. 402-403.

Additional Bibliography 265

Index 269

I

INTRODUCTION

THE BASIC CONCEPTS OF ALGEBRAIC LOGIC

1. Introduction. It has often happened that a theory designed originally as a tool for the study of a physical problem came subsequently to have purely mathematical interest. When that happens, the theory is usually generalized way beyond the point needed for applications, the generalizations make contact with other theories (frequently in completely unexpected directions), and the subject becomes established as a new part of pure mathematics. The part of pure mathematics so created does not (and need not) pretend to solve the physical problem from which it arises; it must stand or fall on its own merits.

Physics is not the only external source of mathematical theories; other disciplines (such as economics and biology) can play a similar role. A recent (and possibly somewhat surprising) addition to the collection of mathematical catalysts is formal logic; the branch of pure mathematics that it has precipitated will here be called *algebraic logic.*

Algebraic logic starts from certain special logical considerations, abstracts from them, places them into a general algebraic context, and, *via* the generalization, makes contact with other branches of mathematics (such as topology and functional analysis). It cannot be overemphasized that algebraic logic is more algebra than logic. Algebraic logic does not claim to solve any of the vexing foundation problems that sometimes occupy logicians. All that is claimed for it is that it is a part of pure mathematics in which the concepts that constitute the skeleton of modern symbolic logic can be discussed in algebraic language. The discussion serves to illuminate and clarify those concepts and to indicate their connection with ordinary mathematics. Whether the subject as a whole will come to be considered sufficiently interesting and sufficiently deep to occupy a place among pure mathematical theories remains to be seen.

The literature of algebraic logic is not yet very extensive, and the few items that are available are highly technical. It is for that reason that this expository paper was written; its main purpose is to kindle interest in a young but promising subject. In such a context it does not seem to be appropriate to burden the reader with the usual scholarly references and assignments of credit. At the very least, however, the names of the principal contributors should be mentioned. Here they are: Curry, Henkin, Rasiowa, Sikorski, and Tarski. Many of the ideas of algebraic logic have been in the air for several years and were known, at least subconsciously, by most logicians. The greatest contributions are those of Tarski; especially relevant is his work (with Jónsson) on Boolean algebras with operators and his theory of cylindric algebras. (Most of the latter material is unfortunately unpublished.) The reader who wishes to study the details will find exact references in two papers on algebraic logic by the present author; the first is in *Compositio Mathematica* (1955), and the second is to appear in *Fundamenta Mathematicae* (1957).[1]

[1] See pp. 71, 72, and 166 of the present text.

2. Boolean algebras. The father of algebraic logic is George Boole; it is appropriate that a discussion of the subject begin with a quick review of the algebras that bear his name. The shortest definition of Boolean algebras is in terms of the theory of rings: a *Boolean algebra* is a ring with unit in which every element is idempotent (*i.e.*, if p is in the ring, then $p^2=p$). The simplest example, and one that plays a vitally important role throughout the theory, is the field of integers modulo 2; this Boolean algebra will be denoted by **O**.

Boolean algebras have an almost embarrassingly rich structure. It is, for instance, an easy consequence of the definition that a Boolean algebra always has characteristic 2 (*i.e.*, $p+p=0$) and that as a ring it is always commutative (*i.e.*, $pq=qp$). In every Boolean algebra there is, moreover, a natural order relation; it is defined by writing $p \leqq q$ if and only if $pq=p$. The algebraic structure and the order structure are as compatible as they can be. The algebraic zero is also the order zero (*i.e.*, the least element of the algebra), and the algebraic unit is also the order unit (*i.e.*, the greatest element of the algebra); in other words, $0 \leqq p \leqq 1$ for every p. With respect to the order, the algebra turns out to be a complemented lattice. The lattice operations can be expressed in terms of the given algebraic operations, as follows: the complement of p, denoted by p', is $1+p$; the infimum of p and q, denoted by $p \wedge q$, is pq; and the supremum of p and q, denoted by $p \vee q$, is $p+q+pq$.

The process of defining many useful operations and relations in a Boolean algebra, in terms of addition and multiplication, is to a large extent reversible. This fact is responsible for the abundance of different axiomatic approaches to the subject. A Boolean algebra can be defined in terms of its partial order, or in terms of complements and suprema, or in terms of complements and infima, and so on and so forth *ad* almost *infinitum*. Thus, for instance, since sums and products can be expressed in terms of complements and infima ($pq=p \wedge q$ and $p+q=(p' \wedge q')' \wedge (p \wedge q)')$, it follows that if a set admits a unary operation and a binary operation satisfying the appropriate conditions, then that set is a Boolean algebra. (It isn't really, but it is close enough to make the distinction pedantic. What should be said is that if addition and multiplication are defined as indicated above, then the set, together with the defined operations, constitutes a Boolean algebra.) The appropriate conditions are simple to describe. They require that the underlying set contain at least two distinct elements, and that

(I 1) $$p \wedge (q \wedge r) = (p \wedge q) \wedge r,$$

(I 2) $$p \wedge q = q \wedge p,$$

(I 3) $$\text{if } p \wedge q' = r \wedge r', \quad \text{then} \quad p \wedge q = p,$$

(I 4) $$\text{if } p \wedge q = p, \quad \text{then} \quad p \wedge q' = r \wedge r',$$

for all elements p, q, and r. (Caution: (I 3) means that if $p \wedge q' = r \wedge r'$ for *some* r, then $p \wedge q = p$, and (I 4) means that if $p \wedge q = p$, then $p \wedge q' = r \wedge r'$ for *all* r.)

Since, inside a fixed non-empty set, set-theoretic complementation and the formation of set-theoretic intersections do satisfy these conditions, any class of sets that is closed under these two operations is a Boolean algebra. The class of all subsets of a non-empty set is the most easily described (but far from the only) example of a Boolean algebra obtained in this way.

[Terminological purists sometimes object to the Boolean use of the word "algebra." The objection is not really cogent. In the first place, the theory of Boolean algebras has not yet collided, and it is not likely to collide, with the theory of linear algebras. In the second place, a collision would not be catastrophic; a Boolean algebra is, after all, a linear algebra over the field of integers modulo 2. The last, but not the least, pertinent comment is a pragmatic one. While, to be sure, a shorter and more suggestive term than "Boolean algebra" might be desirable, the nomenclature is so thoroughly established that to change now would do more harm than good.]

It is amusing and instructive to compare the axiom system (I) with the following, somewhat unorthodox, system of axioms for groups. A group may be defined as a non-empty set with a unary operation ($p \to p^-$) and a binary operation ($(p, q) \to p \times q$) such that

(II 1) $$p \times (q \times r) = (p \times q) \times r,$$

(II 2) $$(p \times q)^- = q^- \times p^-,$$

(II 3) $$\text{if } p \times q = r \times r^-, \quad \text{then} \quad p = q^-,$$

(II 4) $$\text{if } p = q^-, \quad \text{then} \quad p \times q = r \times r^-,$$

for all elements p, q, and r. (Caution: (II 3) means that if $p \times q = r \times r^-$ for *some* r, then $p = q^-$, and (II 4) means that if $p = q^-$, then $p \times q = r \times r^-$ for *all* r.) It is clear that if, in a group, p^- is defined to be the inverse of p, and $p \times q$ is defined to be the product of p and q (in that order), then the conditions (II) are satisfied. The fact that, conversely, the conditions (II) are characteristic of inversion and multiplication in groups is an easy exercise in elementary axiomatics.

[One comment should be made on the independence of the axiom set (II). The fact is that the set is *not* independent; the first three axioms are sufficient to characterize groups, and, in particular, they imply the fourth. The reason (II) is offered in the present form is to emphasize its similarity with the axiom set (I) for Boolean algebras. Each of the axioms (II 1), (II 2), and (II 3) is independent of the other three axioms of the set (II).]

3. Propositional calculi. To understand the connection between Boolean algebra and logic, a good way to begin is to examine how sentences are combined by means of sentential connectives. To ensure an unprejudiced approach to the subject, it is desirable to proceed as abstractly as possible. Suppose, therefore, that there is given an arbitrary non-empty set S; intuitively the elements of S are to be thought of as the basic sentences of some theory that is being studied. Suppose, moreover, that A and N are two distinct objects not contained in S; intuitively A and N are to be thought of as the connectives "and" and "not."

(Warning: in other expositions, A is frequently used for the dual connective "or".) Consider finite sequences whose terms are either elements of S or else A or N. (The empty sequence is not included.) If s is a sequence of length n, say, so that $s_0, \cdots, s_{n-1}$ are elements of $S \cup \{A, N\}$, let Ns be the sequence defined by

$$(Ns)_0 = N, \qquad (Ns)_i = s_{i-1} \qquad (i = 1, \cdots, n);$$

if s and t are sequences (of lengths n and m respectively), let Ast be the sequence defined by

$$(Ast)_0 = A, \quad (Ast)_i = s_{i-1} \ (i = 1, \cdots, n), \quad (Ast)_{n+j} = t_{j-1} \quad (j = 1, \cdots, m).$$

Let S^* be the smallest set of sequences such that (1) if s is a sequence of length 1 whose unique term is in S, then $s \in S^*$, (2) if $s \in S^*$, then $Ns \in S^*$, and (3) if $s \in S^*$ and $t \in S^*$, then $Ast \in S^*$. In other words, S^* is the set of sequences generated from the one-term sequences of S by means of the operations of prefixing N to a sequence and prefixing A to the concatenation of two sequences. Intuitively S^* is to be thought of as the set of sentences generated from the basic sentences by the two basic connectives. The device of writing Ast instead of $s\ A\ t$ (an ingenious Polish invention) is designed to avoid the multiple layers of parentheses that the more intuitive infix notation necessitates.

The set S^* by itself is not quite a proper object of logical study. The trouble is that if, for instance, s and t are in S^*, then Ast and Ats are distinct elements of S^*, whereas common sense seems to demand that if s and t are sentences, then "s and t" and "t and s" should be sentences that "say the same thing" in some sense. Sentences admit, in other words, a natural equivalence relation; such a relation should therefore be introduced into S^*. A little thought about the intuitive interpretation of the elements of S^* will suggest many conditions that equivalence should satisfy. If the assertion that the elements s and t of S^* are equivalent is denoted by $s \equiv t$, then, for all elements s, t, and u of S^*, it should at the very least be required that

(III 1) $\qquad AsAtu \equiv AAstu,$

(III 2) $\qquad Ast \equiv Ats,$

(III 3) $\qquad$ if $AsNt \equiv AuNu$, then $Ast \equiv s,$

(III 4) $\qquad$ if $Ast \equiv s$, then $AsNt \equiv AuNu,$

(III 5) $\qquad$ if $s \equiv t$, then $Ns \equiv Nt,$

(III 6) $\qquad$ if $s \equiv t$, then $Asu \equiv Atu.$

In addition, of course, it is necessary that the concept of equivalence used here be an honest equivalence, *i.e.*, that it be reflexive, symmetric, and transitive.

There are likely to be many equivalence relations satisfying the conditions (III); one such is defined by writing $s \equiv t$ for all s and t. In order to avoid this triviality (and for other reasons), it is desirable to consider the smallest possible equivalence (*i.e.*, the one in which the fewest possible pairs turn out to be equivalent) satisfying these conditions. This makes sense. Indeed, if an equivalence relation is thought of as a certain set of ordered pairs, then the intersection of all the equivalence relations satisfying (III) is an equivalence relation satisfying (III). If, from now on, the symbol $\equiv$ is used to denote this minimal equivalence, then the pair $(S^*, \equiv)$ is one of the simplest non-trivial logical structures; it is usually known as the *propositional* (or *sentential*) *calculus*. There are really many propositional calculi; there is one corresponding to each non-empty set S. It is clear, however, that the only thing that matters (all other differences between the various propositional calculi being essentially notational matters) is the cardinal number of S. It is customary (but not particularly profitable) to assume that S is countably infinite.

4. Axioms and rules. The time has come to make the notation more transparent. While the symbols A and N are technically convenient (no parentheses), it is usually a cumbersome job to decode the sentences involving them and to recapture their intended intuitive content. Accordingly, in what follows, $(s)'$ will be used as an alternative symbol for Ns, and, similarly, $(s) \wedge (t)$ will be used as an alternative symbol for Ast. Thus, for instance, $AsNt$ can be denoted by $(s) \wedge ((t)')$; with the usual mathematical conventions about omitting superfluous parentheses, this becomes $s \wedge t'$. It should now be clear that the conditions (III 1)–(III 4) are only notationally different from (I 1)–(I 4); the conditions (III 5) and (III 6) assert, in customary mathematical language, that $\equiv$ is a congruence relation with respect to the operations of attaching $'$ and infixing $\wedge$.

The equivalence relations that occur in algebra (*e.g.*, the congruence relations in a ring) are usually described by specifying a particular equivalence class (the kernel) and a particular operation (subtraction); it then turns out that two elements are congruent if and only if their difference belongs to the kernel. A similar procedure is available to define the equivalence $\equiv$ described above. In order to motivate the choice of the kernel and the choice of the pertinent subtraction operation, consider first of all an element of S^* that has the form $s \wedge s'$. If the elements of S^* are interpreted as sentences, then there is something obviously undesirable about such an element; in some intuitive sense it is "false." By the same token, the result of attaching $'$ to such an element converts it into something quite laudable; sentences such as that are "true." The kernel that is usually considered is the equivalence class of any particular "true sentence" so obtained. It is pleasant to be able to report that this kernel is independent of the arbitrary choice of its generator. The equivalence class of any element of the form $(s \wedge s')'$ contains all other elements of that form, and, in fact, it consists exactly of all the elements that common sense would declare to be "true." An element of this kernel is called a *tautology* of the propositional calculus. The subtraction operation is the one that associates with two elements s and t of S^* the element $(s \wedge t')' \wedge (s' \wedge t)'$. (In the original notation this reads $ANAsNtNANst$.) The perceptive reader will note that if s and t are interpreted as sentences, then the intuitive interpretation of the proposed "difference" between s and t is the sentence "s if and only if t." This subtraction does what it was tacitly promised it would do; it is true that a necessary and sufficient condition that s be equivalent to t is that the difference of s and t belong to the kernel. (In customary logical language: a necessary and sufficient condition for $s \equiv t$ is that the biconditional of s and t be a tautology. It is a happy circumstance that biconditioning is a commutative operation; the order of s and t is immaterial.)

In principle it is clear how the procedure used above could be reversed. The tautologies could be defined by making a complete list of them, and equivalence could then be defined in terms of tautologies and the formation of biconditionals. The list of tautologies would be infinite, to be sure, but a clever classifier might nevertheless succeed in describing it in a finite number of words. Something like this is what is usually done. In most text-book presentations of the propositional calculus one finds a small list of special tautologies, together with a few easy instructions for manufacturing others; a general tautology is then defined to be a member of the subset of S^* generated from the special tautologies by repeated applications of the instructions. The given special tautologies (whose particular choice is largely a matter of individual taste) are called *axioms*, and the instructions for manufacturing others are called *rules of inference*. This "axiomatic" procedure (which has become traditional) has no particular advantages (or disadvantages) in comparison with the one followed above; its final result is merely an alternative description of the propositional calculus.

[Here, for the sake of completeness, are the details of one of the popular axiomatic approaches to the propositional calculus. Let $s \vee t$ and $s \rightarrow t$ be abbreviations for $(s' \wedge t')'$ and $(s \wedge t')'$ respectively (or, in the original notation, for $NANsNt$ and $NAsNt$ respectively); the axioms are most conveniently, and most understandably, described by means of these abbreviations. The axioms consist of all those elements of S^* that are of one of the following four forms:

(1) $$(s \vee s) \rightarrow s,$$

(2) $$s \rightarrow (s \vee t),$$

(3) $$(s \vee t) \rightarrow (t \vee s),$$

(4) $$(s \rightarrow t) \rightarrow ((u \vee s) \rightarrow (u \vee t)),$$

where s, t, and u are arbitrary elements of S^*. There is only one rule of inference (called *modus ponens*). According to that rule, if s and t are elements of S^* such that both s and $s \rightarrow t$ are tautologies, then t is a tautology.][1]

5. Free Boolean algebras. By whatever method the pair (S^*, $\equiv$) is obtained, once it is at hand it is natural to form the set $\mathbf{A}^*$ of all equivalence classes. In analogy with calling an element of S^* a sentence, an element of $\mathbf{A}^*$ may be called a *proposition*. (The word "proposition" is not always defined this way. The intuitive reasons for using the definition are obvious, but, admittedly, they are not overwhelming.) The set $\mathbf{A}^*$ of propositions possesses, in a natural way, the structure of a Boolean algebra. If $p \in \mathbf{A}^*$, let s be any element of the equivalence class p and write p' for the equivalence class of Ns (*i.e.*, of s'). If both s_1 and s_2

[1] This description of axioms for the propositional calculus contains a subtle error. For the correction, see the paper by Hiz cited in the Additional Bibliography (p. 265).

belong to p, *i.e.*, if $s_1 \equiv s_2$, then, by (III 5), $Ns_1 \equiv Ns_2$, so that the definition of p' is unambiguous. If p and q are in $\mathbf{A}^*$, let s and t be corresponding representative elements (*i.e.*, $s \in p$ and $t \in q$) and write $p \wedge q$ for the equivalence class of Ast (*i.e.*, of $s \wedge t$). The necessary unambiguity argument is based on (III 6) this time. The fact that the conditions (I) are satisfied now follows immediately from a comparison of those conditions with the corresponding conditions (III).

The procedure for arriving at $\mathbf{A}^*$ is familiar to anyone who ever heard of the theory of free groups; the Boolean algebra $\mathbf{A}^*$ is, in fact, isomorphic to the free Boolean algebra generated by the originally given set S. In logical studies it is customary to describe the algebraic properties of $\mathbf{A}^*$ by some rather specialized terminology. Thus, for instance, the fact that $\mathbf{A}^*$ satisfies the first necessary condition for being a Boolean algebra (the possession of at least two distinct elements) is usually described by saying that the propositional calculus is consistent.

In view of what has already been said, it is not at all surprising that the construction of $\mathbf{A}^*$ has a very near parallel in group theory. Given an arbitrary non-empty set S, form the set S^* exactly as before and introduce into S^* the smallest equivalence relation $\equiv$ such that

(IV 1) $$AsAtu \equiv AAstu,$$

(IV 2) $$NAst \equiv ANtNs,$$

(IV 3) $$\text{if } Ast \equiv AuNu, \quad \text{then} \quad s \equiv Nt,$$

(IV 4) $$\text{if } s \equiv Nt, \quad \text{then} \quad Ast \equiv AuNu,$$

(IV 5) $$\text{if } s \equiv t, \quad \text{then} \quad Ns \equiv Nt,$$

(IV 6) $$\text{if } s \equiv t, \quad \text{then} \quad Asu \equiv Atu,$$

for all elements s, t, and u of S^*. The set of all equivalence classes possesses, in a natural way, the structure of a group. The obvious details may safely be omitted; suffice it to say that the proof depends on a comparison of (II) with (IV). The group so obtained is, in fact, isomorphic to the free group generated by the originally given set S.

[The method of sequences-*cum*-equivalence is essentially the only known way of constructing free groups. It is worth noting that for Boolean algebras the following much more elegant method is available. Given an arbitrary (possibly empty) set S, let X be the set of all subsets of S. For each s in S, let $p(s)$ be the set of all those elements of X that contain s. Let $\mathbf{A}^*$ be the Boolean algebra generated by all the sets of the form $p(s)$ with s in S, so that $\mathbf{A}^*$ is a Boolean subalgebra of the Boolean algebra of all subsets of X. Assertion: $\mathbf{A}^*$ is isomorphic to the free Boolean algebra generated by S. The proof of the assertion is an easy exercise in set theory. Unfortunately this method of constructing $\mathbf{A}^*$ is quite provincial; it works like a charm for Boolean algebras, but it does not work for very many algebraic systems, at least not without significant modifications.]

6. Quotients of free algebras. Since the algebraic end-product of the propositional calculus is a free Boolean algebra, that calculus, as it now stands, is too special for most logical purposes. (Freedom for an algebraic system means, roughly speaking, that the only relations among its elements are the trivial ones). The difficulty is visible in the starting point of the construction of the propositional calculus; its source is the fact that the generating set S is a set with no algebraic structure at all. In realistic situations the basic sentences of a theory are likely to be related to each other in various intricate ways, and, in fact, it is usually considered to be one of the chief functions of logic to study such relations and their consequences.

Suppose, for instance, that someone wants to undertake a logical examination of a fragment of the history of English literature. Among the basic sentences of the theory (*i.e.*, among the elements of S) there might occur "Shakespeare wrote Hamlet" and "Bacon wrote Macbeth"; call these sentences s_0 and t_0 respectively. ("Basic sentence" is not the same as "axiom." The sentences s_0 and t_0 need not, at first, be considered either "true" or "false"; they are just considered.) Under these circumstances the investigator would perhaps want to declare the sentence NAs_0t_0 to be an axiom. Since, however, NAs_0t_0 is not a tautology of the propositional calculus, the word "axiom" must be used here in a sense broader than the one discussed above. Such a broadening is perfectly feasible. One way to achieve it is to add another condition to the set (III); the new condition could, for instance, be

(III 7) $$As_0t_0 \equiv As_0Ns_0.$$

The effect of this adjunction is to enlarge the original equivalence relation; the new equivalence relation is not the smallest relation satisfying (III 1)–(III 6), but the smallest relation satisfying (III 1)–(III 7). The same purpose can be accomplished by adjoining to the axioms of the propositional calculus the *extralogical axiom*

$$NAs_0t_0,$$

and then proceeding exactly as before. The result, with either approach, is that each new equivalence class is the union of several old ones. The tautologies, in particular, all belong to the same (new) equivalence class; an element of this equivalence class is usually called a *provable sentence* or a *theorem* of the modified propositional calculus. If there are at least two (new) equivalence classes (in logical language: if the adjoined axioms are consistent), the set **A** of all such classes possesses in a natural way the structure of a Boolean algebra, which, in general, is not free.

The "free" procedure for constructing non-free Boolean algebras has its group-theoretic parallel. One way to ensure that the generators of a group satisfy certain prescribed relations is to build those relations into the equivalence by means of which the quotient system (S^* modulo $\equiv$) is formed. The disadvantage of such a procedure is that it is repetitious; it uses the method of an earlier construction instead of its result. It is more usual, and algebraically more satisfactory, to apply the combinatorial method only once; thereafter, all groups defined by generators and relations are constructed by forming a quotient group of one of the free groups already obtained. This works very smoothly; the main reason it works is that every group is a quotient group of a free group.

Since, similarly, every Boolean algebra is a quotient algebra of a free Boolean algebra, the group-theoretic shortcut is available for Boolean algebras too. Just as in the theory of groups, moreover, the core of the idea is applicable to algebras that are not necessarily free. It makes sense (and it is often useful) to force the elements of a Boolean algebra (or group) to satisfy some relations, even if the algebra (or group) is not free to start with. A well-known example is the process of reducing a group by its commutator subgroup and thereby forcing it to become abelian.

7. Filters and ideals. From the point of view of logic, the reduction of Boolean algebras is connected with the theory of provability. It turns out that from the point of view of algebraic logic the most useful approach to that theory is not to ask "What is a proof?" or "How does one prove something?" but to ask about the structure of the set of provable propositions. Suppose, therefore, that **A** is a Boolean algebra, whose elements are to be thought of, intuitively speaking, as the propositions of some theory, and suppose that a non-empty subset **P** of **A** has been singled out somehow; the elements of **P** are to be thought of as the provable propositions of the theory. Common sense suggests that if s and t are provable sentences, then "s and t" is a provable sentence, and if s is a provable sentence, then "s or t" is a provable sentence, no matter what the sentence t may be. In order to meet these demands of common sense, the set **P** cannot be arbitrary; it must be such that if both p and q belong to **P**, then $p \wedge q$ belongs to **P**, and if p belongs to **P**, then $p \vee q$ belongs to **P** for all q in **A**. If a non-empty subset of a Boolean algebra satisfies these conditions, it is called a *filter*. An illuminating comment is this: a necessary and sufficient condition that a subset **P** of a Boolean algebra be a filter is that $1 \in \mathbf{P}$ (*i.e.*, all tautologies are provable), and that if $p \in \mathbf{P}$ and $(p \rightarrow q) \in \mathbf{P}$ (recall that $(p \rightarrow q) = (p' \vee q)$), then $q \in \mathbf{P}$ (*i.e.*, *modus ponens* is a rule of inference).

A filter is not a commonly encountered mathematical object, but one of its first cousins (namely, an ideal) is known to every mathematician. A non-empty subset **M** of a Boolean algebra is an *ideal* (for occasional emphasis, a *Boolean ideal*) if it contains $p \vee q$ whenever it contains both p and q and if it contains $p \wedge q$ whenever it contains p. Although it looks slightly different, this definition is, in fact, equivalent to the usual one; a Boolean algebra is, after all, a ring, and a Boolean ideal in the present sense is the same as an ordinary algebraic ideal.

Each of the two concepts (filter and ideal) is, in a certain sense, the Boolean dual of the other. (Some authors indicate this relation by using the term *dual-ideal* instead of filter.) Specifically, if **P** is a filter in a Boolean algebra **A**, and if

$\mathbf{M}$ is the set of all those elements p of $\mathbf{A}$ for which $p' \in \mathbf{P}$, then $\mathbf{M}$ is an ideal in $\mathbf{A}$; the reverse procedure (making a filter out of an ideal) works similarly. This comment indicates the logical role of the algebraically more common concept; just as filters arise in the theory of provability, ideals arise in the theory of refutability. (A proposition p is called refutable if its negation p' is provable.) Duality is so ubiquitous in Boolean theory that every development of that theory (unless it is exactly twice as long as it should be) must make an arbitrary choice between two possible approaches. Logic is usually studied from the "1" approach, *i.e.*, the emphasis is on truth and provability, and, consequently, on filters. Since the dual "0" approach uses the algebraically more natural concept of ideal, the remainder of this exposition (addressed to mathematicians, rather than to professional logicians) will be couched in terms of the logically less pleasant concepts of falsehood and refutability.

8. Boolean logics. The preceding discussion was intended to motivate the following definition. A *Boolean logic* is a pair $(\mathbf{A}, \mathbf{M})$, where $\mathbf{A}$ is a Boolean algebra and $\mathbf{M}$ is a Boolean ideal in $\mathbf{A}$. The elements of $\mathbf{A}$ will be called propositions; the elements of $\mathbf{M}$ will be called refutable propositions. The group-theoretic analogue of this concept (the structure consisting of a group and a specified normal subgroup) has not received very much attention, but it would strike most algebraists as a perfectly reasonable object of study.

The concept of a Boolean logic $(\mathbf{A}, \mathbf{M})$ has many similarities with the earlier concept of the propositional calculus $(S^*, \equiv)$. Both objects consist of (1) a set already endowed with some structure, and (2) a congruence relation in that set. (In Boolean theory, as in the rest of algebra, there is a natural one-to-one correspondence between ideals and congruence relations.) It should not be surprising therefore that the "axiomatic" method is the most common way of converting a Boolean algebra into a Boolean logic. In algebraic terms the axiomatic method amounts simply to this: given $\mathbf{A}$, select an arbitrary subset $\mathbf{M}_0$ of $\mathbf{A}$, and let $\mathbf{M}$ be the ideal generated by $\mathbf{M}_0$. (Because $\mathbf{M}$ consists of the refutable propositions, not the provable ones, the elements of $\mathbf{M}_0$ are "anti-axioms"; their negations are axioms in the usual sense.) Even this procedure has its group-theoretic analogue; for an example, recall the usual definition of the commutator subgroup of a group.

Most logical concepts have an algebraic alter-ego definable within the theory of Boolean logics. Two concepts of special importance (consistency and completeness) will be used to illustrate this point.

A Boolean logic $(\mathbf{A}, \mathbf{M})$ is called *consistent* if for no proposition p in $\mathbf{A}$ are both p and p' provable, or, equivalently, if for no p in $\mathbf{A}$ do both p and p' belong to $\mathbf{M}$. Since $\mathbf{M}$ is an ideal, it follows that $(\mathbf{A}, \mathbf{M})$ is consistent if and only if the ideal $\mathbf{M}$ is proper. To say that a Boolean logic is consistent is to say, roughly speaking, that the set of propositions that are refutable in it is not too large.

On intuitive (pragmatic) grounds it is desirable that the set of refutable propositions be not too small. (Recall that there is a natural one-to-one correspondence between refutable propositions and provable propositions.) The simplest way to ensure that the set of refutable propositions is large enough is to insist that, for every proposition p, either p or p' be refutable. A Boolean logic satisfying this condition is called *complete.* In other words, (**A**, **M**) is complete if and only if, for every p in **A**, either $p \in \mathbf{M}$ or $p' \in \mathbf{M}$.

Inconsistent Boolean logics, *i.e.*, logics of the form (**A**, **A**) are not very interesting from either the algebraic or the logical point of view. For a consistent logic (**A**, **M**), the concept of completeness has an elegant algebraic formulation: the logic is complete if and only if the ideal **M** is maximal in the algebra **A**. ("Maximal ideal" in such contexts always means "maximal proper ideal.")

If (**A**, **M**) is a Boolean logic, it is natural to form the quotient system **A**/**M**. A necessary and sufficient condition that **A**/**M** be a Boolean algebra is that it have at least two distinct elements, and a necessary and sufficient condition for that is exactly that **M** be a proper ideal in **A**. On the other hand, a necessary and sufficient condition that (**A**, **M**) be complete is that **A**/**M** have at most two distinct elements; if this is so, and if $\mathbf{M} \neq \mathbf{A}$, then $\mathbf{A}/\mathbf{M} = \mathbf{O}$. (Recall that $\mathbf{O} = \{0, 1\}$. On universal algebraic grounds, the condition that a proper ideal **M** be maximal in an algebra **A** is equivalent to the condition that the quotient algebra **A**/**M** be *simple*, *i.e.*, that **A**/**M** have no non-trivial proper ideals. The only simple Boolean algebra is **O**.) Conclusion: a necessary and sufficient condition that a Boolean logic (**A**, **M**) be both consistent and complete is that $\mathbf{A}/\mathbf{M} = \mathbf{O}$.

For a deeper study of concepts such as consistency and completeness additional logical apparatus is needed; some of it is described below. First, a terminological warning. Consistency and completeness, as defined above, are usually called *simple consistency* and *simple completeness*; perhaps *syntactic* would be a more suggestive adjective here than *simple.* The point is that the propositional calculus, for instance, is usually considered to be a "language," and (simple) consistency and completeness are defined in terms of the intrinsic structure (syntax) of that language. Two related concepts, to be discussed below, are defined in terms of a possible external interpretation (meaning) of the language, and are therefore appropriately called *semantic.* The group-theoretic analog of a semantic concept is one that depends on constructions reaching outside the given group, *i.e.*, typically, on representations of the group. Thus, for instance, "characteristic subgroup" is a syntactic concept, while "character" is a semantic one.

In most of what follows it will be simpler to forget about Boolean logics and to consider Boolean algebras instead. The point is that if (**A**, **M**) is a consistent Boolean logic, then the quotient algebra **A**/**M** can be formed and can be used to study virtually all the logically important properties of (**A**, **M**). Note that if a proposition in **A** is refutable, then its image in **A**/**M** (its equivalence class modulo **M**) is equal to 0, and, similarly, if a proposition in **A** is provable, then its image in **A**/**M** is equal to 1. It is, accordingly, convenient to agree that if p is an element of a Boolean algebra, then "p is refutable" shall be a long way of saying "$p=0$," and, similarly, "p is provable" shall be a long way of saying "$p=1$."

9. Quantifiers. Within the framework of Boolean algebras (or logics) it is easy to give an algebraic formulation of the inference from the premises "Some Greeks are men" and "All men are mortal" to the conclusion "Some Greeks are men." This is not a misprint. Within the framework of Boolean algebras alone it is not possible to formulate the inference that allows, from the same premises, the conclusion "Some Greeks are mortal." The desired inference is justified not by manipulating with propositions as a whole, but by the intrinsic structure of its constituents, and, in particular, by a study of what "some" and "all" mean.

The clue is in the consideration of propositional functions. As their name indicates, propositional functions are functions whose values are propositions. If, for instance, for every natural number x, $p(x)$ is the sentence "x is even" and $q(x)$ is the sentence "$2x=1$," then p and q are propositional functions. (The present discussion is only heuristic, of course, but even so the contrasting use of "sentence" and "proposition" needs a little justification. The justification is a very common one in mathematics; it is convenient, though incorrect, to identify an equivalence class with a representative element. Recall that a proposition was defined above as an equivalence class of sentences. The solecism is the same as the one every analyst commits when he speaks of an element of L_2 as a function.) From the algebraic point of view, a propositional function is a function defined on an arbitrary non-empty set with values in a Boolean algebra.

To single out a particular theory for logical examination means, algebraically, to fix a Boolean algebra $\mathbf{B}$. If, in addition, a certain non-empty set X is selected, then the ground is prepared for a discussion of propositional functions. (Intuitively the elements of X may be thought of as the objects that the propositions of the theory talk about.) The set $\mathbf{A}$ of all functions from X to $\mathbf{B}$ is in a natural way a Boolean algebra (pointwise operations), and the same is true of many of its subsets. The constant functions (*i.e.*, the functions p obtained by selecting an element p_0 in $\mathbf{B}$ and writing $p(x)=p_0$ for every x in X) constitute a subalgebra of the algebra $\mathbf{A}$; that subalgebra is obviously isomorphic to $\mathbf{B}$. In other words, the propositions of a theory are (or, rather, may be identified with) particular propositional functions of that theory.

What is the effect of "some" in a sentence such as "for some x, x is even"? (It is understood here that the underlying set X is the set of natural numbers; the value-algebra $\mathbf{B}$ has not been, and need not be, specified). The most striking effect it has is to convert a propositional function into a constant, or, *via* the identification convention of the preceding paragraph, into a constant function. This effect is analogous to the effect of "sup" in "$\sup_x f(x)$" and to the effect of "$\int_0^1 \cdots dx$" in "$\int_0^1 f(x)dx$" (where the function f is, say, a real-valued continuous function defined on the closed unit interval). Accordingly, the algebraic analogue of "some" ought to be a mapping that sends a certain Boolean algebra (propositional functions) into a certain subalgebra (constant functions). Such a mapping (usually denoted by $\exists$) is indeed at the basis of the theory of propositional functions.

It is now wise to abstract from the motivation. Consider a perfectly arbitrary Boolean algebra **A**, and let $\exists$ be a mapping of **A** into itself. If **A** were an algebra of propositional functions, and if $\exists$ were the operator "some," then, presumably, $\exists$ would satisfy some rather special conditions; the problem now is to find a reasonable list of conditions that characterize such a special $\exists$. This is easy; here are some of them:

(Q 1) $$\exists 0 = 0,$$

(Q 2) $$p \leqq \exists p,$$

(Q 3) $$\exists(p \vee q) = \exists p \vee \exists q,$$

(Q 4) $$\exists\exists p = \exists p.$$

(In these conditions, and also in (Q 5) below, p and q are arbitrary elements of **A**). The intuitive grounds for the conditions are easy to see. Suppose, for instance, that $q(x)$ is "$2x = 1$"; here, once more, **A** is to be thought of as an algebra of propositional functions whose domain is the set of natural numbers. Since in all reasonable theories of the arithmetic of natural numbers the sentence "$2x = 1$" is refutable, each value of the propositional function q is refutable, and therefore (in accordance with the agreement concerning the reduction of Boolean logics to Boolean algebras) the propositional function q is equal to the constant function 0. The same considerations apply to the sentence "for some x, $2x = 1$" and thus serve to complete the illustration of (Q 1). The remaining conditions (Q 2), (Q 3), and (Q 4) are illustrated similarly. In intuitive terms, (Q 2) says that each value of p implies that "for some x, $p(x)$"; the construction of analogous readings of (Q 3) and (Q 4) is left as an exercise to the reader.

It is amusing to observe that (except for notation) the conditions (Q 1)–(Q 4) are well-known to most mathematicians (but in a very different context). They are, all but verbatim, the Kuratowski axioms for a closure operator on a topological space. They do not, however, serve to characterize the "some" operator; in technical language, the theory of closure algebras is not co-extensive with the monadic functional calculus. Another glance at the conditions as they now stand should arouse the suspicion that something is missing; the trouble is that they relate $\exists$ to $\vee$ only (0 and $\leqq$ can be defined in terms of $\vee$), and say nothing about the relation of $\exists$ to either $\wedge$ or $'$. The missing condition is this:

(Q 5) $$\exists(\exists p)' = (\exists p)'.$$

In intuitive terms, (Q 5) serves as at least a partial reminder of the fact that the constant functions form a Boolean algebra (so that, in particular, they are closed under complementation), and that "some" applied to a constant function has no effect.

[Two remarks are in order. (1) The conditions (Q 1)–(Q 5) are equivalent to a shorter set, namely to (Q 1), (Q 2), and

(Q 6) $$\exists(p \wedge \exists q) = \exists p \wedge \exists q.$$

The proof of this equivalence is just axiom-chopping. The longer set was selected for presentation here because of its greater intuitive content. (2) The motivation of (Q 1)–(Q 5), and also of (Q 6), could have been, and sometimes is, based on quasi-geometric considerations, involving the formation of (possibly infinite) Boolean suprema. The didactic danger of this motivation is its emphasis on the intuitive idea that "some" is just an infinite "or." The idea is not wholly unsound, but real progress in algebraic logic was achieved only after the realization that $\exists$ is really a unary operation, not an infinitary one.]

The way is now clear to a precise definition: an *existential quantifier* is a mapping $\exists$ of a Boolean algebra into itself satisfying the conditions (Q 1)–(Q 5). Dually, a *universal quantifier* is a mapping $\forall$ of a Boolean algebra into itself satisfying the conditions

(Q′ 1) $$\forall 1 = 1,$$

(Q′ 2) $$\forall p = p,$$

(Q′ 3) $$\forall(p \wedge q) = \forall p \wedge \forall q,$$

(Q′ 4) $$\forall\forall p = \forall p,$$

(Q′ 5) $$\forall(\forall p)' = (\forall p)'.$$

It is easy to see that $\forall$ bears the same relation to the intuitive "all" as $\exists$ bears to "some."

There is a very close connection between existential quantifiers and universal ones: if $\exists$ is an existential quantifier on, say, a Boolean algebra **A**, and if a mapping $\forall$ of **A** into itself is defined by

$$\forall p = (\exists p')',$$

then $\forall$ is a universal quantifier on **A**; if, in reverse, a universal quantifier $\forall$ is given, and if $\exists$ is defined by

$$\exists p = (\forall p')',$$

then $\exists$ is an existential quantifier. ("Always p" is the same as "not sometimes not p," and "sometimes p" is the same as "not always not p"). The thoroughgoing symmetry of this situation justifies an asymmetric treatment; anything that can be said about an $\exists$ has an obvious dual about an $\forall$, and it is therefore sufficient to discuss in detail only one of these two objects. In what follows $\exists$ will be given preferred treatment, and, in fact, the word "quantifier" will be used from now on in the sense of "existential quantifier." (This is in line with the algebraists preference for ideals instead of filters; the dual of an algebraist would select $\forall$ for preferred treatment.) Universal quantifiers, whenever they must be considered, will always be given their full name.

10. Monadic algebras. Once the concept of a quantifier is at hand, it is immediately possible to generalize the concept of a Boolean algebra in a manner adapted to an algebraic answer to the question of why some Greeks are mortal. Technically the generalized algebras (called monadic algebras) play an intermediate role; they are slightly more useful than Boolean algebras, but not nearly so useful as certain even more general algebras (the polyadic algebras that will be described a little later). Psychologically and historically, however, the intermediate generalization has some value. Its psychological value is that it exhibits in a simplified form some of the curious properties of its generalized version; its historical value is that it puts into a modern algebraic context the oldest known systematic treatment of logic, namely Aristotle's syllogistics.

A *monadic* (Boolean) *algebra* is a pair ($\mathbf{A}$, $\exists$), where $\mathbf{A}$ is a Boolean algebra and $\exists$ is a quantifier on $\mathbf{A}$. (The word "monadic" serves as a reminder of the one additional operation that distinguishes monadic algebras from Boolean algebras.) The elementary theory of monadic algebras is a routine matter; subalgebras, homomorphisms, ideals, and similar universal algebraic concepts (with a qualifying "monadic" when clarity demands it) are defined in a completely unsurprising manner. Free monadic algebras can be defined and studied (if desired) by the techniques used in the study of free Boolean algebras; the analogue of the propositional calculus is called the *monadic functional calculus.*

The concept of a monadic logic arises naturally in connection with the theory of provability and refutability in the monadic functional calculus: a *monadic logic* is a pair ($\mathbf{A}$, $\mathbf{M}$), where $\mathbf{A}$ is a monadic algebra (with quantifier $\exists$, say) and $\mathbf{M}$ is a monadic ideal in $\mathbf{A}$. (It is convenient here, as in other parts of algebra, to be mildly forgetful of the completely rigorous definitions, and, accordingly, to identify a monadic algebra with the underlying Boolean algebra. If this were not done, a monadic logic would have to be denoted by a symbol such as (($\mathbf{A}$, $\exists$), $\mathbf{M}$).) The fact that a monadic ideal is a Boolean ideal invariant under the application of $\exists$ corresponds to the logical fact that if p is a refutable propositional function, then $\exists p$ is also refutable.

To discuss the (simple, or syntactic) consistency and completeness of Boolean logics, it is desirable to introduce a new term: an element p of a monadic algebra will be called *closed* if $\exists p = p$. Intuitively, closed propositions correspond to what were called constant functions before, or, in other words, to propositions instead of propositional functions. There is a natural way of associating a Boolean logic ($\mathbf{A}_0$, $\mathbf{M}_0$) with every monadic logic ($\mathbf{A}$, $\mathbf{M}$); the algebra $\mathbf{A}_0$ is the set of all closed elements of $\mathbf{A}$ and the ideal $\mathbf{M}_0$ is the intersection of $\mathbf{M}$ with $\mathbf{A}_0$. A monadic logic ($\mathbf{A}$, $\mathbf{M}$) is called syntactically consistent (or complete) if the associated Boolean logic ($\mathbf{A}_0$, $\mathbf{M}_0$) is consistent (or complete).

Why is it necessary to modify the Boolean definitions of consistency and of completeness for the monadic situation? Consistency could conceivably be defined by the requirement that for no p in $\mathbf{A}$ should both p and p' belong to $\mathbf{M}$, and, similarly, completeness could be defined by the requirement that for every p in $\mathbf{A}$ either p or p' should belong to $\mathbf{M}$. What is wrong with these definitions? The answer is that there is nothing wrong as far as consistency is concerned, and very much is wrong as far as completeness is concerned. The alternative definition of consistency is equivalent to the one officially adopted above, but the alternative definition of completeness is out of harmony with both the official definition and common sense. If, as in some previous examples, $p(x)$ is "x is even," then it is not at all reasonable to demand of a logic sufficient power to settle the provability or refutability of the propositional function p. The sentences "for some x, x is even" and "for all x, x is even" are closed; a reasonable logical theory of arithmetic should declare each of them to be either provable or refutable. The function p, however, is not a sentence; common sense demands that both it and its negation should fail to be either provable or refutable.

11. Syllogisms. The discussion of monadic algebras and logics up to this point was merely an adaptation of the corresponding discussion of Boolean facts. Some progress has been made, nevertheless; monadic logics (unlike Boolean logics) contain the general theory of syllogisms. For an example, consider again the premises "Some Greeks are men" and "All men are mortal," and consider the desired conclusion "Some Greeks are mortal." To make an algebraic model of the situation, let X be an appropriate set and let $\mathbf{B}$ be an appropriate Boolean algebra of propositions about the elements of X. The set X, for instance, could be the set of all animals, and $\mathbf{B}$ could be an algebra containing, for each x in X, the propositions "x is Greek," "x is a man," and "x is mortal." (This description is, of course, much too colloquial for complete precision, but there is no difficulty at all in converting it into honest mathematics.) If these propositions are denoted by $p(x)$, $q(x)$, and $r(x)$, respectively, and if $\mathbf{A}$ is the algebra of all propositional functions such as p, q, and r, then $\mathbf{A}$ (with the intuitively obvious "some" quantifier in the role of $\exists$) is a monadic algebra suitable for the study of the inference described above. The first premise is $\exists(p \wedge q)$, the second premise is $\forall(q \rightarrow r)$, and the conclusion is $\exists(p \wedge r)$. (The universal quantifier $\forall$ here is the natural dual of the given existential quantifier $\exists$.) The algebraic justification of the inference is that if $\mathbf{A}$ is the monadic algebra of a monadic logic such that both premises belong to the filter of provable propositions, then the conclusion also belongs to that filter.

There is a special aspect of the theory of syllogisms that still remains to be converted into algebra; it is exemplified by the classical premises "Socrates is a man" and "All men are mortal," together with the conclusion "Socrates is mortal." In highly informal language, the trouble with this syllogism is that the algebraic theory (so far) is equipped to deal with generalities only, and is unable to say anything concrete; Socrates, however, is a concrete individual entity. The preceding paragraph shows that a monadic algebra can be taught to say "All men are mortal." The reason is that "manhood" can easily be thought of as an element of a monadic algebra, since it is the obvious abstraction of the propositional function whose value at each point x of some set is "x is a man." Socrates, on the other hand, is a "constant," and there is no immediately apparent way of pointing to him. (The use of the word "constant" here and below is quite different from its earlier use in the phrase "constant function." The important concept now is what logicians call an individual constant.) A classical artifice, designed to deal with just this difficulty, is to promote Socrates to a propositional function, *i.e.*, to identify him with the function whose value at x is

"x is Socrates." This procedure is both intuitively and algebraically artificial; Socrates should not be a propositional function but a possible argument of such functions.

To find the proper algebraization of the concept of a constant it is necessary only to recall that the elements of a monadic algebra are abstractions of the concept of a propositional function, and to determine the algebraic effect of replacing the argument of such a function by some fixed element of its domain. If p is a propositional function with domain X, and if $x_0 \in X$, then $p(x_0)$ is a proposition, or, *via* the obvious identification convention, a propositional function with only one value. The mapping $p \rightarrow p(x_0)$ (the evaluation map induced by x_0) clearly preserves the Boolean operations (*i.e.*, suprema, infima, and complements). If the function p itself has only one value (equivalently, if $p = \exists q$ for some q), then that value is $p(x_0)$, *i.e.*, the mapping leaves the range of $\exists$ elementwise fixed. If, on the other hand, $\exists$ is applied to the (constant) function $p(x_0)$, the result is the same function, *i.e.*, $\exists$ leaves the range of the mapping elementwise fixed. These considerations motivate the following general definition: a *constant* of a monadic algebra $\mathbf{A}$ is a Boolean endomorphism c of $\mathbf{A}$ such that

$$c\exists = \exists \quad \text{and} \quad \exists c = c.$$

This definition is applicable to the mortality of Socrates. If, as before, q is manhood and r is mortality, and if Socrates is taken to be a constant, say c, of a monadic algebra containing q and r, then the algebraic justification of the classical syllogism is this: if $\mathbf{A}$ is the monadic algebra of a monadic logic such that both $\forall(q \rightarrow r)$ and cq belong to the filter of provable propositions, then cr also belongs to that filter.

Constants are much more important than their more or less casual introduction above might indicate; the concept of a constant (suitably generalized to the polyadic situation) is probably the most important single concept in algebraic logic. This should not be too surprising; in the intuitive interpretation, the constants of a theory constitute, after all, the subject matter that the propositions of the theory talk about. Algebraically constants play a crucial role in the representation theory of monadic (and polyadic) algebras.

12. Quantifier algebras. The theory of propositional functions of only one variable (and of their abstract algebraic counterparts) is as insufficient for the understanding of logic and its applications as the theory of ordinary numerical functions of one variable is insufficient for the understanding of calculus. Mathematics contains not only sentences of the form "x is positive," but also sentences such as "x is less than y" and "x is between y and z." Although a mathematician with no training in modern logic is likely to be suspicious when he is told that syllogisms are not enough for mathematics, the basis of that assertion is nothing more profound or esoteric than the need to consider propositional functions of several variables. This is all that DeMorgan meant when he said that the scholastics, after two millennia of Aristotelean tradition, were still unable to prove that if a horse is an animal, then a horse's tail is an animal's tail.

Suppose, to begin modestly, that one wants to study propositional functions of three variables. The most immediately apparent new phenomenon is the possibility of partial quantification: the theory must be able to treat sentences of the form "there is an x_1 and there is an x_3, such that $p(x_1, x_2, x_3)$." (Here, as before, it is enough to discuss existential quantifiers; the corresponding theory of universal quantifiers is obtained by a simple dualization.) Another way of expressing the phenomenon is to say that the appropriate algebraic system must possess not one quantifier but many. What appears to be going on is this: if I is the set of relevant indices, so that $I = \{1, 2, 3\}$ in the present example, then there is a quantifier (*i.e.*, an existential quantifier) corresponding to each subset (*e.g.*, to $\{1, 3\}$) of the set I. If $\exists(J)$ is the quantifier corresponding to the subset J of I, then, of course, the way that $\exists(J)$ depends on J should be specified. The example shows the way. If J is empty, then prefixing $\exists(J)$ to p should produce no change in p, and if both J and K are subsets of I, then prefixing $\exists(J)$ and $\exists(K)$ to p, in either order, should have the same effect as prefixing the quantifier corresponding to the union of J and K.

The preceding paragraph suggests the definition of an algebraic system, which, however, turns out to fall far short of what is needed. It is worth a brief look anyway. Call a ***quantifier algebra*** a triple $(\mathbf{A}, I, \exists)$, where $\mathbf{A}$ is a Boolean algebra, I is a set, and $\exists$ is a function from subsets of I to quantifiers on $\mathbf{A}$ such that

$$(\exists\ 1) \qquad \exists(\varnothing)p = p$$

whenever $p \in \mathbf{A}$, and

$$(\exists\ 2) \qquad \exists(J)\exists(K) = \exists(J \cup K)$$

whenever J and K are subsets of I. In analogy with logical usage, an element of the set I will be called a *variable* (or, in more detail, an *individual variable*) of the quantifier algebra $(\mathbf{A}, I, \exists)$. Caution: a variable, in this sense, does not vary at all; it merely serves as a reminder of the place into which a "variable," in the intuitive sense, could be substituted, if, that is, the elements of $\mathbf{A}$ were propositional functions and not elements of a quite abstract Boolean algebra.

The concept of a quantifier algebra is a proper generalization of the concept of a monadic algebra; a quantifier algebra $(\mathbf{A}, I, \exists)$ for which the set I of variables consists of exactly one element may be identified with the monadic algebra $(\mathbf{A}, \exists(I))$. The *degree* of a quantifier algebra is defined to be the cardinal number of the set of its variables; the last comment says that quantifier algebras of degree 1 are essentially the same as monadic algebras. It is crucial for applications to permit the consideration of quantifier algebras of infinite degree. While, to be sure, each particular propositional function that is likely to arise has only finitely many arguments, the number of arguments in a reasonably extensive theory (*e.g.*, in mathematics) is not likely to be bounded. To keep the generalization from running away with itself, it is often advisable to restrict the study of quantifier algebras to the locally finite case. The quantifier algebra $(\mathbf{A}, I, \exists)$ is called *locally finite* if to every element p of **A** there corresponds a finite subset J of I such that $\exists(I-J)p=p$; this is the abstract formulation of the concrete promise that no propositional function under consideration will actually depend on infinitely many variables.

13. Polyadic algebras. The reason that quantifier algebras are an inefficient logical tool is that they do not allow the discussion of transformations of variables. Consider, for instance, a sentence of the form "if $p(x_1, x_2)$ and $p(x_2, x_1)$, then $x_1=x_2$." Such sentences usually occur in the definition of a partial order; they indicate the need for discussing the passage from $p(x_1, x_2)$ to $p(x_2, x_1)$. For a more complicated example of the same type consider the problem of converting $p(x_1, x_2, x_3, x_4)$ into $p(x_4, x_4, x_1, x_3)$. There is no way of describing such conversions in terms of Boolean operations and quantifications alone; the best way of incorporating them into the general theory is by postulating them outright. What appears to be going on is this: to every transformation τ on I (*i.e.*, to every mapping of the set I into itself) there corresponds a mapping, say $\mathbf{S}(\tau)$, of the set of propositional functions under consideration into itself. The mappings $\mathbf{S}(\tau)$ preserve the Boolean operations (*i.e.*, they are Boolean endomorphisms). If τ is the identity transformation on I (to be denoted by δ), then $\mathbf{S}(\tau)$ leaves every p invariant; if both σ and τ are transformations on I, then the effect of $\mathbf{S}(\sigma)$ on $\mathbf{S}(\tau)p$ is the same as the effect of $\mathbf{S}(\sigma\tau)$ on p.

In analogy with quantifier algebras, it is possible to define a *transformation algebra* as a triple (**A**, I, **S**), where **A** is a Boolean algebra, I is a set, and **S** is a function from transformations on I to Boolean endomorphisms on **A**, such that

(**S** 1) $$\mathbf{S}(\delta)p = p$$

whenever $p \in \mathbf{A}$, and

(**S** 2) $$\mathbf{S}(\sigma)\mathbf{S}(\tau) = \mathbf{S}(\sigma\tau)$$

whenever σ and τ are transformations on I.

Neither quantifier algebras nor transformation algebras are of much logical significance; the important concept is one that possesses both structures at the same time. It is a matter of universal algebraic experience, however, that it is not enough to impose two different structures on the same set; in order to get a usable theory, it is necessary to describe the compatibility conditions that relate the two structures to each other. (A topological group is not merely a set that is simultaneously a group and a topological space.) The appropriate conditions are discovered by experimentation with the special structures that the theory is intended to generalize; their final justification is their success. To examine here the experimentation that led to the definition of polyadic algebras would be more boring than profitable; suffice it to report the results. A *polyadic* (Boolean) *algebra* is a quadruple (**A**, I, **S**, $\exists$) subject to the following conditions: (i) ($\exists$ 1) and ($\exists$ 2) hold, or, more precisely, (**A**, I, $\exists$) is a quantifier algebra, (ii) (**S** 1) and (**S** 2) hold, or, more precisely, (**A**, I, **S**) is a transformation algebra, and (iii)

$$(\exists \mathbf{S}\ 1) \qquad \mathbf{S}(\sigma)\exists(J) = \mathbf{S}(\tau)\exists(J)$$

whenever J is a subset of I and σ and τ are transformations on I such that $\sigma i = \tau i$ for all i in $I - J$, and

$$(\exists \mathbf{S}\ 2) \qquad \exists(J)\mathbf{S}(\tau) = \mathbf{S}(\tau)\exists(\tau^{-1}J)$$

whenever J is a subset of I and τ is a transformation on I such that two distinct elements of I are never mapped by τ onto the same element of J.

The conditions ($\exists \mathbf{S}$ 1) and ($\exists \mathbf{S}$ 2) were recorded here for the sake of completeness only; no technical use will be made of them in this merely descriptive report. It is perhaps worth remarking though that they (or, rather, the provisos that accompany them) are not so complicated as they seem on first sight. What they amount to is a condensation of the usual and intuitively obvious relations between quantifications and transformations. Suppose, for example, that i and j are distinct elements of I; let J be the singleton $\{i\}$ and let τ be the transformation that maps i onto j and everything else (including j itself) onto itself. Since τ agrees with δ outside J, it follows from ($\exists \mathbf{S}$ 1) that $\mathbf{S}(\tau)\exists(J) = \exists(J)$; this equation corresponds to the familiar fact that once a variable has been quantified, the replacement of that variable by another one has no further effect. To get another example, note, for the same τ and J, that $\tau^{-1}J = \varnothing$. It follows from ($\exists \mathbf{S}$ 2) that $\exists(J)\mathbf{S}(\tau) = \mathbf{S}(\tau)$; this equation corresponds to the familiar fact that once a variable has been replaced by another one, a quantification on the replaced variable has no further effect.

[Polyadic algebras stand in the same relation to the so-called *pure* first-order functional calculus as do Boolean algebras to the propositional calculus. The simplest example of an *applied* functional calculus is one that is equipped to discuss the concept of equality. By a suitable adaptation of standard logical methods, such applied calculi can be treated within the framework of polyadic algebras; the details are of no relevance here. What should be mentioned, however, is that there is another way of algebraizing the functional calculus with equality, namely, *via* Tarski's concept of a cylindric algebra. Roughly speaking, a cylindric algebra is a quantifier algebra together with certain distinguished elements; the distinguished elements play the role of sentences that assert equations among variables. The theory of cylindric algebras is an efficient algebraic tool for studying calculi with equality. The exact relations between polyadic algebras and cylindric algebras are of considerable technical interest; they are still in the process of being clarified. It is already known, however, that in most important cases the two concepts are equivalent. The way that transformations enter into the theory of cylindric algebras is an ingenious trick; the idea is that to say "$p(x_1, x_1)$" is the same as to say "there is an x_2 such that $p(x_1, x_2)$ and $x_2 = x_1$."]

14. Semantic concepts. The discussion of polyadic logics and their syntactic consistency and completeness proceeds in a straightforward manner. A *polyadic logic* is a pair (**A**, **M**), where **A** is a polyadic algebra and **M** is a polyadic ideal in

A. An element p of a polyadic algebra **A** is called *closed* if $\exists(I)p=p$ (where I, of course, is the set of variables of **A**). Just as for monadic logics, there is a natural way of associating a Boolean logic ($\mathbf{A}_0$, $\mathbf{M}_0$) with every polyadic logic (**A**, **M**); the algebra $\mathbf{A}_0$ is the set of all closed elements of **A** and the ideal $\mathbf{M}_0$ is the intersection of **M** with $\mathbf{A}_0$. A polyadic logic (**A**, **M**) is called syntactically consistent (or complete) if the associated Boolean logic ($\mathbf{A}_0$, $\mathbf{M}_0$) is consistent (or complete).

All this is pretty routine stuff by now. The element of novelty is in the study of the "semantic" theory of polyadic logics, *i.e.*, in the study of "interpretations" of a logic in a "model" and in the study of "truth" and "validity." (It must be admitted that these concepts could have been introduced in connection with Boolean logics and monadic logics also. Since, however, in those simple situations, they exhibit the deceptive simplicity of a degenerate case, their premature study would have been more confusing than helpful.) The natural habitat of semantic theories is situated somewhere near a polyadic *logic*, but the machinery for examining such a theory is easier for a polyadic *algebra*. The reduction of (syntactically consistent) logics to algebras is an easy step: if $\mathbf{M}\neq\mathbf{A}$, then $\mathbf{A}/\mathbf{M}$ can be formed and can be used to study (**A**, **M**). (This kind of thing was done before; *cf.* the corresponding discussion in the Boolean case.) Some concepts disappear in the passage from (**A**, **M**) to $\mathbf{A}/\mathbf{M}$ and others take on an interesting new guise. Thus, for instance, to say that a polyadic algebra is (syntactically) consistent is to say merely that it is there; on the other hand, to say that a polyadic algebra is (syntactically) complete is to say that it is simple. (Just as for Boolean logics, it turns out that a polyadic logic (**A**, **M**), with $\mathbf{M}\neq\mathbf{A}$, is syntactically complete if and only if the polyadic ideal **M** is a maximal ideal in the polyadic algebra **A**.)

The usual way to describe a model is this: take a non-empty set X, interpret the variables (*i.e.*, the elements of I) as variables (in the intuitive sense) varying over X, and interpret the elements of **A** as propositional functions on an appropriate Cartesian power of X with values that are either true or false. This is somewhat vague; a slight change in language converts it, however, into an algebraically precise and clean definition. If X is a non-empty set and if I is an arbitrary set, the set of all functions from the Cartesian product X^I into the Boolean algebra **O** possesses in a natural way the structure of a polyadic Boolean algebra; a *model* is, by definition, a polyadic subalgebra of such a total functional algebra.

Once the concept of a model is known, the definitions of the remaining semantic concepts are automatic. The purpose of the following paragraphs is first to define the usual semantic concepts and then to examine their algebraic significance.

An *interpretation* of a polyadic algebra (**A**, I, **S**, $\exists$) in a model is a polyadic homomorphism from **A** onto the model. An element p of **A** is *true* in an interpretation f if $fp=1$; if $fp=0$, then p is *false* in the interpretation. (Caution: fp is a function from X^I into **O** and is therefore not necessarily either 0 or 1. To say that p is true for f means that the image fp of p under f is the constant function whose value is 1.) An element p of **A** is *valid* if it is true in every interpretation; if p is false in every interpretation, then p is called *contravalid*. An element p of **A** is *satisfiable* if there is an interpretation for which it is not false (*i.e.*, if p is not contravalid).

The algebra **A** is *semantically consistent* if it has an interpretation in a model. The definition of semantic completeness requires a little more motivation. It is a trivial consequence of the definition of Boolean homomorphism (sometimes even built in as part of the definition) that the unit element of **A** is always valid, and, dually, that the zero element is always contravalid; in logical terms this says that provable propositions are valid, and refutable ones are contravalid. (The assumption of syntactic consistency is usually, and quite properly, made explicit at this point. In the present discussion, dealing with algebras instead of logics, the role of that assumption is played by the fact that **A** is an algebra at all, or, in more detail, that **A** has a unit element distinct from zero.) Semantic completeness requires the converse assertions; the algebra **A** is *semantically complete* if the unit is the only valid element, or, equivalently, if every non-zero element is satisfiable. In logical terms these requirements say that everything valid is provable, or, equivalently, that everything that is not refutable is satisfiable.

All the preceding definitions are based on one of them, namely, on the definition of a model. The situation can be better understood if it is first generalized and then specialized in a new direction. A model was defined as an element of a certain particular class of polyadic algebras, but, in the subsequent definitions, no special properties of that class were used. Those definitions, in other words, are relative to the prescribed class $\mathfrak{M}$ of models, and to nothing else; a significant generalization of them is obtained if $\mathfrak{M}$ is replaced by an arbitrary class $\mathcal{C}$ of algebras. The concepts so obtained may be referred to by using $\mathcal{C}$ as a prefix, so that expressions such as "$\mathcal{C}$-interpretation" and "$\mathcal{C}$-valid" now make sense. (Only in case $\mathcal{C} = \mathfrak{M}$ may the prefix be omitted.) An interesting specialization is obtained by letting the class $\mathcal{S}$ of all simple algebras play the role of $\mathcal{C}$. The discussion that follows will be in terms of $\mathcal{S}$ rather than $\mathfrak{M}$; the vitally important relation between $\mathcal{S}$ and $\mathfrak{M}$ will be treated afterward.

To say that an algebra **A** is semantically $\mathcal{S}$-consistent means, by definition, that there exists a homomorphism from **A** onto a simple algebra. This, in turn, is equivalent, on universal algebraic grounds, to the existence of a maximal ideal in **A**. The problem of the existence of maximal ideals is well known in algebra; the cases in which the solution is affirmative can usually be settled by a straightforward application of Zorn's lemma. The theory of polyadic algebras is one of these pleasant cases; Zorn's lemma applies and proves that maximal ideals always do exist. In other words, every polyadic algebra is (semantically) $\mathcal{S}$-consistent. The situation is reminiscent of syntactic consistency, and, in fact, it turns out that the two concepts are the same. It is profitable to return (but just for a moment) to logics instead of algebras. Recall that a logic (**A**, **M**) is syntactically consistent if and only if **M** is a proper ideal in **A**. The general definition of semantic $\mathcal{C}$-consistency for a logic (**A**, **M**) requires the existence of an algebra **C** in the class $\mathcal{C}$ and the existence of a homomorphism f from **A** onto **C** such that $fp=0$ whenever $p \in$ **M**. If $\mathcal{C}$ is $\mathcal{S}$, then this is equivalent to the requirement that **M** be included in some maximal ideal. Since Zorn's lemma can be used to show not only that maximal ideals exist, but that, in fact, every proper ideal is included in some maximal ideal, syntactic consistency and semantic $\mathcal{S}$-consistency are the same whether they are approached *via* logics (**M** is a proper ideal) or *via* algebras (**A** is a non-degenerate algebra).

To say that an algebra **A** is semantically $\mathcal{C}$-complete means, by definition, that for every non-zero element p of **A** there exists an algebra **C** in the class $\mathcal{C}$ and there exists a homomorphism f from **A** onto **C** such that $fp\neq 0$. In this formulation $\mathcal{C}$-completeness may not look familiar even to a professional algebraist; an easy argument serves, however, to prove that the definition is equivalent to one that is very well known indeed. The fact is that **A** is $\mathcal{C}$-complete if and only if **A** is a *subdirect sum* of algebras belonging to the class $\mathcal{C}$. The argument (and even the definition of subdirect sum) will be omitted here; they are mentioned only to show that the concept ($\mathcal{C}$-completeness for an arbitrary $\mathcal{C}$) does fit into the framework of algebra.

In view of the relation between maximal ideals on the one hand and homomorphisms onto simple algebras on the other hand, the definition of semantic $\mathcal{S}$-completeness for an algebra **A** reduces to this: to every non-zero element of **A** there corresponds a maximal ideal not containing it. Equivalently: the intersection of all the maximal ideals of **A** is the singleton $\{0\}$. This too is a well-known algebraic phenomenon. In analogy with some other parts of algebra, a polyadic algebra will be called *semisimple* if the intersection of all its maximal ideals is trivial; the result is that a polyadic algebra is $\mathcal{S}$-complete if and only if it is semisimple, or, equivalently, if and only if it is a subdirect sum of simple algebras.

15. The Gödel theorems. The title of this paper promised a discussion of the basic concepts of algebraic logic, and indeed the definitions of the basic *concepts* were emphasized much more than the basic *results* concerning them. As a partial rectification of this unbalanced state of affairs the present (concluding) section is devoted to the statement of two of the deepest theorems of the field. The original formulations and proofs of these justly celebrated theorems are due to Gödel; the results are known as the Gödel completeness theorem and the Gödel incompleteness theorem.

The word "completeness" in "completeness theorem" refers to semantic completeness. The algebraic statement of the theorem is short and elegant: *every locally finite polyadic algebra of infinite degree is semantically complete* (or, equivalently, $\mathfrak{M}$-complete). The crux of the proof is a characterization of simple polyadic algebras. It is very easy to see that every model is a simple algebra; the hard thing to prove is that, conversely, every simple, locally finite polyadic algebra of infinite degree is (isomorphic to) a model. The proof makes use of the polyadic generalization of what was called a "constant" of a monadic algebra.

In terms of the symbols used above, the easy fact is that $\mathfrak{M} \subset \mathcal{S}$ and the hard one is that (in the presence of local finiteness and infinite degree) $\mathcal{S} \subset \mathfrak{M}$. In view of the equation $\mathcal{S} = \mathfrak{M}$, semantic completeness (*i.e.*, $\mathfrak{M}$-completeness) is the same as $\mathcal{S}$-completeness, and, consequently, the Gödel completeness theorem implies that every locally finite polyadic algebra of infinite degree is semisimple. In fact, *every* polyadic algebra is semisimple, and this assertion is sometimes taken to be the algebraic version of the Gödel completeness theorem. The proof of semisimplicity is quite easy; the crucial fact is not that something is semisimple but that in the most important cases semisimplicity is the same as semantic completeness.

The word "incompleteness" in "incompleteness theorem" refers to syntactic incompleteness. The result here is not so general as the completeness theorem; the subject matter is not all polyadic algebras, but only relatively few of them. The algebras covered by the theorem are usually described by saying that they are adequate for elementary arithmetic. Because a precise explanation of what this means would involve a rather long and technical detour, no such explanation will be presented here, and, consequently, the incompleteness theorem will be described rather than stated. The striking qualities of the theorem are sufficiently great to remain visible even under such cavalier treatment.

Even if the definition of the class of algebras under consideration is not made explicit, it is convenient to have a short phrase in which to refer to them; in what follows "Peano algebra" will be used instead of "polyadic algebra that is adequate for elementary arithmetic." The Gödel incompleteness theorem asserts the existence of "undecidable propositions." Since in the passage from logics to algebras the statement that an element p is refutable (or provable) was identified with the statement that $p=0$ (or $p=1$), and since "undecidable" means "neither refutable nor provable," the assertion reduces to the existence of a (closed) element different from both 0 and 1. The ideal generated by such an element is a non-trivial proper ideal. Conversely, every non-trivial proper ideal contains an undecidable (closed) element. These facts indicate that the Gödel incompleteness theorem asserts the existence, in Peano algebras, of non-trivial proper ideals; the definition of syntactic completeness shows now that the theorem does indeed assert that something is (syntactically) incomplete. The usual, logical, formulation explicitly makes the assumption that the underlying logic is consistent; the treatment of algebras instead of logics makes it unnecessary to mention such an assumption here.

The Gödel theorem does not assert that every Peano algebra is syntactically incomplete. It asserts, instead, that the definition of Peano algebras is not a faithful algebraic transcription of all intuitive facts about elementary arithmetic. In algebraic terms this means that while some Peano algebras may be syntactically complete, there definitely exist others that are not. The situation is analogous to the one in the theory of groups. A class of polyadic algebras could be defined that is adequate for a discussion of elementary group theory. For this class the assertion that every two elements of a group commute would be undecidable, in the obvious sense that the assertion is true for some groups (and, therefore, corresponds to the unit element of some of the polyadic algebras under consideration) and false for others.

What has been said so far makes the Gödel incompleteness theorem take the following form: not every Peano algebra is syntactically complete. In view of the algebraic characterization of syntactic completeness this can be rephrased thus: not every Peano algebra is a simple polyadic algebra. This is the description that was promised above. What follows is another rephrasing of this description; the rephrasing, possibly of some mnemonic value, makes its point by making a pun. Consider one of the systems of axiomatic set theory that is commonly accepted as a foundation for all extant mathematics. There is no difficulty in constructing polyadic algebras with sufficiently rich structure to mirror that axiomatic system in all detail. Since set theory is, in particular, an adequate foundation for elementary arithmetic, each such algebra is a Peano algebra. The elements of such a Peano algebra correspond in a natural way to the propositions considered in mathematics; it is stretching a point, but not very far, to identify such an algebra with mathematics itself. Some of these "mathematics" may turn out to possess no non-trivial proper ideals, *i.e.*, to be syntactically complete; the Gödel theorem implies that some of them will certainly be syntactically incomplete. The conclusion is that the crowning glory of modern logic (algebraic or not) is the assertion: *mathematics is not necessarily simple.*

II

GENERAL THEORY

ALGEBRAIC LOGIC, I
MONADIC BOOLEAN ALGEBRAS

Preface. The purpose of the sequence of papers here begun is to make algebra out of logic. For the propositional calculus this program is in effect realized by the existing theory of Boolean algebras. An indication of how the program could be realized for the first-order functional calculus was published recently (3). In this paper the details will be carried out for the so-called first-order monadic functional calculus.

While the projected sequel will, in part at least, supersede some of the present discussion, the major part of this paper is an indispensable preliminary to that sequel. In order to be able to understand the algebraic versions of the intricate substitution processes that give the polyadic calculi their characteristic flavor, it is necessary first to understand the algebraic version of the logical operation of quantification. The latter is the subject matter of this paper, which, accordingly, could have been subtitled: "An algebraic study of quantification."

The theory of what will presently be called monadic (Boolean) algebras is discussed here not as a possible tool for solving problems about the foundations of mathematics, but as an independently interesting part of algebra. A knowledge of symbolic logic is unnecessary for an understanding of this theory; the language and the techniques used are those of modern algebra and topology.

It will be obvious to any reader who happens to be familiar with the recent literature of Boolean algebras that the results that follow lean heavily on the works of M. H. Stone and A. Tarski. Without the inspiration of Stone's representation theory and without Tarski's subsequent investigations of various Boolean algebras with operators, the subject of algebraic logic could not have come into existence. The present form of the paper was strongly influenced by some valuable suggestions of Mr. A. H. Kruse and Mr. B. A. Galler.

PART 1

Algebra

1. Functional monadic algebras. There is no novelty nowadays in the observation that propositions, whatever they may be, tend to band together and form a Boolean algebra. On the basis of this observation it is natural to interpret the expression "propositional function" to mean a function whose values are in a Boolean algebra. Accordingly, we begin our algebraic study of quantification with the consideration of a non-empty set X (the *domain*) and a Boolean algebra $\mathbf{B}$ (the *value-algebra*). The set $\mathbf{B}^X$ of all functions from X to $\mathbf{B}$ is itself a Boolean algebra with respect to the pointwise operations. Explicitly, if p and q are in $\mathbf{B}^X$, then the supremum $p \vee q$ and the complement p' are defined by

$$(p \vee q)(x) = p(x) \vee q(x) \text{ and } p'(x) = (p(x))'$$

for each x in X; the zero and the unit of $\mathbf{B}^X$ are the functions that are constantly equal to 0 and to 1, respectively.

The chief interest of $\mathbf{B}^X$ comes from the fact that it is more than just a Boolean algebra, What makes it more is the possibility of associating with each element p of $\mathbf{B}^X$ a subset $\mathbf{R}(p)$ of $\mathbf{B}$, where

$$\mathbf{R}(p) = \{p(x) : x \in X\}$$

is the range of the function p. With the set $\mathbf{R}(p)$, in turn, there are two obvious ways of associating an element of $\mathbf{B}$: we may try to form the supremum and the infimum of $\mathbf{R}(p)$. The trouble is that unless $\mathbf{B}$ is complete (in the usual lattice-theoretic sense), these extrema need not exist, and, from the point of view of the intended applications, the assumption that $\mathbf{B}$ is complete is much too restrictive. The remedy is to consider, instead of $\mathbf{B}^X$, a Boolean subalgebra $\mathbf{A}$ of $\mathbf{B}^X$ such that (i) for every p in $\mathbf{A}$ the supremum $\bigvee \mathbf{R}(p)$ and the infimum $\bigwedge \mathbf{R}(p)$ exist in $\mathbf{B}$, and (ii) the (constant) functions $\exists p$ and $\forall p$, defined by

$$\exists p(x) = \bigvee \mathbf{R}(p) \text{ and } \forall p(x) = \bigwedge \mathbf{R}(p)$$

belong to $\mathbf{A}$. Every such subalgebra $\mathbf{A}$ will be called a *functional monadic algebra*, or, to give it its full title, a $\mathbf{B}$-valued functional monadic algebra with domain X. The reason for the word "monadic" is that the concept of a monadic algebra (to be defined in appropriate generality below) is a special case of the concept of a polyadic algebra; the special case is characterized by the superimposition on the Boolean structure of exactly one additional operator.

A simple example of a functional monadic algebra is obtained by assuming that **B** is finite (or, more generally, complete), and letting $\mathbf{B}^X$ itself play the role of **A**. An equally simple example, in which **A** is again equal to $\mathbf{B}^X$, is obtained by assuming that X is finite. An example with **B** and X unrestricted is furnished by the set of all those functions from X to **B** that take on only a finite number of values.

If **B** happens to be the (complete) Boolean algebra of all subsets of a set Y, and if y is a point in Y, then a value $p(x)$ of a function p in **A** ($= \mathbf{B}^X$) corresponds in a natural way to the proposition "y belongs to $p(x)$." Since supremum in **B** is set-theoretic union, it follows that each value of $\exists p$ corresponds to "there is an x such that y belongs to $p(x)$," and, dually, each value of $\forall p$ corresponds to "for all x, y belongs to $p(x)$." For this reason, the operator $\exists$ on a functional monadic algebra is called a *functional existential quantifier*, and the operator $\forall$ is called a *functional universal quantifier*.

It is frequently helpful to visualize the example in the preceding paragraph geometrically. If X and Y are both equal to the real line, then **A** is naturally isomorphic to the algebra of all subsets of the Cartesian plane, via the isomorphism that assigns to each p in **A** the set $\{(x, y) : y \in p(x)\}$. The set that corresponds to $\exists p$ under this isomorphism is the union of all horizontal lines that pass through some point of the set corresponding to p; the set that corresponds to $\forall p$ is the union of all horizontal lines that are entirely included in the set corresponding to p.

In the definition of a functional monadic algebra it is not necessary to insist that for every p in **A** *both* $\exists p$ and $\forall p$ exist and belong to **A**: either one alone is sufficient. The reason for this is the validity of the identities

$$\forall p = (\exists p')' \text{ and } \exists p = (\forall p')'.$$

In more detail: if **A** is a Boolean subalgebra of $\mathbf{B}^X$ such that, for every p in **A**, the supremum $\bigvee \mathbf{R}(p)$ exists and the function

$\exists p$, whose value at every point is that supremum, belongs to **A**, then, for every p in **A**, the infimum $\bigwedge \mathbf{R}(p)$ also exists and the function $\forall p$, whose value at every point is that infimum, also belongs to **A**. The converse of this assertion is, in an obvious sense, its dual, and is also true. The perfect duality between $\exists$ and $\forall$ justifies the asymmetric treatment in what follows; we shall study $\exists$ alone and content ourselves with an occasional comment on the behavior of $\forall$.

The functional existential quantifier $\exists$ on a functional monadic algebra **A** is *normalized, increasing,* and *quasi-multiplicative.* In other words

$$(Q_1) \qquad \exists 0 = 0,$$

$$(Q_2) \qquad p \leqq \exists p,$$

$$(Q_3) \qquad \exists(p \wedge \exists q) = \exists p \wedge \exists q,$$

whenever p and q are in **A**. The assertions (Q_1) and (Q_2) are immediate consequences of the definition of $\exists$. The proof of (Q_3) is based on the following distributive law (true and easy to prove for every Boolean algebra **B**): if $\{p_i\}$ is a family of elements of **B** such that $\bigvee_i p_i$ exists, then, for every q in **B**, $\bigvee_i (p_i \wedge q)$ exists and is equal to $(\bigvee_i p_i) \wedge q$. The corresponding assertions for a functional universal quantifier are obtained from (Q_1)—(Q_3) upon replacing $\exists$, 0, $\leqq$, and $\wedge$ by $\forall$, 1, $\geqq$, and $\vee$, respectively.

2. Quantifiers. A general concept of quantification that applies to any Boolean algebra is obtained by abstraction from the functional case. In the process of abstraction the domain X and the value-algebra **B** disappear. What remains is the following definition: a *quantifier* (properly speaking, an *existential quantifier*) is a mapping $\exists$ of a Boolean algebra into itself, satisfying the conditions (Q_1)—(Q_3). The concept of an existential quantifier occurs implicitly in a brief announcement of some related work of Tarski and Thompson (9). The concept of a *universal quantifier* is defined by an obvious dualization, or, if preferred, via the equation $\forall p = (\exists p')'$. Since we have agreed to refer to universal quantifiers only tangentially, the adjective "existential" will usually be omitted.

The following examples show that the conditions (Q_1)—(Q_3)

are independent of each other. For (Q_1): **A** is arbitrary and $\exists p = 1$ for all p in **A**. For (Q_2): **A** is arbitrary and $\exists p = 0$ for all p in **A**. For (Q_3): **A** is the class of all subsets of a topological space that includes a non-closed open set and $\exists p$ is the closure of p for all p in **A**.

It is worth while to look at some quantifiers that are at least *prima facie* different from the functional examples of the preceding section. (i) The identity mapping of a Boolean algebra into itself is a quantifier; this quantifier will be called *discrete*. (ii) The mapping defined by $\exists 0 = 0$ and $\exists p = 1$ for all $p \neq 0$ is a quantifier; this quantifier will be called *simple*. (The reason for the terms "discrete", borrowed from topology, and "simple", borrowed from algebra, will become apparent later; cf. sections 3 and 5, respectively.) (iii) Suppose that **A** is the class of all subsets of some set and that G is a group of one-to-one transformations of that set onto itself. If $\exists p = \cup_{g \in G}\, gp$ for all p in **A**, then $\exists$ is a quantifier; $\exists p$ is the least set including p that is invariant under G. (Examples of this type are of some importance in ergodic theory.) (iv) Suppose that **A** is the algebra of all subsets of, say, the real line, modulo sets of Lebesgue measure zero. (Generalizations to other measure spaces are obvious.) If, for all p in **A**, $\exists p$ is the measurable cover of p (modulo sets of Lebesgue measure zero, of course), then $\exists$ is a quantifier.

To obtain insight into the algebraic properties of a quantifier, it is now advisable to derive certain elementary consequences of the definition. Several of these consequences are almost trivial and are stated formally for convenience of reference only. The important facts are that a quantifier is *idempotent* (Theorem 1) and *additive* (Theorem 2). Throughout the following statements it is assumed that **A** is a Boolean algebra and that $\exists$ is a quantifier on **A**.

LEMMA 1. $\exists 1 = 1$.

PROOF. Put $p = 1$ in (Q_2).

Theorem 1. $\exists\exists = \exists$.

PROOF. Put $p = 1$ in (Q_3) and apply Lemma 1.

LEMMA 2. *A necessary and sufficient condition that an element p of* **A** *belong to the range of $\exists$, i.e., that $p \in \exists(\mathbf{A})$, is that $\exists p = p$.*

PROOF. If $p \in \exists(\mathbf{A})$, say $p = \exists q$, then $\exists p = \exists\exists q = \exists q$ (by Theorem 1), so that, indeed, $\exists p = p$. This proves necessity;

sufficiency is trivial.

LEMMA 3. *If* $p \leqq \exists q$, *then* $\exists p \leqq \exists q$.

PROOF. By assumption $p \wedge \exists q = p$; it follows from (Q_3) that $\exists p = \exists(p \wedge \exists q) = \exists p \wedge \exists q$, so that, indeed, $\exists p \leqq \exists q$.

LEMMA 4. *A quantifier is monotone; i.e., if* $p \leqq q$, *then* $\exists p \leqq \exists q$.

PROOF. Note that $q \leqq \exists q$ by (Q_2) and apply Lemma 3.

LEMMA 5. $\exists(\exists p)' = (\exists p)'$.

PROOF. Since $(\exists p)' \wedge \exists p = 0$, it follows that

$$\begin{aligned} 0 &= \exists((\exists p)' \wedge \exists p) \qquad &&(\text{by } (Q_1)),\\ &= \exists(\exists p)' \wedge \exists p \qquad &&(\text{by } (Q_3)), \end{aligned}$$

and hence that $\exists(\exists p)' \leqq (\exists p)'$. The reverse inequality is immediate from (Q_2).

LEMMA 6. *The range* $\exists(\mathbf{A})$ *of the quantifier* $\exists$ *is a Boolean subalgebra of* $\mathbf{A}$.

PROOF. If p and q are in $\exists(\mathbf{A})$, then (by Lemma 2) $p = \exists p$ and $q = \exists q$, and consequently (by (Q_3)) $p \wedge q = \exists p \wedge \exists q = \exists(p \wedge \exists q)$. This proves that $\exists(\mathbf{A})$ is closed under the formation of infima. If $p \in \exists(\mathbf{A})$, then (again by Lemma 2) $p = \exists p$, and therefore (by Lemma 5) $p' = (\exists p)' = \exists(\exists p)'$. This proves that $\exists(\mathbf{A})$ is closed under the formation of complements.

Theorem 2. $\exists(p \vee q) = \exists p \vee \exists q$.

PROOF. Since $p \leqq p \vee q$ and $q \leqq p \vee q$, it follows from Lemma 4 that $\exists p \leqq \exists(p \vee q)$ and $\exists q \leqq \exists(p \vee q)$, and hence that $\exists p \vee \exists q \leqq \exists(p \vee q)$. To prove the reverse inequality, observe first that both $\exists p$ and $\exists q$ belong to $\exists(\mathbf{A})$ and that therefore (by Lemma 6) $\exists p \vee \exists q$ belongs to $\exists(\mathbf{A})$. It follows from Lemma 2 that $\exists(\exists p \vee \exists q) = \exists p \vee \exists q$. Since $p \leqq \exists p \vee \exists q$ (by (Q_2)), and, similarly, $q \leqq \exists p \vee \exists q$, so that $p \vee q \leqq \exists p \vee \exists q$, Lemma 4 implies that $\exists(p \vee q) \leqq \exists(\exists p \vee \exists q)$; this, together with what was just proved about $\exists(\exists p \vee \exists q)$, completes the proof of the theorem.

It is sometimes necessary to know the relation between quantification and relative complementation (where the relative complement of q in p is defined by $p - q = p \wedge q'$) and the relation between quantification and Boolean addition (where the Boolean sum, or symmetric difference, of p and q is defined by $p + q = (p - q) \vee (q - p)$). The result and its proof are simple.

LEMMA 7. $\exists p - \exists q \leqq \exists(p - q)$ and $\exists p + \exists q \leqq \exists(p + q)$.

PROOF. Since $p \vee q = (p - q) \vee q$, it follows (by Theorem 2) that $\exists p \vee \exists q = \exists(p - q) \vee \exists q$. Forming the infimum of both sides of this equation with $(\exists q)'$, we obtain

$$\exists p - \exists q = \exists(p - q) - \exists q \leqq \exists(p - q).$$

The result for Boolean addition follows from two applications of the result for relative complementation.

3. Closure operators. A *closure operator* is a normalized, increasing, idempotent, and additive mapping of a Boolean algebra into itself; in other words, it is an operator $\exists$ on a Boolean algebra $\mathbf{A}$, such that the conditions stated in (Q_1), (Q_2), Theorem 1, and Theorem 2 are satisfied. The first systematic investigation of the algebraic properties of closure operators was carried out by McKinsey and Tarski (5). A typical example of a closure operator is obtained by taking $\mathbf{A}$ to be the class of all subsets of a topological space and defining $\exists p$ to be the closure of p for every p in $\mathbf{A}$. Included among the results of the preceding section is the fact that every quantifier is a closure operator. In the converse direction, the only obvious thing that can be said is that the closure operator on a discrete topological space is a quantifier. It is, in fact, a discrete quantifier; this is the reason for the use of the word "discrete" in connection with quantifiers.

Despite the apparently promising connection between quantification and topology, it turns out that the topological point of view is almost completely valueless in the study of quantifiers. Not only is it false that every closure operator is a quantifier, but, in fact, the discrete (and therefore topologically uninteresting) closure operators are essentially the only ones that are quantifiers. The precise statement of the facts is as follows. The closure operator on a topological space is a quantifier if and only if, in that space, every open set is closed, or, equivalently, every closed set is open. (The proof is an easy application of Lemma 5.) In such a space the relation R, defined by writing $x \, R \, y$ whenever x belongs to the closure of the one-point set $\{y\}$, is an equivalence whose associated quotient space is discrete. Conversely, every space with this latter property has a quantifier for its closure operator. It follows that such spaces are as nearly discrete as a space not satisfying any separation axioms can ever be; in particular the T_1-spaces among them are discrete. Since these results border on pathology, and are of no importance for the theory of quantification, the details are omitted.

Nevertheless, closure operators play a useful role in quantifier algebra. The point is that it is frequently necessary to define a Boolean operator by certain algebraic constructions, and then to prove that the operator so constructed is a quantifier. It is usually easy to prove that the construction leads to a closure operator; the proof of quasi-multiplicativity, however, is likely to be more intricate. For this reason, it is desirable to have at hand a usable answer to the question: when is a closure operator a quantifier?

Theorem 3. *If* $\exists$ *is a closure operator on a Boolean algebra* **A**, *then the following conditions are mutually equivalent.*

(i) $\exists$ *is a quantifier.*

(ii) *The range of* $\exists$ *is a Boolean subalgebra of* **A**.

(iii) $\exists(\exists p)' = (\exists p)'$ *for all* p *in* **A**.

PROOF. The implication from (i) to (ii) is the statement of Lemma 6. To derive (iii) from (ii), note first (cf. Lemma 2) that $p \in \exists(\mathbf{A})$ if and only if $\exists p = p$. It follows that (iii) is equivalent to the assertion that $(\exists p)' \in \exists(\mathbf{A})$ for all p, and this in turn is an immediate consequence (via (ii)) of the fact that $\exists p \in \exists(\mathbf{A})$ for all p. It remains only to prove that if (iii) is satisfied, then $\exists$ is quasi-multiplicative.

Since $p \wedge \exists q \leqq p \leqq \exists p$, it follows that $\exists(p \wedge \exists q) \leqq \exists\exists p = \exists p$. (The reasoning here depends on the fact that an additive operator, and hence in particular a closure operator, is monotone; cf. Lemma 4.) Similarly, since $p \wedge \exists q \leqq \exists q$, it follows that $\exists(p \wedge \exists q) \leqq \exists q$ and hence that $\exists(p \wedge \exists q) \leqq \exists p \wedge \exists q$. To prove the reverse inequality, note that

$$p = (p \wedge \exists q) \vee (p \wedge (\exists q)') \leqq (p \wedge \exists q) \vee (\exists q)'$$

and that, therefore, $\exists p \leqq \exists(p \wedge \exists q) \vee (\exists q)'$. (This is where (iii) is used.) Forming the infimum of both sides of this relation with $\exists q$, we obtain

$$\exists p \wedge \exists q \leqq \exists(p \wedge \exists q) \wedge \exists q = \exists(p \wedge \exists q),$$

and the proof is complete.

4. Relative completeness. There are certain similarities between Boolean homomorphisms and quantifiers. A homomorphism is a mapping, satisfying certain algebraic conditions, from one Boolean algebra into another; a quantifier is a mapping, satisfying certain algebraic conditions, from a Boolean algebra into itself. A homomorphism uniquely determines a subset of its domain (namely, the kernel). The homomorphism theorem can be viewed as a characterization of kernels; it asserts that a subset of a Boolean algebra is the kernel of a homomorphism if and only if it is a proper ideal. Similarly, a quantifier uniquely determines a subset of its domain (namely, the range). The purpose of this section is to point out that the range uniquely determines the quantifier and to characterize the possible ranges of quantifiers.

A Boolean subalgebra **B** of a Boolean algebra **A** will be called *relatively complete* if, for every p in **A**, the set $\mathbf{B}(p)$, defined by

$$\mathbf{B}(p) = \{q \in \mathbf{B}\colon\ p \leqq q\},$$

has a least element (and therefore, *a fortiori*, an infimum). Relatively complete subalgebras are the objects whose relation to quantifiers is the same as the relation of proper ideals to homomorphisms.

Theorem 4. *If* $\exists$ *is a quantifier on a Boolean algebra* **A** *and if* **B** *is the range of* $\exists$, *then* **B** *is a relatively complete subalgebra of* **A**, *and, moreover, if* $\mathbf{B}(p) = \{q \in \mathbf{B}\colon\ p \leqq q\}$, *then* $\exists p = \bigwedge \mathbf{B}(p)$ *for every* p *in* **A**.

Proof. The fact that **B** is a Boolean subalgebra of **A** is already known from Lemma 6. If $q \in \mathbf{B}(p)$, then, of course, $p \leqq q$, and therefore $\exists p \leqq \exists q = q$. Since $\exists p \in \mathbf{B}(p)$, it follows that $\mathbf{B}(p)$ does indeed have a least element and that, moreover, that least element is equal to $\exists p$.

Theorem 5. *If* **B** *is a relatively complete subalgebra of a Boolean algebra* **A**, *then there exists a unique quantifier on* **A** *with range* **B**.

Proof. Write, for each p in **A**, $\exists p = \bigwedge \mathbf{B}(p)$; it is to be proved that $\exists$ is a quantifier on **A** and that $\exists(\mathbf{A}) = \mathbf{B}$.

(i) If $p = 0$, then $\mathbf{B}(p) = \mathbf{B}$ and therefore $\exists 0 = 0$.

(ii) Since $p \leqq q$ whenever $q \in \mathbf{B}(p)$, it follows that $p \leqq \bigwedge \mathbf{B}(p) = \exists p$.

(iii) If $p \in \mathbf{B}$, then $p \in \mathbf{B}(p)$ and therefore $\exists p = \bigwedge \mathbf{B}(p) \leqq p$. It follows from (ii) that (iiia) $\exists p = p$ whenever $p \in \mathbf{B}$. Since $\exists p \in \mathbf{B}$ for all p in **A**, this result in turn implies that (iiib) $\exists \exists p = \exists p$ whenever $p \in \mathbf{A}$.

(iv) If p_1 and p_2 are in **A**, then $\exists p_1$ and $\exists p_2$ are in **B** and therefore, since **B** is a Boolean algebra, $\exists p_1 \vee \exists p_2 \in \mathbf{B}$. Since, by (ii), $p_1 \vee p_2 \leqq \exists p_1 \vee \exists p_2$, so that $\exists p_1 \vee \exists p_2 \in \mathbf{B}(p_1 \vee p_2)$, it follows that $\exists(p_1 \vee p_2) = \bigwedge \mathbf{B}(p_1 \vee p_2) \leqq \exists p_1 \vee \exists p_2$. On the other hand, since $\exists(p_1 \vee p_2) \in \mathbf{B}(p_1 \vee p_2)$, it follows that $p_1 \vee p_2 \leqq \exists(p_1 \vee p_2)$ and hence that $p_1 \leqq \exists(p_1 \vee p_2)$ and $p_2 \leqq \exists(p_1 \vee p_2)$. The definition of $\exists$ implies that $\exists p_1 \leqq \exists(p_1 \vee p_2)$ and $\exists p_2 \leqq \exists(p_1 \vee p_2)$, and hence that $\exists p_1 \vee \exists p_2 \leqq \exists(p_1 \vee p_2)$.

(v) The range of $\exists$ is included in $\mathbf{B}$ by the definition of $\exists$. The result (iiia) implies the reverse inclusion, so that $\exists(\mathbf{A}) = \mathbf{B}$.

In (i)—(iv) we saw that $\exists$ is a closure operator. Since, by (v), the range of $\exists$ is a Boolean algebra, it follows from Theorem 3 that $\exists$ is a quantifier. The existence proof is complete; uniqueness is an immediate consequence of Theorem 4.

5. Monadic algebras. A *monadic algebra* is a Boolean algebra $\mathbf{A}$ together with a quantifier $\exists$ on $\mathbf{A}$. The elementary algebraic theory of monadic algebras is similar to that of every other algebraic system, and, consequently, it is rather a routine matter. Thus, for example, a subset $\mathbf{B}$ of a monadic algebra $\mathbf{A}$ is a *monadic subalgebra* of $\mathbf{A}$ if it is a Boolean subalgebra of $\mathbf{A}$ and if it is a monadic algebra with respect to the quantifier on $\mathbf{A}$. In other words, a Boolean subalgebra $\mathbf{B}$ of $\mathbf{A}$ is a monadic subalgebra of $\mathbf{A}$ if and only if $\exists p \in \mathbf{B}$ whenever $p \in \mathbf{B}$. The central concept is, as usual, that of a homomorphism; a *monadic homomorphism* is a mapping f from one monadic algebra into another, such that f is a Boolean homomorphism and $f\exists p = \exists f p$ for all p. Associated with every homomorphism f is its kernel $\{p: fp = 0\}$. The kernel of a monadic homomorphism is a *monadic ideal*; i.e., it is a Boolean ideal $\mathbf{I}$ in $\mathbf{A}$ such that $\exists p \in \mathbf{I}$ whenever $p \in \mathbf{I}$. The adjective "monadic" will be used with "subalgebra," "homomorphism," etc., whenever it is advisable to emphasize the distinction from other kinds of subalgebras, homomorphisms, etc. — e.g., from the plain Boolean kind. Usually, however, the adjective will be omitted and the context will unambiguously indicate what is meant.

The homomorphism theorem (every proper ideal is a kernel) and the consequent definition of monadic quotient algebras work as usual. If $\mathbf{A}$ is a monadic algebra and $\mathbf{I}$ is a monadic ideal in $\mathbf{A}$, form the Boolean quotient algebra $\mathbf{B} = \mathbf{A}/\mathbf{I}$, and consider the natural Boolean homomorphism f from $\mathbf{A}$ onto $\mathbf{B}$. There is a unique, natural way of converting $\mathbf{B}$ into a monadic algebra so that f becomes a monadic homomorphism (with kernel **I**, of course). Indeed, if p_1 and p_2 are in **A**, and if $fp_1 = fp_2$, then $f(p_1 + p_2) = 0$, or, equivalently, $p_1 + p_2 \in \mathbf{I}$. Since **I** is a monadic ideal, it follows that $\exists(p_1 + p_2) \in \mathbf{I}$. By Lemma 7, $\exists p_1 + \exists p_2 \in \mathbf{I}$, or, equivalently, $f\exists p_1 = f\exists p_2$. This conclusion justifies the following procedure: given q in **B**, find p in **A** so that $fp = q$, and define $\exists$ on q by $\exists q = f\exists p$. The preceding argument shows that the definition is unambiguous; a straightforward verification shows that $\exists$ is a quantifier on **B**.

A monadic algebra is *simple* if $\{0\}$ is the only proper ideal in it. A monadic ideal is *maximal* if it is a proper ideal that is not a proper subset of any other proper ideal. The connection between maximal ideals and simple algebras is an elementary part of universal algebra: the kernel of a homomorphism is a maximal ideal if and only if its range is a simple algebra.

LEMMA 8. *A monadic algebra is simple if and only if its quantifier is simple.*

PROOF. If **A** is simple and if $p \in \mathbf{A}$, $p \neq 0$, write $\mathbf{I} = \{q : q \leqq \exists p\}$. Since, clearly, **I** is a non-trivial monadic ideal, it follows that $\mathbf{I} = \mathbf{A}$, and hence, in particular, that $1 \in \mathbf{I}$. This implies that $\exists p = 1$ whenever $p \neq 0$. Suppose, conversely, that $\exists p = 1$ whenever $p \neq 0$, and suppose that **I** is a monadic ideal in **A**. If $p \in \mathbf{I}$, then $\exists p \in \mathbf{I}$; if, moreover, $p \neq 0$, this implies that $1 \in \mathbf{I}$ and hence that $\mathbf{I} = \mathbf{A}$. In other words, every non-trivial ideal in **A** is improper; this proves that **A** is simple.

The only simple *Boolean* algebra is the two-element algebra, to be designated throughout the sequel as **O**. This Boolean algebra is a subalgebra, and, what is more, a relatively complete subalgebra, of every Boolean algebra. Lemma 8 asserts that a monadic algebra is simple if and only if the relatively complete subalgebra associated with its quantifier is equal to **O**, or, in other words, if and only if the range of its quantifier is a simple Boolean algebra.

The connection between simple Boolean algebras and simple monadic algebras is even closer than that indicated in the preceding paragraph; it turns out that the simplest examples of monadic algebras (in both the popular and the technical sense of "simple") are the **O**-valued functional algebras.

Theorem 6. *A monadic algebra is simple if and only if it is (isomorphic to) an* **O**-*valued functional monadic algebra.*

PROOF. If **A** is an **O**-valued functional monadic algebra with domain X, and if p is a non-zero element of A, then $p(x_0) = 1$ for some point x_0 in X. It follows that $1 \in \mathbf{R}(p)$ and hence that $\bigvee \mathbf{R}(p) = 1$. The definition of functional quantification implies that $\exists p = 1$. Since this proves that $\exists p = 1$ whenever $p \neq 0$, i.e., that $\exists$ is simple, the desired result follows from Lemma 8.

The converse is just as easy to prove, but the proof makes use of a relatively deep fact, namely Stone's theorem on the representation of Boolean algebras (7).

If **A** is a simple monadic algebra, then **A** is, in particular, a Boolean algebra, to which Stone's theorem is applicable. It follows that there exist (i) a set X, (ii) a Boolean subalgebra **B** of $\mathbf{O}^X$, and (iii) a Boolean isomorphism f from **A** onto **B**. Since, by Lemma 8, the quantifier of **A** is simple, and since, by Lemma 8 and the first part of this proof, the quantifier of **B** is simple, it follows that f preserves quantification, i.e., that f is automatically a monadic isomorphism between the monadic algebras **A** and **B**.

6. Monadic logics. In the usual logical treatment of Boolean algebras and their generalizations, certain elements of the appropriate Boolean algebra are singled out and called "provable". From the algebraic point of view, the definition of provability in any particular case is irrelevant; what is important is the algebraic structure of the set of all provable elements. It is convenient, in the examination of that structure, to dualize, i.e., to consider not provability but refutability. There is an obvious relation between the two concepts; clearly p should be called refutable if and only if p' is provable.

Suppose accordingly that **A** is a monadic algebra whose elements, for heuristic purposes, are thought of as propositions, or, rather, as propositional functions. What properties does it seem reasonable to demand of a subset **I** of **A** in order that its elements deserve to be called refutable? Clearly if p and q are refutable, then $p \vee q$ should also be refutable, and if p is refutable, then $p \wedge q$ should be refutable no matter what q may be. In other words, **I** should be, at least, a Boolean ideal in **A**. That is not enough, however;

I should also bear the proper relation to quantification. If, in other words, p is refutable (and here it is essential that p be thought of as a propositional function, and not merely as a proposition), then $\exists p$ should also be refutable. The requirement (satisfied by the set of refutable elements of the usual logical algebras) converts **I** into a monadic ideal.

The following definition is now adequately motivated: a *monadic logic* is a pair (**A**, **I**), where **A** is a monadic algebra and **I** is a monadic ideal in **A**. The elements p of **I** are the *refutable* elements of the logic; if $p' \in \mathbf{I}$, then p is called *provable.*

For monadic logics, as for most other mathematical systems, representation theory plays an important role. Representation theory proceeds, as always, by selecting a class of particularly simple and "concrete" monadic logics, and asking to what extent every monadic logic is representable by means of logics of that class. For intuitively obvious reasons there is universal agreement on which logics should be called "concrete". The technical term for a concrete monadic logic is *model*; a model is, by definition, a monadic logic (**A**, **I**), where **A** is an **O**-valued functional monadic algebra and **I** is the trivial ideal $\{0\}$. Note that since an **O**-valued functional monadic algebra is simple (Theorem 6), **I** could only be $\{0\}$ or **A**, and the latter choice is obviously uninteresting.

An *interpretation* of a monadic logic (**A**, **I**) in a model (**B**, $\{0\}$) is a monadic homomorphism f from **A** into **B** such that $fp = 0$ whenever $p \in \mathbf{I}$. A convenient way of expressing the condition on the homomorphism is to say that every refutable element is *false* in the interpretation. If, in other words, an element p of **A** is called *universally invalid* whenever it is false in every interpretation, then, by definition, every refutable element is universally invalid. There could conceivably be elements in **A** that are not refutable but that are nevertheless universally invalid. If there are no such elements, i.e., if every universally invalid element is refutable, the logic is said to be *semantically complete.* This definition sounds a little more palatable in its dual form: a logic is semantically complete if every universally valid element is provable. (The element p is *universally valid* if $fp = 1$ for every interpretation f, i.e., if p is *true* in every interpretation.) Elliptically but suggestively, semantic completeness can be described by saying that everything true is provable.

7. Semisimplicity. Semantic completeness demands of a logic (**A**, **I**) that the ideal **I** be relatively large. If, in particular, **I** is very large, i.e., **I** $=$ **A**, then the logic is semantically complete, simply because every element of **A** is refutable. (The fact that in this case there are no interpretations is immaterial.) If **I** $\neq$ **A**, then the quotient algebra **A**/**I** may be formed, and the problem of deciding whether or not the *logic* (**A**, **I**) is semantically complete reduces to a question about the *algebra* **A**/**I**.

Since every interpretation of (**A**, **I**) in a model (**B**, $\{0\}$) induces in a natural way a homomorphism from **A**/**I** into **B**, and since (by Theorem 6) the only restriction on **B** is that it be simple, the question becomes the following one. Under what conditions on a monadic algebra **A** is it true that whenever an element p of **A** is mapped on 0 by every homomorphism from **A** into a simple algebra, then $p = 0$? Since every monadic subalgebra of an **O**-valued functional monadic algebra is an algebra of the same kind, it follows from Theorem 6 that every monadic subalgebra of a simple monadic algebra is also simple, and, consequently, that the difference between "into" and "onto" is not essential here. Because of the correspondence between homomorphisms with simple ranges and maximal ideals, the question could also be put this way: under what conditions on a monadic algebra **A** is it true that whenever an element p of **A** belongs to all maximal ideals, then $p = 0$? In analogy with other parts of algebra, it is natural to say that a monadic algebra **A** is *semisimple* if the intersection of all maximal ideals in **A** is $\{0\}$. The question now becomes: which monadic algebras are semisimple? The answer is quite satisfying.

Theorem 7. *Every monadic algebra is semisimple.*

Remark. Since monadic algebras constitute a generalization of Boolean algebras, Theorem 7 asserts, in particular, that every Boolean algebra is semisimple. This consequence of Theorem 7 is well known: it is an immediate consequence of Stone's representation theorem, and it is often presented as the most important step in the proof of that theorem. The proof of the present generalization can be carried out by a monadic imitation of any one of the usual proofs of its Boolean special case. The proof below adopts the alternative procedure of deducing the generalization from the special case.

PROOF. It is to be proved that if **A** is a monadic algebra and if p_0 is a non-zero element of **A**, then there exists a monadic maximal ideal **I** in **A** such that $p_0 \epsilon' \mathbf{I}$. It follows from the known Boolean version of the theorem that there exists a Boolean maximal ideal $\mathbf{I}_0$ in **A** such that $p_0 \epsilon' \mathbf{I}_0$. If **I** is the set of all those elements p in **A** for which $\exists p \in \mathbf{I}_0$, then it is trivial to verify that **I** is a monadic ideal and that $p_0 \epsilon' \mathbf{I}$. The proof of semisimplicity can be completed by showing that **I** is maximal. Suppose therefore that **J** is a monadic ideal properly including **I**. It follows that **J** contains an element p such that $\exists p \epsilon' \mathbf{I}_0$. Since **J** is a monadic ideal, $\exists p \in \mathbf{J}$. Since $\exists p \epsilon' \mathbf{I}_0$, and since $\mathbf{I}_0$ is a Boolean maximal ideal, $(\exists p)' \in \mathbf{I} \subset \mathbf{J}$. The last two sentences together imply that $1 \in \mathbf{J}$, so that $\mathbf{J} = \mathbf{A}$; the proof is complete.

It should be remarked that there are natural examples of Boolean algebras with operators (4), easy generalizations of monadic algebras, that are not semisimple. Thus, for instance, the semisimplicity of the closure algebra of a topological space appears to depend on which separation axioms the space satisfies.

PART 2

Topology

8. Hemimorphisms and Boolean relations. The algebraic theory of Boolean algebras is better understood if their topological theory is taken into account; the same is true of monadic algebras. In what follows we shall therefore make use of the topological version of Stone's theorem, in the following form: there is a one-to-one correspondence between Boolean algebras **A** and Boolean spaces X such that each algebra **A** is isomorphic to the algebra of all clopen subsets of the corresponding space X (8). Explanation of terms: a *clopen* set in a topological space is a set that is simultaneously closed and open, and a *Boolean space* is a totally disconnected compact Hausdorff space, i.e., a compact Hausdorff space in which the clopen sets form a base. The one-to-one nature of the correspondence must of course be interpreted with the usual algebraic-topological grain of salt: X uniquely determines **A** to within an isomorphism, and **A** uniquely determines X to within a homeomorphism. The algebra **A** corresponding to a space X will be called the *dual algebra* of X, and the space X corresponding to an algebra **A** will be called the *dual space* of **A**.

There is a natural isomorphism between the dual algebra **A** of a Boolean space X and the set of all continuous functions from X into **O**. In order to interpret this assertion, we must, of course, endow **O** with a topology; this is done, once and for all, by declaring open every one of the four subsets of **O**, so that **O** itself becomes a (discrete) Boolean space. The isomorphism mentioned just above assigns to each clopen subset of X its characteristic function. It is algebraically convenient to identify **A** with the algebra of all continuous functions from X into **O**. In view of this identification, every Boolean algebra that occurs below is to be regarded as identical with the algebra of all **O**-valued continuous functions on its dual space. Thus, for example, if **A** is a Boolean algebra with dual space X, and if $p \in \mathbf{A}$ and $x \in X$, then $p(x)$ makes sense; it has the value 1 or 0 according as x belongs or does not belong to that clopen subset of X which corresponds to the element p of **A**.

The duality theory of monadic algebras is conveniently studied at a slightly more general level than might appear relevant at first sight. The point is that it is possible to treat simultaneously both the old theory of homomorphisms and the newer theory of quantifiers. The appropriate general concept is that of a *hemimorphism*, defined as a mapping f from a Boolean algebra **A** into a Boolean algebra **B**, such that $f0 = 0$ and $f(p \vee q) = fp \vee fq$ for all p and q in **A**. The reason for the name is that, roughly speaking, a hemimorphism preserves half the structure of a Boolean algebra. Hemimorphisms occur, under the name of "normal and additive functions", in the work of Jónsson and Tarski (4). Two elementary consequences of the definition are that a hemimorphism is *monotone* ($fp \leqq fq$ whenever $p \leqq q$) and *submultiplicative* ($f(p \wedge q) \leqq fp \wedge fq$). It is clear that every homomorphism (from a Boolean algebra **A** to a Boolean algebra **B**) and every quantifier (from a Boolean algebra **A** to itself) is a hemimorphism.

The proper topological concept is that of a Boolean relation. A *relation* between two sets Y and X (or, more accurately, from Y to X) is, as always, a subset φ of the Cartesian product $Y \times X$; the assertion $(y, x) \in \varphi$ is conveniently abbreviated as $y\varphi x$. If Q is a subset of Y, the *direct image* of Q under φ, in symbols φQ, is the set of all those points x in X for which there exists a point

y in Q such that $y\varphi x$. If φ^{-1} denotes the *inverse* of the relation φ, i.e., φ^{-1} is the set of all those pairs (x, y) in $X \times Y$ for which $y\varphi x$, then the *inverse image* of a subset P of X under φ, in symbols $\varphi^{-1}P$, is the direct image of P under φ^{-1}, or, equivalently, it is the set of all those points y in Y for which there exists a point x in P such that $y\varphi x$. If $y \in Y$ and $\{y\}$ is the set whose only element is y, then $\varphi\{y\}$ is also denoted by φy; similarly if $x \in X$, then $\varphi^{-1}\{x\}$ is also denoted by $\varphi^{-1}x$. These concepts are well known; they are explicitly mentioned here only in order to establish the notation. Several other related concepts (e.g., *equivalence relation* and *relation product*) will be used below without explicit definition.

A *Boolean relation* is a relation φ from a Boolean space Y to a Boolean space X, such that the inverse image of every clopen set in X is a clopen set in Y and such that the direct image of every point in Y is a closed set in X. It is easy to verify that a function from a Boolean space Y to a Boolean space X is a Boolean relation if and only if it is continuous. It is also pertinent to remark that the two conditions in the definition of a Boolean relation are independent of each other. Indeed, if X is a Boolean space with a single cluster point $\bar{x}$, if $Y = X$, and if $\varphi = (Y \times X) - \{(\bar{x}, \bar{x})\}$, then $\varphi^{-1}P = Y$ for every non-empty clopen subset P of X, so that the inverse image of every clopen set is clopen, but $\varphi\bar{x} = X - \{\bar{x}\}$, so that the direct image of a point is not always closed. Any discontinuous function from a Boolean space Y to a Boolean space X furnishes an example of a relation for which the direct image of every point is closed, but the inverse image of a clopen set is not always clopen.

9. Duality. Suppose that **A** and **B** are Boolean algebras, with respective dual spaces X and Y. If f is a hemimorphism from **A** into **B**, its dual, denoted by f^*, is the relation from Y to X defined by

$$f^* = \bigcap_{p \in \mathbf{A}}\{(y, x)\colon\ p(x) \leqq fp(y)\};$$

in other words, yf^*x if and only if $p(x) \leqq fp(y)$ for all p in **A**. (Symbols such as $fp(y)$ will be used frequently in what follows. They are to be interpreted, in every case, by first performing all the indicated functional operations and then evaluating the resulting function at the indicated point. Thus, explicitly, $fp(y) = (fp)(y)$.)

If φ is a Boolean relation from Y to X, its dual, denoted by φ^*, is the mapping that assigns to every element p of $\mathbf{A}$ a function $\varphi^* p$ from Y to $\mathbf{O}$, defined by

$$\varphi^* p(y) = \bigvee \{p(x) \colon y\varphi x\}.$$

The following theorem is the principal result of the theory of Boolean duality.

Theorem 8. *If f is a hemimorphism, then f^* is a Boolean relation, and $f^{**} = f$. If φ is a Boolean relation, then φ^* is a hemimorphism and $\varphi^{**} = \varphi$. If f and φ are each other's duals, then*

$$(*) \qquad \{y \colon fp(y) = 1\} = \varphi^{-1}\{x \colon p(x) = 1\}$$

for every p in $\mathbf{A}$.

Remark. A similar but weaker theorem has been published by Jónsson and Tarski (4), and the same comment is true for some of the results obtained in sections 10 and 11. The essential difference between the present approach and that of Jónsson and Tarski can be described as follows: since they do not have the concept of a Boolean relation, they are unable to state which relations between Boolean spaces can occur as the duals of hemimorphisms.

Proof. Since

$$f^* y = \bigcap_{p \in \mathbf{A}} \{x \colon p(x) \leqq fp(y)\},$$

so that $f^* y$ is obviously closed, in order to prove that f^* is a Boolean relation, it is sufficient to prove that $(*)$ holds with f^* in place of φ. Given p_0 in $\mathbf{A}$, write

$$P_0 = \{x \colon p_0(x) = 1\} \text{ and } Q_0 = \{y \colon fp_0(y) = 1\};$$

it is to be proved that $f^{*-1}P_0 = Q_0$. If $p_0 = 0$, then P_0 is empty; since, by the definition of a hemimorphism, $f0 = 0$, it follows that Q_0 is empty. In what follows, it is therefore permissible to assume that $p_0 \neq 0$ and hence that P_0 is not empty.

If $y \in f^{*-1}P_0$, then there exists a point x in P_0 such that yf^*x, i.e., such that $p(x) \leqq fp(y)$ for all p in $\mathbf{A}$. It follows in particular, with $p = p_0$, that $1 = p_0(x) \leqq fp_0(y)$, so that $y \in Q_0$. This proves that $f^{*-1}P_0 \subset Q_0$; the reverse inclusion lies slightly deeper.

It is to be proved that if $y \in' f^{*-1}P_0$, then $fp_0(y) = 0$. To say that $y \in' f^{*-1}P_0$ means that the assertion yf^*x is false for every

x in P_0. This, in turn, means that to every x in P_0 there corresponds an element p_x of $\mathbf{A}$ such that the assertion $p_x(x) \leqq fp_x(y)$ is false, i.e., such that $p_x(x) = 1$ and $fp_x(y) = 0$. Since

$$P_0 \subset \cup_{x \in P_0}\{z\colon\ p_x(z) = 1\},$$

the fact that P_0 is closed (and therefore compact), together with the fact that each set $\{z\colon\ p_x(z) = 1\}$ is open, implies that there exists a finite subset $\{x_1, \ldots, x_n\}$ of P_0 such that

$$P_0 \subset \cup_{j=1}^{n}\{z\colon\ p_{x_j}(z) = 1\}.$$

The assumption that P_0 is not empty is reflected here in the fact that $n \neq 0$, i.e., that the finite set is not empty.

Write $\bar{p} = \vee_{j=1}^{n} p_{x_j}$. If $z \in P_0$, then $p_{x_j}(z) = 1$ for at least one j, and, therefore, $\bar{p}(z) = 1$; it follows that $p_0 \leqq \bar{p}$. Since f is a hemimorphism, $f\bar{p} = \vee_{j=1}^{n} fp_{x_j}$; since $fp_x(y) = 0$ for all x in P_0, it follows that $f\bar{p}(y) = 0$. Since a hemimorphism is monotone, it follows, finally, that $f\bar{p}_0(y) = 0$; this completes the proof that f^* satisfies (*) and is, consequently, a Boolean relation.

It still remains to be shown that $f^{**} = f$. Since yf^*x implies that $p(x) \leqq fp(y)$ for all p, the inequality $\vee\{p(x)\colon\ yf^*x\} \leqq fp(y)$ is obvious. The reverse inequality amounts to the assertion that if $p(x) = 0$ whenever yf^*x, then $fp(y) = 0$. But to say that $p(x) = 0$ whenever yf^*x is equivalent to saying that y does not belong to $f^{*-1}\{x\colon\ p(x) = 1\}$. It follows from (*) that $\vee\{p(x)\colon\ yf^*x\} = fp(y)$, and hence, by the definition of the dual of a Boolean relation, that $f^{**} = f$.

Suppose now that φ is a Boolean relation from Y to X. It follows from the definition of φ^* that $\varphi^*p(y) = 1$ if and only if there is a point x such that $p(x) = 1$ and such that $y\varphi x$; in other words, (*) holds with φ^* in place of f. Since the inverse image under φ of a clopen set is clopen, it follows that φ^* maps $\mathbf{A}$ into $\mathbf{B}$. The verification that φ^* is a hemimorphism is a matter of trivial routine.

It remains only to verify that $\varphi^{**} = \varphi$. If $y\varphi x$, then, for every p in $\mathbf{A}$, $p(x)$ is one of the terms whose supremum is $\varphi^*p(y)$, and, consequently, $p(x) \leqq \varphi^*p(y)$. On the other hand, if the assertion $y\varphi x$ is false, then (since φy is closed) the Boolean nature of X implies that there exists an element p_0 in $\mathbf{A}$ such that $p_0(x) = 1$ and such that $p_0(z) = 0$ whenever $y\varphi z$. It follows that $\varphi^*p_0(y) = 0$

and hence that the assertion that $p(x) \leqq \varphi^* p(y)$ for all p is also false. Conclusion: $y\varphi x$ if and only if $p(x) \leqq \varphi^* p(y)$ for all p, and hence, by the definition of the dual of a hemimorphism, $\varphi^{**} = \varphi$. The proof of Theorem 8 is complete.

COROLLARY. *Under a Boolean relation, the inverse image of every point is closed.*

PROOF. If φ is a Boolean relation and $f = \varphi^*$, then

$$\varphi^{-1}x = \cap_{p \in \mathbf{A}} \{y\colon p(x) \leqq fp(y)\}$$

for every x in X.

10. The dual of a homomorphism. In view of the duality theorem, it is possible, in principle, to translate every algebraic property of hemimorphisms into a topological property of Boolean relations, and vice versa. The purpose of this section is to carry out the translation for some of the properties of importance in the theory of monadic algebras.

LEMMA 9. *If X and Y are Boolean spaces and if φ and ψ are Boolean relations from Y to X such that $\varphi^{-1}P = \psi^{-1}P$ for every clopen subset P of X, then $\varphi = \psi$.*

PROOF. If x and y are such that $y\varphi x$ and if P is a clopen subset of X such that $x \in P$, then $y \in \varphi^{-1}P$ and therefore, by assumption, $y \in \psi^{-1}P$. This implies that $P \cap \psi y$ is not empty. In other words: every neighborhood of x meets the set ψy. Since ψy is closed, it follows that $x \in \psi y$, i.e., that $y\psi x$. This proves that $\varphi \subset \psi$; the reverse inclusion follows by symmetry.

Theorem 9. *If* **A**, **B**, *and* **C** *are Boolean algebras, with respective dual spaces X, Y, and Z, and if f and g are hemimorphisms from* **A** *into* **B** *and from* **B** *into* **C** *respectively, then $(gf)^* = f^*g^*$.*

PROOF. An application of (*) (section 9) to gf shows that

$$(gf)^{*-1}\{x\colon p(x) = 1\} = \{z\colon gfp(z) = 1\}$$

for all p in **A**. An application of (*) to g shows that

$$g^{*-1}\{y\colon q(y) = 1\} = \{z\colon gq(z) = 1\}$$

for all q in **B**; replacing q by fp and applying (*) to f, we deduce that

$$g^{*-1}f^{*-1}\{x\colon p(x) = 1\} = \{z\colon gfp(z) = 1\}$$

for all p in **A**. Together the two equations involving gfp imply that

$$(gf)^{*-1}P = g^{*-1}f^{*-1}P$$

for every clopen subset P of X. Since it is an elementary fact about relations that $g^{*-1}f^{*-1} = (f^*g^*)^{-1}$, the desired result follows[1] from Lemma 9.

LEMMA 10. *If a hemimorphism f (from* **A** *to* **B**) *and a Boolean relation φ (from Y to X) are each other's duals, then a necessary and sufficient condition that f be multiplicative ($f(p \wedge q) = fp \wedge fq$) is that φ be a function.*

PROOF. If φ is a function, then $y\varphi x$ means that $\varphi y = x$ and therefore

$$fp(y) = \bigvee \{p(x)\colon\ y\varphi x\} = p(\varphi y).$$

The multiplicativity of f is the result of a straightforward computation. If φ is not a function, then there exists a point y in Y and there exist distinct points x_0 and x_1 in X so that $y\varphi x_0$ and $y\varphi x_1$. If p is an element of **A** such that $p(x_0) = 0$ and $p(x_1) = 1$, then (since $p'(x_0) = 1$) $fp'(y) = fp(y) = 1$ and therefore $fp' \wedge fp \neq 0$. Since $f(p \wedge p') = f0 = 0$, it follows that f is not multiplicative.

In the following two lemmas, as in Lemma 10, f is a hemimorphism from **A** to **B** and φ is the corresponding Boolean relation from Y to X.

LEMMA 11. *A necessary and sufficient condition that $f1 = 1$ is that φ have Y for its domain.*

PROOF. It follows from (*), with $p = 1$, that

$$\{y\colon\ f1(y) = 1\} = \varphi^{-1}X,$$

and hence that $f1 = 1$ if and only if $\varphi^{-1}X = Y$.

LEMMA 12. *A necessary and sufficient condition that f be a homomorphism ($fp' = (fp)'$) is that φ be a function with domain Y.*

PROOF. It is a well-known and easily proved fact about Boolean homomorphisms that a hemimorphism f is a homomorphism if and only if it is multiplicative and satisfies the equation $f1 = 1$. The desired conclusion now follows from Lemmas 10 and 11.

[1] See (a) on p. 165.

At this point the generalized version of Boolean duality makes contact with the standard version. The duality between homomorphisms f (from $\mathbf{A}$ to $\mathbf{B}$) and continuous functions φ (from Y to X) is known. It is known also that f is a monomorphism (i.e., an isomorphism into) if and only if φ maps Y onto X, and f is an epimorphism (i.e., a homomorphism onto) if and only if φ is one-to-one (8); these facts will be used below.

11. The dual of a quantifier. Suppose, throughout this section, that $\mathbf{A}$ is a Boolean algebra, with dual X, and that f is a hemimorphism from $\mathbf{A}$ into itself, with dual φ. A useful auxiliary concept in this case is the relation $\hat{\varphi}$ in X; by definition

$$\hat{\varphi} = \cap_{p \in \mathbf{A}} \{(y, x) \colon fp(x) = fp(y)\}.$$

In other words, $y\hat{\varphi}x$ if and only if $fp(x) = fp(y)$ for all p in $\mathbf{A}$. It is obvious that the relation $\hat{\varphi}$ is an equivalence with domain X (and, therefore, with range X).

LEMMA 13. *A necessary and sufficient condition that f be a quantifier is that $\varphi = \hat{\varphi}$.*

PROOF. If f is a quantifier, then, of course, f is increasing. It follows that if $y\hat{\varphi}x$, then $p(x) \leqq fp(x) = fp(y)$ for all p in $\mathbf{A}$, so that $y\varphi x$; in other words, $\hat{\varphi} \subset \varphi$. To prove the reverse inclusion, suppose that $y\varphi x$, i.e., that $p(x) \leqq fp(y)$ for all p in $\mathbf{A}$. This inequality, applied first to fp and then to $(fp)'$ in place of p, yields

$$\text{(i)} \qquad fp(x) \leqq ffp(y) = fp(y)$$

(cf. Theorem 1), and

$$\text{(ii)} \qquad (fp)'(x) \leqq f(fp)'(y) = (fp)'(y)$$

(cf. Lemma 5). Since (ii) is equivalent to the reverse of (i), it follows that $fp(x) = fp(y)$ for all p in $\mathbf{A}$, and hence that $y\hat{\varphi}x$. This means that $\varphi \subset \hat{\varphi}$ and therefore completes the proof of the necessity of the condition.

For sufficiency, it is to be proved that if $\varphi = \hat{\varphi}$, then f is increasing and quasi-multiplicative. The assertion that f is increasing means that $p(x) \leqq fp(x)$ whenever $p \in \mathbf{A}$ and $x \in X$, and, consequently, it is equivalent to the assertion that φ is reflexive. If $\varphi = \hat{\varphi}$, or, more generally, if φ is known to be an equivalence with domain X, then φ is reflexive and therefore f is increasing. (This comment will be used again in the proof of Lemma 14 below.) To prove that f is quasi-multiplicative, note first that

$$f(p \wedge fq)(y) = \bigvee\{(p \wedge fq)(x)\colon y\varphi x\} = \bigvee\{p(x) \wedge fq(x)\colon y\varphi x\}.$$

Since $\varphi = \hat{\varphi}$, so that, in particular, $\varphi \subset \hat{\varphi}$, the condition $y\varphi x$ implies that $fq(x) = fq(y)$. It follows that

$$f(p \wedge fq)(y) = \bigvee\{p(x) \wedge fq(y)\colon y\varphi x\} = \bigvee\{p(x)\colon y\varphi x\} \wedge fq(y) = \\ = fp(y) \wedge fq(y);$$

this completes the proof of sufficiency.

LEMMA 14. *A necessary and sufficient condition that $\varphi = \hat{\varphi}$ is that φ be an equivalence with domain X.*

PROOF. Necessity is trivial. To prove sufficiency, assume that φ is an equivalence with domain X. It was already remarked that this implies that f is increasing.

It follows, exactly as in the corresponding part of the proof of Lemma 13, that $\hat{\varphi} \subset \varphi$. To prove the reverse inclusion, assume that $y\ \varphi\ x$. Since

$$fp(x) = \bigvee\{p(z)\colon x\varphi z\} \text{ and } fp(y) = \bigvee\{p(z)\colon y\varphi z\},$$

and since $x\varphi z$ is equivalent to $y\varphi z$, it follows that $fp(x) = fp(y)$ for all p, i.e., that $\varphi \subset \hat{\varphi}$.

Theorem 10. *A necessary and sufficient condition that f be a quantifier is that φ be an equivalence with domain X.*

PROOF. Obvious from Lemmas 13 and 14.

The two extreme quantifiers are sometimes of interest.

LEMMA 15. *If $\exists$ is the discrete quantifier on* **A**, *then $\exists^*$ is the identity (i.e., $y\exists^* x$ if and only if $y = x$); if $\exists$ is the simple quantifier on* **A**, *then $\exists^* = X \times X$ (i.e., $y\exists^* x$ for all x and y).*

PROOF. If $\exists$ is discrete, then, by definition, $\exists p = p$ for all p. It follows that $y\exists^* x$ if and only if $p(x) = p(y)$ for all p (cf. Lemma 13), and hence if and only if $x = y$. (Observe also that the discrete quantifier on **A** coincides with the identity homomorphism from **A** onto itself; it follows from the duality theory of homomorphisms that its dual is the identity mapping from X onto itself.) If $\exists$ is simple, then $\exists p = 1$ whenever $p \neq 0$; it follows *a fortiori* that $p(x) \leqq \exists p(y)$ whenever $p \neq 0$ (for all x and y). Since the inequality is still true when $p = 0$, the proof is complete.

PART 3

Representation

12. Boolean mappings. Just as the ordinary duality theory shows that the study of Boolean algebras is equivalent to the study of Boolean spaces, the duality theory of quantifiers shows that the study of monadic algebras is equivalent to the study of Boolean spaces equipped with a Boolean equivalence relation. An equivalence relation φ in a topological space X determines a quotient space X/φ; the first purpose of the present section is to use this fact in order to describe a topological version of monadic algebra in terms of objects that are more familiar than Boolean equivalence relations. In the approach it is convenient not to use the theory of quotient spaoes but to use, instead, methods more directly relevant to Boolean theory. The appropriate concept is that of a *Boolean mapping*, defined as a continuous and open mapping from one Boolean space onto another. The principal result is that the study of quantifiers is equivalent to the study of Boolean mappings.

LEMMA 16. *If* $\exists$ *is a quantifier on a Boolean algebra* **A**, *then there exists a Boolean mapping* π, *from the dual space* X *of* **A** *to a Boolean space* Y, *such that* $x_1 \exists^* x_2$ *is equivalent to* $\pi x_1 = \pi x_2$.

PROOF. Let **B** be the range of $\exists$, so that **B** is a Boolean subalgebra of **A**, and let Y be the dual space of **B**. Despite the identification convention of section 8, it is not permissible to identify **B** with the algebra of all continuous functions from Y to **O**; the trouble is that, in view of that identification convention, **B** is already concretely given as an algebra of continuous functions from X to **O**. All that it is possible to say is that there is an isomorphism, say f, from **B** onto the algebra of all continuous functions from Y to **O**, and, fortunately, this is all that is needed.

The identity mapping from **B** into **A** is a homomorphism; its dual is a continuous function π from X onto Y. The assertion that π is the dual of the embedding of **B** into **A** means that the result of evaluating a function q in **B** at a point x in X is the same as the value of the image function fq on Y at the point πx of Y; in other words,

$$q(x) = fq(\pi x)$$

for all q and all x.

In view of Lemma 13, the condition $x_1 \exists^* x_2$ is equivalent to the validity of $\exists p(x_1) = \exists p(x_2)$ for all p in **A**; since $\exists$ maps **A** onto **B**, this, in turn, is equivalent to the validity of $q(x_1) = q(x_2)$ for all q in **B**. It follows from the definition of π that $x_1 \exists^* x_2$ if and only if $fq(\pi x_1) = fq(\pi x_2)$ for all q in **B**; since, finally, f is an isomorphism, the condition is equivalent to $\pi x_1 = \pi x_2$.

One consequence of the preceding paragraph is that $\exists^* Q = \pi^{-1}\pi Q$ for every subset Q of X. Indeed, if $x_0 \in \exists^* Q$, then $x_0 \exists^* x$ for some x in Q, and therefore $\pi x_0 = \pi x$ for some x in Q. The last assertion means that $\pi x_0 \in \pi Q$, or, equivalently, that $x_0 \in \pi^{-1}\pi Q$. This proves that $\exists^* Q \subset \pi^{-1}\pi Q$; the reverse inclusion follows by retracing the steps of the argument in the reverse order.

Since $\exists^*$ is a Boolean relation, the equation $\exists^* Q = \pi^{-1}\pi Q$ implies that $\pi^{-1}\pi Q$ is clopen whenever Q is clopen. Since it is an elementary fact about mappings between compact Hausdorff spaces, i.e., mappings such as π, that if $\pi^{-1}P$ is open (or closed), then P is open (or closed), it follows that πQ is clopen whenever Q is clopen. The Boolean nature of X implies now that π is open; the proof is complete.

Lemma 17. *If π is a Boolean mapping from a Boolean space X to a Boolean space Y, then there exists a quantifier $\exists$ on the dual algebra* **A** *of X such that $x_1 \exists^* x_2$ is equivalent to $\pi x_1 = \pi x_2$.*

Proof. Define a relation φ in X by writing $x_1 \varphi x_2$ if and only if $\pi x_1 = \pi x_2$. Clearly φ is an equivalence relation with domain X, and $\varphi Q = \pi^{-1}\pi Q$ for every subset Q of X. The Boolean nature of π implies that the (inverse) image under φ of a clopen set is clopen and that the (direct) image under φ of a point is closed. (Since φ is an equivalence, the distinction between direct and inverse images is purely verbal.) In other words, φ is a Boolean equivalence with domain X; the desired result follows from Theorem 10.

It is not difficult to verify that the correspondences described in Lemmas 16 and 17 are in an obvious sense dual to each other. If, in other words, $\exists$ is a quantifier and π is the Boolean mapping that corresponds to $\exists$ via Lemma 16, then the quantifier that corresponds to π via Lemma 17 is the same as $\exists$, to within an isomorphism, and a similar statement holds with the order of π and $\exists$ reversed. Since the proof of this assertion involves absolutely no conceptual difficulties, it may safely be omitted.

13. Constants. The first-order monadic functional calculus is often described as the modern version of Aristotelean (syllogistic) logic. Correspondingly, the algebraic formulation of Aristotelean logic is to be found in the theory of monadic algebras; the fact that a certain syllogism is valid can be described by asserting that the ideal generated by two specified elements of a monadic algebra contains a third specified element.

The point is worth a little closer examination. It turns out that the logically relevant concept is not that of an ideal (cf. section 6) but the dual concept of a filter. A (Boolean) *filter*, by definition, is a non-empty subset $\mathbf{F}$ of a Boolean algebra $\mathbf{A}$, such that if p and q are in $\mathbf{F}$, then $p \wedge q \in \mathbf{F}$, and if $p \in \mathbf{F}$, then $p \vee q \in \mathbf{F}$ for all q in $\mathbf{A}$. A *monadic filter* is a subset $\mathbf{F}$ of a monadic algebra $\mathbf{A}$, such that $\mathbf{F}$ is a Boolean filter in $\mathbf{A}$ and such that $\forall p \in \mathbf{F}$ whenever $p \in \mathbf{F}$. The validity of the syllogism *Barbara* can now be described in the following terms: if p, q, and r are elements of a monadic algebra, then the monadic filter generated by $\forall(p' \vee q)$ and $\forall(q' \vee r)$ contains $\forall(p' \vee r)$.

If the preceding example is examined with a view to specializing it to the well-known syllogism concerning the mortality of Socrates, an intuitively unsatisfactory aspect of the situation emerges. In highly informal language, the trouble is that the theory is equipped to deal with generalities only, and is unable to say anything concrete. Thus, for instance, by an appropriate choice of notation, a monadic algebra might be taught to say "all men are mortal", but difficulties are encountered in trying to teach it to say "Socrates is a man." The point is that "manhood" and "mortality" can easily be thought of as elements of a monadic algebra, since they are the obvious abstractions of the propositional functions whose value at each x of some set is, respectively, "x is a man" and "x is mortal". Socrates, on the other hand, is a "constant", and there is no immediately apparent way of pointing to him. A classical artifice, designed to avoid this situation, is to promote Socrates to a propositional function, namely the function whose value at x is "x is Socrates". This procedure is both intuitively and algebraically artificial; Socrates is not a proposition, but an entity about which propositions may be made.

Intuitively, constants play the same role in algebraic logic as distinguished elements (e.g., the unit element of a group) play in ordinary algebra. What is desired is to single out certain points of a given domain X and to build them into the theory. Since, however, the ultimate objects of the theory are not points of X, but abstract elements suggested by functions on X, the desideratum becomes that of finding an algebraic description of what it means to replace the argument of a function by a constant.

The correct description is quickly suggested by the study of functional algebras. Suppose, to be specific and to avoid irrelevant, non-algebraic difficulties, that **B** is a complete Boolean algebra, and that **A** is the functional monadic algebra of all functions from some domain X into **B**. A convenient way to describe the act of replacing the argument x of a function p in **A** by a fixed element x_0 of X is to introduce the mapping c that associates with p the function cp, where

$$cp(x) = p(x_0)$$

for all x in X. It is clear that c is a Boolean endomorphism on **A**. Since, moreover, $\exists p$ is always a constant function, its value at x_0 is the same as its constant value, so that $c\exists p = \exists p$. Since, similarly, cp is always a constant function, an application of the quantifier leaves it unchanged, so that $\exists cp = cp$.

The following definition is now adequately motivated: a *constant* of a monadic algebra **A** is a Boolean endomorphism c on **A**, such that

$$c\exists = \exists \text{ and } \exists c = c.$$

It follows from the definition that (i) c is the identity on the range of $\exists$, and (ii) the range of c is included in the range of $\exists$. Conversely, (iii) if c is a Boolean endomorphism satisfying (i) and (ii), then c is a constant. (iv) A constant is idempotent ($c^2 = c$). (v) If c is a constant of **A**, then $cp \leqq \exists p$ for all p in **A**. If c is a constant of **A**, then (vi) $c\forall = \forall$, and (vii) $\forall c = c$, and, conversely, (viii) if c is a Boolean endomorphism satisfying (vi) and (vii), then c is a constant. In view of (vi), (vii), and (viii), all true assertions about constants have true duals, and, in particular, (ix) if c is a constant of **A**, then $\forall p \leqq cp$ for all p in **A**.

With all the machinery at hand, it is easy to formulate the algebraic version of the syllogism about the mortality of Socrates. It asserts that if p and q are elements of a monadic algebra **A**, and if c is a constant of **A**, then the filter generated by $\forall(p' \vee q)$ and cp contains cq.

The introduction of constants was motivated above by logical considerations. It turns out that constants are also of great algebraic importance, and that, in fact, they play a central role in the proof of the fundamental representation theorem for monadic algebras. A final pertinent comment is this: although the concept of a constant is a purely algebraic one, in the discussion of the existence of constants it is convenient to make use of the topological theory of duality; it is for that reason that the definition was not given before.

14. Cross sections. A *cross section* of a continuous mapping π from a topological space X onto a topological space Y is a continuous mapping σ from Y into X such that $\pi\sigma y = y$ for all y in Y. Cross sections do not always exist, not even if X and Y are Boolean spaces. Example: let X be a Boolean space with exactly two cluster points, let Y be obtained from X by identifying the two cluster points, and let π be the identification mapping. One reason this example works is that π is not a Boolean mapping. Unfortunately, however, even Boolean mappings do not always have a cross section. Examples to show this are not trivial to construct; a suitable one has been constructed by J. L. Kelley and is described in the work of Arens and Kaplansky (1).

Cross sections of Boolean mappings are intimately related to constants of monadic algebras; the relation depends, naturally, on the correspondence between Boolean mappings and quantifiers. Suppose, indeed, that **A** is a monadic algebra, with quantifier $\exists$ and dual space X, and let π be a Boolean mapping from X onto a Boolean space Y such that $x_1 \exists^* x_2$ is equivalent to $\pi x_1 = \pi x_2$ (Lemma 16).

LEMMA 18. *There is a one-to-one correspondence between all constants c of* **A** *and all cross sections σ of π, such that*

$$cp(x) = p(\sigma\pi x)$$

for all p in **A** *and all x in X.*

Proof. Suppose that c is a constant and let γ be its dual; in other words, γ is a continuous mapping from X into X such that $cp(x) = p(\gamma x)$ for all p and all x. The cross section σ corresponding to c is defined for a point $y = \pi x$ of X by writing $\sigma y = \gamma x$. It must, of course, be proved that this definition is unambiguous; i.e., that if $\pi x_1 = \pi x_2$, than $\gamma x_1 = \gamma x_2$. Indeed: if $\pi x_1 = \pi x_2$, then $x_1 \exists^* x_2$, so that $\exists p(x_1) = \exists p(x_2)$ for all p. This relation, applied to cp in place of p, yields the conclusion that $cp(x_1) = cp(x_2)$ for all p, and hence that $p(\gamma x_1) = p(\gamma x_2)$ for all p; the equation $\gamma x_1 = \gamma x_2$ now follows immediately. Clearly σ maps Y into X, and, if $p \in \mathbf{A}$, $x \in X$, and $y = \pi x$, then

$$cp(x) = p(\gamma x) = p(\sigma y) = p(\sigma\pi x).$$

It remains to prove that σ is continuous. Suppose, for this purpose, that $p \in \mathbf{A}$. The set $\{x: p(x) = 1\}$ is a typical clopen set in X; it is to be proved that its inverse image under σ is open. Since $y \in \sigma^{-1}\{x: p(x) = 1\}$ if and only if $y = \pi x$ with $p(\gamma x) = 1$, i.e., if and only if $y \in \pi\{x: cp(x) = 1\}$, the desired result follows from the fact that π is open.[1]

Suppose next that σ is a cross section of π; the constant c corresponding to σ is defined for an element p of **A** by writing $cp(x) = p(\sigma\pi x)$ for all x in X. Clearly c is a Boolean endomorphism of **A**. Since $\pi\sigma y = y$ for all y, so that $\pi\sigma\pi x = \pi x$ for all x, it follows from the relation between π and $\exists$ that $\sigma\pi x \exists^* x$ for all x. This implies that

$$c\exists p(x) = \exists p(\sigma\pi x) = \exists p(x)$$

for all p and all x, and hence that $c\exists = \exists$. Finally, if $x_1 \exists^* x_2$, then $\pi x_1 = \pi x_2$, so that $p(\sigma\pi x_1) = p(\sigma\pi x_2)$, or $cp(x_1) = cp(x_2)$; this implies that

$$\exists cp(x_0) = \bigvee\{cp(x): x \exists^* x_0\} = cp(x_0)$$

for all p and all x_0, and hence that $\exists c = c$. The proof of the lemma is complete.

[1] See (b) on p. 166.

From Lemmas 17 and 18 we can conclude that there exist monadic algebras that possess no constants. The assertion of the existence of a constant for a monadic algebra **A**, whenever it is true, can be regarded as an extension theorem. Indeed, a Boolean endomorphism of **A** is a constant of **A** if and only if it is an extension (to a homomorphism from **A** to $\exists(\mathbf{A})$) of the identity mapping (from $\exists(\mathbf{A})$ to $\exists(\mathbf{A})$). These considerations make contact with a theorem of Sikorski (6). The reason Sikorski's theorem is not available to prove the existence of constants is that Sikorski needs the added assumption that the prescribed range algebra is complete.

15. Rich algebras. A constant c of a monadic algebra **A** with quantifier will be called a *witness* to an element p of **A** if $\exists p = cp$; we shall also say that c is a witness to p *with respect to* **A**, or, more simply, *in* **A**. If **A** and $\mathbf{A}^+$ are monadic algebras such that **A** is a monadic subalgebra of $\mathbf{A}^+$ and such that every element of **A** has a witness in $\mathbf{A}^+$, we shall say that $\mathbf{A}^+$ is a *rich extension* of **A**, or, more simply, that $\mathbf{A}^+$ is rich *for* **A**. A *rich algebra* is one that is rich for itself.

LEMMA 19. *If p_0 is an arbitrary element of a monadic algebra* **A**, *then there exists a monadic algebra $\mathbf{A}_0^+$ including* **A** *as a monadic subalgebra and such that* (i) *there is a witness c_0 to p_0 in $\mathbf{A}_0^+$, and* (ii) *every constant of* **A** *has an extension to a constant of $\mathbf{A}_0^+$.*

PROOF. Let X be the dual space of **A** and let π be a Boolean mapping from X onto a Boolean space Y, such that $x_1 \exists^* x_2$ is equivalent to $\pi x_1 = \pi x_2$ (Lemma 16). If $X^+ = X \times X$, $Y^+ = X \times Y$, and $\pi^+(x, y) = (x, \pi y)$ whenever $(x, y) \in X^+$, then π^+ is a Boolean mapping from X^+ onto Y^+. To this Boolean mapping there corresponds a quantifier $\exists^+$ on the dual algebra $\mathbf{A}^+$ of X^+ such that $(x_1, y_1) \exists^{+*} (x_2, y_2)$ is equivalent to $\pi^+(x_1, y_1) = \pi^+(x_2, y_2)$ (Lemma 17). It follows from the definition of π^+ that if $p(x, y) = q(x) \wedge r(y)$, where q and r are in **A**, then

$$\begin{aligned}\exists^+ p(x_0, y_0) = \bigvee\{p(x, y)\colon x_0 = x,\ y_0 \exists^* y\} = q(x_0) \wedge \bigvee\{r(y)\colon y_0 \exists^* y\} \\ = q(x_0) \wedge \exists r(y_0).\end{aligned}$$

If we write $p^+(x, y) = p(y)$ for every p in **A**, it follows from what was just proved about $\exists^+$ that the mapping $p \to p^+$ (which is obviously a Boolean isomorphism from **A** into $\mathbf{A}^+$) is a monadic isomorphism. We may therefore regard **A** as a monadic subalgebra of $\mathbf{A}^+$ whenever it is convenient to do so.

We observe next that every constant c of **A** has a natural extension to a constant c^+ of $\mathbf{A}^+$. To say that c^+ is an extension of c means, of course, that $c^+p^+ = (cp)^+$ whenever $p \in \mathbf{A}$. To prove this, let σ be a cross section of π such that $cp(x) = p(\sigma\pi x)$ (Lemma 18) and write $\sigma^+(x, y) = (x, \sigma y)$ whenever $(x, y) \in Y^+$. It is easy to verify that σ^+ is a cross section of π^+ and that the constant c^+ corresponding to the cross section σ^+ via Lemma 18 is the desired extension.

The algebra $\mathbf{A}^+$ has a simple Boolean endomorphism c_0^+ that is almost a constant of $\mathbf{A}^+$; by definition, $c_0^+p(x, y) = p(x, x)$. It is trivial to verify that c_0^+ is an idempotent Boolean endomorphism and that $\exists^+ c_0^+ = c_0^+$; the only reason c_0^+ is not a constant is that the equation $c_0^+\exists^+ = \exists^+$ need not hold. (The equation does not hold, because, for instance, if $p \in \mathbf{A}$, then $\exists^+ p^+(x, y) = \exists p(y)$, whereas $c_0^+ \exists^+ p^+(x, y) = \exists p(x)$.) We shall force c_0^+ to become a constant by identifying $c_0^+ \exists^+ p$ with $\exists^+ p$ for every p in $\mathbf{A}^+$. Precisely speaking, we shall consider in $\mathbf{A}^+$ the monadic ideal $\mathbf{I}^+$ generated by all elements of the form $\exists^+ p + c_0^+ \exists^+ p$, and we shall form the quotient algebra $\mathbf{A}^+/\mathbf{I}^+$. We shall prove that during the reduction modulo $\mathbf{I}^+$ the subalgebra **A** is not disturbed, and that after that reduction the endomorphism c_0^+ becomes a constant.

An element of $\mathbf{A}^+$ belongs to $\mathbf{I}^+$ if and only if it is dominated by the supremum of a finite set of generators. If $q \in \mathbf{I}^+$, so that

$$q \leqq \bigvee_{i=1}^{n}(\exists^+ p_i + c_0^+ \exists^+ p_i),$$

then, applying c_0^+ to both sides, we obtain $c_0^+ q = 0$. If, in particular, $q \in \mathbf{A}$ and $q^+ \in \mathbf{I}^+$, then $c_0^+ q^+ = 0$, and, since $c_0^+ q^+(x, y) = q(x)$, it follows that $q = 0$. If f^+ is the natural homomorphism from $\mathbf{A}^+$ onto $\mathbf{A}^+/\mathbf{I}^+$, the result we just obtained implies that f^+ is one-to-one on **A**, i.e., that we may regard **A** as a monadic subalgebra of $\mathbf{A}^+/\mathbf{I}^+$ whenever it is convenient to do so.

Since, as shown in the preceding paragraph, c_0^+ maps $\mathbf{I}^+$ onto 0, it follows that if p and q are in $\mathbf{A}^+$ and if $f^+p = f^+q$, then $c_0^+p = c_0^+q$. This implies that c_0^+ maps cosets of $\mathbf{I}^+$ onto cosets of $\mathbf{I}^+$ and hence that c_0^+ may be regarded as a mapping of $\mathbf{A}^+/\mathbf{I}^+$ into itself. A routine verification shows that c_0^+ is, in fact, a constant of $\mathbf{A}^+/\mathbf{I}^+$, as promised.

An argument similar to the one just given shows that every constant of $\mathbf{A}^+$ can be transferred to $\mathbf{A}^+/\mathbf{I}^+$. Suppose indeed that c^+ is a constant of $\mathbf{A}^+$. If p and q are in $\mathbf{A}^+$ and if $f^+p = f^+q$, then $p + q \in \mathbf{I}^+$. Since $c^+(p + q) \leqq \exists^+(p + q)$ and $\exists^+(p + q) \in \mathbf{I}^+$, it follows that $c^+p + c^+q \in \mathbf{I}^+$ and hence that $f^+c^+p = f^+c^+q$. This implies that c^+ maps cosets of $\mathbf{I}^+$ onto cosets of $\mathbf{I}^+$ and hence that c^+ may be regarded as a mapping of $\mathbf{A}^+/\mathbf{I}^+$ into itself. The fact that c^+ is a constant of $\mathbf{A}^+/\mathbf{I}^+$ follows from the fact that f^+ is a monadic homomorphism. From this result and from what we said earlier about extending constants from $\mathbf{A}$ to $\mathbf{A}^+$, it follows that every constant of $\mathbf{A}$ has a natural extension to a constant of $\mathbf{A}^+/\mathbf{I}^+$. The proof of the transferability of c^+ to $\mathbf{A}^+/\mathbf{I}^+$ did not use any special properties of $\mathbf{A}^+$ and $\mathbf{I}^+$; we have proved, in fact, that a constant can always be transferred to a quotient algebra. Since the final algebra $\mathbf{A}_0^+$ that we shall construct will be a quotient algebra of $\mathbf{A}^+/\mathbf{I}^+$, as soon as we prove that the natural homomorphism from $\mathbf{A}^+/\mathbf{I}^+$ onto $\mathbf{A}_0^+$ does not disturb $\mathbf{A}$ (i.e., is one-to-one on $\mathbf{A}$), it will follow automatically that $\mathbf{A}_0^+$ has property (ii).

To construct $\mathbf{A}_0^+$ out of $\mathbf{A}^+/\mathbf{I}^+$, we simply force the identification of $c_0^+p_0^+$ with $\exists^+p_0^+$. Precisely speaking, we shall consider in $\mathbf{A}^+/\mathbf{I}^+$ the monadic ideal $\mathbf{I}_0^+$ generated by $\exists^+p_0^+ - c_0^+p_0^+$ (or, rather, by the coset of $\mathbf{I}^+$ containing that element), and we shall form the quotient algebra $\mathbf{A}_0^+ = (\mathbf{A}^+/\mathbf{I}^+)/\mathbf{I}_0^+$. In accordance with what was said above all that remains to be proved is that if an element q of $\mathbf{A}$ belongs to $\mathbf{I}_0^+$, then $q = 0$. An element of $\mathbf{A}^+/\mathbf{I}^+$ belongs to $\mathbf{I}_0^+$ if and only if it is dominated by the coset of $\exists^+p_0^+ - c_0^+p_0^+$. What we must prove reduces therefore to the following implication: if $q \in \mathbf{A}$ and if

$$q^+ \leqq \exists^+p_0^+ - c_0^+p_0^+,$$

then $q^+ \in \mathbf{I}^+$ (i.e., q^+ is equal to 0 modulo $\mathbf{I}^+$).[1] In fact we can conclude that $q^+ = 0$. If $p_0 = 0$, this is trivial, since $q^+ \leqq \exists^+p_0^+$. If $p_0 \neq 0$, let x_0 be a point of X such that $p_0(x_0) = 1$, and form the infimum of both sides of the assumed inequality with $c_0^+p_0^+$. The result is that $q^+ \wedge c_0^+p_0^+ = 0$ and hence that $q(y) \wedge p_0(x) = 0$ for all x and y; the desideratum follows by setting x equal to x_0. The proof of Lemma 19 is complete.

[1] See (c) on pp. 165-66.

Theorem 11. *Every monadic algebra is a subalgebra of a rich algebra.*

PROOF. Repeated applications of Lemma 19 show that if $\mathbf{A}_0$ is a monadic algebra, then there exists a monadic algebra $\mathbf{A}_1$ including $\mathbf{A}_0$ as a monadic subalgebra and such that (i) $\mathbf{A}_1$ is a rich extension of $\mathbf{A}_0$ and (ii) every constant of $\mathbf{A}_0$ has an extension to a constant of $\mathbf{A}_1$. The applications of Lemma 19 have to repeated rather often, to be sure; precisely speaking, what is involved is transfinite induction. The witnesses to the elements of $\mathbf{A}_0$ are introduced one by one. The constants obtained at each stage are carried along to all subsequent stages, and, finally, to the union of the chain of algebras so obtained. The details are automatic.

The proof of the theorem consists of repeated applications of the result of the preceding paragraph. The applications have to be repeated countably often in this case; what is involved is elementary mathematical induction. Given $\mathbf{A}_n$, we denote by $\mathbf{A}_{n+1}$ a rich extension of it with the constant-extension-property. Thus the constants obtained at each stage are carried along to all subsequent stages, and, finally, to the union of the increasing sequence of algebras so obtained. Since that union is obviously a rich algebra, the proof of Theorem 11 is complete.

Constants play the same role in the theory of monadic algebras as homomorphisms into $\mathbf{O}$ play in the theory of Boolean algebras; in a (somewhat vague) sense a constant is "locally" a homomorphism into $\mathbf{O}$. Theorem 11 is the monadic substitute for Stone's theorem on the existence of sufficiently many maximal ideals; the representation theorem of the next section is based on Theorem 11 almost the same way as Stone's representation theorem is based on the maximal ideal theorem.

16. Representation. It is a routine application of universal algebraic techniques (2) to put together the simplicity theorem (Theorem 6) and the semisimplicity theorem (Theorem 7) to obtain a representation theorem that exhibits every monadic algebra as a subdirect union of $\mathbf{O}$-valued functional monadic algebras. The purpose of this final section is to discuss a stronger and more useful representation theorem that asserts, in effect, that the functional algebras with which we began the theory of monadic algebras exhaust all possible cases.

Theorem 12. *If* **A** *is a monadic algebra, then there exists a set* X *and there exists a Boolean algebra* **B**, *such that* (i) **A** *is isomorphic to a* **B**-*valued functional algebra* $\tilde{\mathbf{A}}$ *with domain* X, *and* (ii) *for every element* $\tilde{p}$ *of* $\tilde{\mathbf{A}}$ *there exists a point* x *in* X *with* $\tilde{p}(x) = \exists \tilde{p}(x)$.

PROOF. The conclusions of the theorem are such that if they are valid for an algebra, then they are automatically valid for all its subalgebras. It follows from this comment and from Theorem 11 that there is no loss of generality in assuming that **A** is rich. This means that for each element p of **A** there exists at least one constant $c = c_p$ of **A** such that $cp = \exists p$; let X be a set of constants containing at least one such c for each p. Let the Boolean algebra **B** be the range $\exists(\mathbf{A})$ of the quantifier $\exists$ on **A**. Define a mapping f from **A** into $\mathbf{B}^X$, i.e., associate a function $\tilde{p} = fp$, from X into **B**, with every element p of **A**, by writing $\tilde{p}(c) = cp$. Since $\exists cp = cp$, the value $\tilde{p}(c)$ is indeed in **B** for every c in X, so that $\tilde{p} \in \mathbf{B}^X$. Since each c in X is a Boolean endomorphism on **A**, and since the Boolean operations in $\mathbf{B}^X$ are defined pointwise, a routine verification shows that f is a Boolean homomorphism. If $fp = 0$, i.e., if $\tilde{p}(c) = 0$ for all c in X, then, in particular, $\exists p = c_p p = \tilde{p}(c_p) = 0$, and therefore, $p = 0$; this proves that the homomorphism f is one-to-one.

Let $\tilde{\mathbf{A}}$ be the range of f, so that $\tilde{\mathbf{A}}$ is a Boolean subalgebra of $\mathbf{B}^X$; it is to be proved that $\tilde{\mathbf{A}}$ is a functional monadic algebra and that f is a monadic isomorphism between **A** and $\tilde{\mathbf{A}}$. If $\tilde{p} = fp \in \tilde{\mathbf{A}}$, the range $\mathbf{R}(\tilde{p})$ of the function $\tilde{p}$ contains, in particular, the element $\exists p = c_p p = \tilde{p}(c_p)$ of **B**; since $cp \leqq \exists p$ for every constant c, it follows that $\mathbf{R}(\tilde{p})$ has a largest element, and therefore a supremum, namely $\exists p$. This proves that $\exists \tilde{p}$ exists and has the value $\exists p$ at each c in X. On the other hand, $f\exists p(c) = c\exists p = \exists p$ for all c, so that $\exists fp = \exists \tilde{p} = f\exists p$; the proof of the theorem is complete.

It is instructive to observe that the simplicity and semisimplicity theorems can be recaptured from this general representation theorem. The purpose of the following considerations is to show how that can be done.

LEMMA 20. *If* **A** *is a* **B**-*valued functional monadic algebra with domain* X, *such that for every* p *in* **A** *there exists a point* x *in* X *with* $p(x) = \exists p(x)$, *and if* f_0 *is a Boolean homomorphism from* **B** *into* **O**, *then the mapping* f, *from* **A** *into* $\mathbf{O}^X$, *defined by* $fp(x) = f_0(p(x))$, *is a monadic homomorphism.*

Proof. The fact that f is a Boolean homomorphism is an easily verified consequence of the fact that f_0 is such, and of the fact that the Boolean operations in $\mathbf{O}^X$ are defined pointwise. The proof of the fact that f is a monadic homomorphism is also similar to the corresponding part of the proof of Theorem 12. Indeed, if $p_0 \in \mathbf{A}$, and if x_0 is a point such that $p(x_0) = \exists p(x_0)$, then $\mathbf{R}(fp)$ contains $f_0(p(x_0))$ as its largest element, and therefore $\exists fp(x) = f_0(p(x_0))$ for all x. Since, on the other hand, $\exists p$ is a constant function, so that

$$f\exists p(x) = f_0(\exists p(x)) = f_0(\exists p(x_0)) = f_0(p(x_0)),$$

it follows that $\exists fp = f\exists p$, as desired.

It is now an easy matter to give an alternative proof of the deeper half of the simplicity theorem. In view of Theorem 12, there is no loss of generality in restricting attention to a functional monadic algebra $\mathbf{A}$ that satisfies the condition of Lemma 20. If $\mathbf{A}$ is simple, then select the f_0 in Lemma 20 arbitrarily; since, by simplicity, the homomorphism f so obtained has a trivial kernel, it follows that f is an isomorphism.

Once the simplicity theorem is known, the assertion of semisimplicity takes the following form: if $p \in \mathbf{A}$ and $p \neq 0$, then there exists a monadic homomorphism f from $\mathbf{A}$ into an $\mathbf{O}$-valued functional algebra, such that $fp \neq 0$. For the proof, find x in X so that $p(x) \neq 0$, and (by the semisimplicity theorem for ordinary Boolean algebras) find f_0 so that $f_0(p(x)) \neq 0$; the f obtained from Lemma 20 satisfies the stated condition.

University of Chicago.

REFERENCES

R. F. Arens and I. Kaplansky

[1] *Topological representations of algebras*, Trans. Amer. Math. Soc., 63 (1948) 457—481.

G. Birkhoff

[2] *Lattice theory*, New York, 1948; p. 91.

P. R. Halmos

[3] *Polyadic Boolean algebras*, Proc. Nat. Acad. Sci., U.S.A., 40 (1954) 296—301.

B. Jónsson and A. Tarski

[4] *Boolean algebras with operators*, Amer. J. Math., 73 (1951) 891—939.

J. C. C. McKinsey and A. Tarski

[5] *The algebra of topology*, Ann. Math., 45 (1944) 141—191.

R. Sikorski

[6] *A theorem on extension of homomorphisms*, Ann. Soc. Polon. Math., 21 (1948) 332—335.

M. H. Stone

[7] *The theory of representations for Boolean algebras*, Trans. Amer. Math. Soc., 40 (1936) 37—111.

M. H. Stone

[8] *Applications of the theory of Boolean rings to general topology*, Trans. Amer. Math. Soc., 41 (1937) 321—364.

A. Tarski and F. B. Thompson

[9] *Some general properties of cylindric algebras*, Bull. Amer. Math. Soc., 58 (1952) 65.

(Oblatum 12-5-54).

III

REPRESENTATION

THE REPRESENTATION OF MONADIC BOOLEAN ALGEBRAS

Introduction. A monadic (Boolean) algebra is a Boolean algebra A together with an operator $\exists$ on A (called an existential quantifier, or, simply, a quantifier) such that

$$\exists 0 = 0,$$

$$p \leq \exists p,$$

$$\exists (p \wedge \exists q) = \exists p \wedge \exists q$$

whenever p and q are in A. A systematic study of monadic algebras appears in [3]; operators such as $\exists$ had occurred before in, for instance, Lewis' studies of modal logic and Tarski's studies of algebraic logic. One motivation for studying monadic algebras is the desire to understand certain aspects of mathematical logic; the connection with logic is also the source of much of the terminology and notation used in the theory. There are, however, many other ways of motivating the study. Monadic algebras arise naturally in, for instance, set theory, pure measure theory, and ergodic theory.

The main purpose of this paper is to present a new proof of the fundamental representation theorem for monadic algebras, much simpler than the original one. At the same time it is pertinent to call attention to some hitherto unnoticed and quite amusing relations among four results: the monadic representation theorem, Sikorski's extension theorem for Boolean homomorphisms [6], Gleason's characterization of "projective" compact spaces [2], and Michael's theorem on cross sections of mappings with a zero-dimensional range [5].

Representation. A typical way to manufacture monadic algebras is this. Let C be an arbitrary Boolean algebra, let k be a positive integer, and let A be a set of ordered k-tuples of elements of C (not necessarily all of them) such that (i) if p and q are in A, then p' and $p \vee q$ are in A (where the Boolean complements and suprema are defined coordinate-wise), and (ii) if p is in A, $p = (p_1, \cdots, p_k)$, and if $\bar{p} = \bigvee_{n=1}^{k} p_n$, then $(\bar{p}, \cdots, \bar{p})$ [k coordinates] is in A. If $\exists p$ is defined to be $(\bar{p}, \cdots, \bar{p})$, then A becomes a monadic algebra.

An infinite generalization of this construction is useful. Let C be an arbitrary Boolean algebra, let K be an arbitrary non-empty set, and let A be a set of functions from K to C such that (i) A is a Boolean algebra with respect to the pointwise Boolean operations, and (ii) if p is in A, then the range of p (a subset of C) has a supremum $\bar{p}$ in C, and the function that takes the value $\bar{p}$ at each point of K is in A. If $\exists p$ is defined to be that function, then A becomes

Received May 28, 1958. Research supported in part by a contract with the U.S. Air Force.

a monadic algebra. Every monadic algebra obtained in this way is called a C-valued *functional algebra* with domain K.

An element p of a monadic algebra A (functional or not) is called *closed* if $\exists p = p$. The set of closed elements coincides with the range $\exists A$ of the quantifier $\exists$; it always constitutes a Boolean subalgebra of A. If A happens to be a C-valued functional algebra, then the elements of $\exists A$ are just the functions whose values are all the same. The mapping that sends each such function onto its unique value is a Boolean monomorphism from $\exists A$ into C; it may well fail to be an epimorphism.

The representation theorem for monadic algebras says (among other things) that every monadic algebra A is isomorphic to a functional one. The proof of such an assertion must at some stage exhibit the domain K and the value-algebra C that are to be used. It is natural to try to define C as $\exists A$; the moral of the preceding paragraph, however, is that the attempt is not likely to succeed.

Another difficulty in the theory of monadic algebras (already visible in the very definition of a functional algebra) is that the various Boolean algebras that enter the discussion need not be complete. (Completeness here means lattice-theoretic completeness, i.e., the requirement that every set of elements have both a supremum and an infimum.) Fortunately, however, every Boolean algebra can be embedded into a complete Boolean algebra (in general in many ways). One way to do this is via the Stone theory of Boolean duality. Every Boolean algebra is isomorphic to the set of all clopen (closed and open) subsets of an essentially unique Boolean space (totally disconnected compact Hausdorff space); the Boolean algebra of *all* subsets of that space is a complete Boolean algebra.

It turns out that the completability of Boolean algebras overcomes both the obstacles encountered so far. If, to be more precise, A is a monadic algebra, then (it is proved below that) A is isomorphic to a C-valued functional algebra, where C is an arbitrary complete Boolean algebra that includes $\exists A$.

The next problem is to find an algebraic characterization of a domain that can be used in the proof of the representation theorem. The way to do that is to analyze the algebraic role played by the points of the domain for an algebra that is already known to be functional. The result of such an analysis (see [3, § 13]) is the following definition. If A is a monadic algebra (not necessarily functional) and if C is a Boolean algebra (not necessarily complete) that includes the algebra $\exists A$ of closed elements of A, then a *C-constant* of A is a Boolean homomorphism that maps A into C and is equal to the identity on $\exists A$. This definition includes the extreme case in which $C = \exists A$; in that case a C-constant is called simply a *constant*. (The definition of constants in [3] is easily seen to be equivalent to the present one. The terminology is suggested by the use of the phrase "individual constant" in mathematical logic.)

Constants (and, more generally, C-constants) solve the representation problem completely, provided sufficiently many of them can be found. The way to say this precisely is to use the following definition. A monadic algebra A is *C-rich* (where C, here and below, is a Boolean algebra including $\exists A$) if to each element p_0 of A there corresponds a C-constant b (not necessarily unique) such that $\exists p_0 = bp_0$ (i.e., b is a "witness" to $\exists p_0$). If $C = \exists A$, a C-rich algebra is called simply *rich*.

REPRESENTATION THEOREM. *If K is the set of all C-constants of a C-rich monadic algebra A, then A is isomorphic to a C-valued functional monadic algebra with domain K.*

The proof of the theorem is essentially the same as the proof of Theorem 12 in [3]; a hint will suffice here. Assign to each p in A the function p_K in C^K (the set of all functions from K to C) defined for each b in K by $p_K(b) = bp$. The correspondence $p \rightarrow p_K$ is a Boolean homomorphism; C-richness implies that it is a monomorphism. The assumption of C-richness implies also that the range of p_K contains a greatest element, namely $\exists p$, and hence that the correspondence $p \rightarrow p_K$ is a monadic as well as a Boolean homomorphism.

Extension. It is not obvious a priori (and, in fact, it is not true) that C-constants always exist. As long as nothing is known about the existence of such constants, the applicability of the representation theorem is rather limited. What follows is first the statements of the two powerful positive results that can be obtained along these lines, and afterward the proofs of those statements.

THEOREM 1. *If C is a complete Boolean algebra that includes the Boolean algebra of all closed elements of a monadic algebra A, then A is C-rich.*

Since every Boolean algebra can be embedded into a complete one in a canonical manner (the Stone completion), Theorem 1 and the representation theorem together imply that every monadic algebra is isomorphic to a functional one. Varsavsky [7] mentions that essentially the same approach to the main representation theorem of [3] has also been suggested by Monteiro.

In an important special case the conclusion of Theorem 1 can be improved spectacularly.

THEOREM 2. *Every countable monadic algebra is rich.*

These are the main facts; next come the proofs.

EXTENSION LEMMA. *Suppose that a Boolean algebra B is generated by a Boolean subalgebra B_0 and an element p_0. Suppose, moreover, that a_0 is a homomorphism from B_0 into a Boolean algebra C and that p^- and p^+ are elements of C with the following properties: if $p_0 \geq q \in B_0$, then $p^- \geq a_0 q$, and if $p_0 \leq r \in B_0$, then $p^+ \leq a_0 r$. If p is any element of C such that $p^- \leq p \leq p^+$, then there exists a unique homomorphism a from B to C such that a agrees with a_0 on B_0, and such that $ap_0 = p$.*

This technicality is a very slight generalization of a step in a proof in Sikorski's work [6]. The idea is this: since B_0 and p_0 generate B, every element of B has a (not necessarily unique) representation in the form

$$(q \wedge p_0) \vee (r \wedge p_0')$$

with q and r in B_0 ; the mapping a that sends each such element onto

$$(a_0 q \wedge p) \vee (a_0 r \wedge p')$$

is unambiguously defined and does everything that is required of it.

The main application of the extension lemma is the extension theorem for Boolean homomorphisms.

THEOREM 3. (Sikorski) *If B is a Boolean subalgebra of a Boolean algebra A, and if a is a homomorphism from B into a complete Boolean algebra C, then a can be extended to a homomorphism b from A into C.*

The proof consists of transfinitely many applications of the extension lemma (or, equivalently, of one application together with Zorn's lemma). The completeness of B is used to show the existence of elements such as p^- and p^+.

Everything is now ready for the proof of Theorem 1. It is to be proved that for each element p_0 of A there exists a C-constant b such that $\exists p_0 = bp_0$. Write $B_0 = \exists A$, let B be the Boolean algebra generated by B_0 and p_0 , and let a_0 be the identity mapping of B_0 into C. The extension lemma can be applied with $p^- = \forall p_0 \ (= (\exists p_0')')$, $p^+ = \exists p_0$, and $p = \exists p_0$; the conclusion is that there exixts a homomorphism a from B into C such that a maps p_0 onto $\exists p_0$ and is equal to the identity on $\exists A$. (The completeness of C is not needed for this part of the argument; $\exists A$ itself could have played the role of C). The last step is to apply Theorem 3; the conclusion is exactly the existence of a constant b (i.e., a homomorphism) of the desired sort.

It is perhaps worth noting that there are other ways of simplifying the proof of the representation theorem given in [3]; each new proof offers new hope for simplifying (or in some cases attaining) the proof of a corresponding theorem in one of the various generalizations of the subject. Two alternative proofs were offered by Leon LeBlanc (in a course on algebraic logic at the University of Chicago in 1957). In outline form they run as follows. (I) On universal algebraic grounds every monadic algebra is isomorphic to a subdirect sum of simple algebras. Each direct summand is easy to represent as a 2-valued functional algebra [3, Theorem 6]. (The symbol "2" denotes, here and below, the two-element Boolean algebra.) The Cartesian product of the domains used in these representations is a suitable alternative domain for each one of them. Since $(2^K)^N$ is identifiable with $(2^N)^K$ (where K is a suitable simultaneous domain for each direct summand, and N is the index-set of the direct sum decomposition), the direct sum is isomorphic to a 2^N-valued functional monadic algebra. (II) A constant of a simple monadic algebra is nothing more than a 2-valued Boolean homomorphism; see [3, Lemma 8]. The existence of sufficiently many such homomorphisms implies that every simple monadic algebra is rich. Since direct sums of constants are constants, it follows easily that a direct sum of rich algebras is rich. This implies (via the universal algebraic fact mentioned above) that every monadic algebra is isomorphic to a subalgebra of a rich algebra. At this point the proof makes contact with [3, Theorem 11]; the rest of the proof of the representation theorem can now proceed (easily and algebraically) as in [3].

It remains to prove Theorem 2. Here too the principal tool is the extension lemma, or, rather, the following consequence of it.

COROLLARY 1. *If C is a complete Boolean algebra that includes the algebra $\exists A$ of all closed elements of a monadic algebra A, and if a_0 is a C-constant of a monadic subalgebra B_0 of A, then a_0 can be extended to a C-constant b of A.*

To prove the corollary, choose first an arbitrary element p_0 of $\exists A$. It is easy to see that the Boolean algebra B generated by B_0 and p_0 is a monadic subalgebra of A. Since C includes $\exists A$, the element p_0 belongs to C, and it makes sense to apply the extension lemma with $p^- = p^+ = p = p_0$. The conclusion is the existence of a homomorphism a that maps B into C and is equal to the identity on $B \cap \exists A$ (i.e., on $\exists B$). In other words, a_0 can be extended to a C-constant a of B. The construction can be repeated with B and a in place of B_0 and a_0. Transfinite induction (or argument à la Zorn) yields finally a monadic subalgebra of A (call it B again) and a C-constant of it (call it a again); the difference between the new B and the old B_0 is that B includes all $\exists A$. The proof of the corollary is completed by an application of Theorem 3.

Countable monadic algebras can now be approached by approximation. The climb up the inductive ladder is achieved by means of the following consequence of Corollary 1.

COROLLARY 2. *Every constant of a monadic subalgebra of a finite monadic algebra can be extended to a constant of the entire algebra.*

The proof is immediate since finiteness implies completeness. It should be noted that transfinite methods are not needed to prove Corollary 2; for finite algebras the extension theorem and Corollary 1 can be obtained by finite proofs.

A final preparatory comment is this: a finitely generated monadic algebra is finite. The comment is not profound, but neither is it completely trivial; the first attempt at a proof is not likely to be successful. The result follows from the work of Bass on free monadic algebras [1]; for an alternative treatment see [4].

Suppose now that A is a countable monadic algebra; it is to be proved that for each element p_0 of A there exists a constant b such that $\exists p_0 = bp_0$. Let A_0 be the monadic subalgebra of A generated by p_0. The algebra A_0 is finite, and it can be made the first term of an ascending sequence $\{A_n\}$ of finite monadic algebras whose union is A. (Adjoin the elements of A one after another.) The algebra A_0 is rich; this is a trivial special case of Theorem 1. (Transfinite methods to prove this special case are even more wasteful now than before, but it is of some interest to see how the finite cases fit into the general context. The algebra A_0 has at most four atoms.) Let b_0 be constant of A_0 such that $\exists p_0 = b_0 p_0$. By an inductive application of Corollary 2, there exists a sequence $\{b_n\}$ of constants (where b_n is a constant of A_n of course) each of which (after b_0) is an extension of its predecessor. The unique common extension b of all the b_n is the constant whose existence is needed to complete the proof of Theorem 2.

Topologization. Topological concepts enter into the study of Boolean algebras via the Stone duality theory. The dual of a Boolean algebra A is a Boolean space X (the set of all 2-valued homomorphisms on A with the product-space topology). The duals of Boolean homomorphisms are continuous transformations, and, as usual, onto and one-to-one are each other's duals.

A monadic algebra A is uniquely determined by its Boolean structure together with the Boolean subalgebra $\exists A$ of closed elements; see [3, § 4]. A Boolean subalgebra that is fit to be the range of a quantifier on a Boolean algebra A is called a relatively complete subalgebra of A. (For present purposes the definition of relative completeness is irrelevant.) In other words, a monadic algebra may be characterized as a triple (A, B, c), where A and B are Boolean algebras and c is a monomorphism from B onto a relatively complete subalgebra of A. The dual of such an object is a triple (X, Y, h), where X and Y are Boolean spaces and h is a continuous and open mapping from X onto Y. (The open character of h is what corresponds to relative completeness; see [3, § 12].)

The elements p of A are (correspond to) clopen subsets of X; the action of the quantifier is described in topological terms by

$$\exists p = h^{-1}hp.$$

Constants correspond to cross sections of the mapping h. A *cross section* of h is a continuous mapping g from Y into X such that hg is equal to the identity on Y; the action of the corresponding constant b is described in topological terms by

$$bp = g^{-1}p.$$

To describe the more general C-constants, let Z be the dual space of C, and let a be the identity mapping from B into C. In algebraic terms a C-constant is a homomorphism b from A into C such that the diagram

$$\begin{array}{ccc} & A & \\ {}^{c}\nearrow & & \searrow^{b} \\ B & \underset{a}{\longrightarrow} & C \end{array}$$

is commutative. The corresponding topological concept is a continuous mapping

g from Z into X such that the diagram

$$\begin{array}{ccc} & X & \\ {}^{h}\swarrow & & \nwarrow^{g} \\ Y & \underset{f}{\leftarrow} & Z \end{array}$$

is commutative; here f, of course, is the dual of the injection a, so that f is a continuous mapping of Z onto Y.

The principal tool used in the proof of Theorem 1 is Theorem 3 (Sikorski's extension theorem). In both theorems an important role is played by the assumption that the algebra C is complete. The topological equivalent of that assumption says that the space Z is *extremally disconnected* (i.e., that the closure of every open subset of Z is open). In view of this equivalence the dual of Theorem 3 can be formulated as follows.

THEOREM 4. *If h is a continuous mapping from a Boolean space X onto a Boolean space Y, and if f is a continuous mapping from an extremally disconnected Boolean space Z into Y, then there exists a continuous mapping g from Z into X such that $hg = f$.*

It turns out that Theorem 4 is a special case of a theorem proved by Gleason [2]. The general version says that the same conclusion holds even if X and Y are not required to be Boolean spaces, but are allowed to be arbitrary compact Hausdorff spaces. Gleason has proved also that if Z is a compact Hausdorff space such that the conclusion of Theorem 4 holds for all compact Hausdorff spaces X and Y and all continuous mappings f and h (with h onto), then Z must be extremally disconnected; in this sense Gleason's generalized version of Theorem 4 is the best possible result of its kind. (A large part of the clarification of the relation between Sikorski's theorem and Gleason's was obtained in the course of a conversation with F. B. Wright.)

Both Theorem 1 and Theorem 2 assert the existence of many homomorphisms (constants) satisfying certain conditions. A glance at the proofs shows that the main difficulty in both cases is the existence of one such homomorphism. To arrange for that homomorphism to satisfy a prescribed equation (be a witness to something) is a matter of relatively easy algebraic technique. Accordingly, the heart of Theorem 1 is the weakened statement obtained by concluding the existence of just one C-constant. That weakened statement, in turn, is a rather special case of Sikorski's theorem, and hence, in dual form, of Gleason's theorem. A similar procedure (weaken, generalize, dualize) leads from Theorem 2 to the following result.

THEOREM 5. *A continuous and open mapping from a Boolean space with countable base onto a Boolean space has a cross section.*

Two remarks are in order. First, the assumption that the mapping h be open was generalized out of Theorem 4; in the present case it is indispensable.

Second, the assumption of a countable base (i.e., Hausdorff's second axiom of countability) is just the topological version (the dual) of the countability assumption in Theorem 2.

It turns out that Theorem 5 is a special case of a theorem proved by Michael [5]. The general version says that the same conclusion holds even if the domain of the mapping is required to be merely a complete metric space and the range a paracompact zero-dimensional space. Roughly speaking, Michael stands in the same relation to Theorem 2 as Gleason to Theorem 1.

References

1. Hyman Bass, *Finite monadic algebras*, Proceedings of the American Mathematical Society, vol. 9(1958), pp. 258–268.
2. Andrew M. Gleason, *Projective topological spaces*, Illinois Journal of Mathematics, vol. 2(1958), pp. 482–489.
3. Paul R. Halmos, *Algebraic logic*, I. *Monadic Boolean algebras*, Compositio Mathematica, vol. 12(1955), pp. 217–249.
4. Paul R. Halmos, *Free monadic algebras*, Proceedings of the American Mathematical Society, vol. 10(1959), pp. 219–227.
5. E. Michael, *Selected selection theorems*, American Mathematical Monthly, vol. 63(1956), pp. 233–238.
6. Roman Sikorski, *A theorem on extensions of homomorphisms*, Annales Polonici Mathematici, vol. 21(1948), pp. 332–335.
7. O. Varsavsky, *Quantitiers and equivalence relations*, Revista Matemática Cuyana, to appear.

University of Chicago
and
Institute for Advanced Study

IV

FREEDOM

FREE MONADIC ALGEBRAS[1]

A monadic (Boolean) algebra is a Boolean algebra A together with an operator $\exists$ on A (called an existential quantifier, or, simply, a quantifier) such that $\exists 0=0$, $p \leqq \exists p$, and $\exists(p \wedge \exists q)=\exists p \wedge \exists q$ whenever p and q are in A. Most of this note uses nothing more profound about monadic algebras than the definition. The reader interested in the motivation for and the basic facts in the theory of monadic algebras may, however, wish to consult [**2**].

Every Boolean algebra can be converted into a monadic algebra, usually in several ways. (One way is to write $\exists p=p$ for all p; another is to write $\exists p=0$ or 1 according as $p=0$ or $p\neq 0$. These special operators are known as the discrete and the simple quantifier, respectively.) It follows, a fortiori, that every Boolean algebra can be embedded into a monadic algebra, and it is clear, on grounds of universal algebra, that among the monadic extensions of a Boolean algebra there is one that is "as free as possible."

To be more precise, let us say that a monadic algebra A is a *free monadic extension* of a Boolean algebra B if

(i) *B is a Boolean subalgebra of A,*

(ii) *A is (monadically) generated by B,*

(iii) *every Boolean homomorphism g that maps B into an arbitrary monadic algebra C has a (necessarily unique) extension to a monadic homomorphism f that maps A into C.*

The statement that a monadic extension "as free as possible" always exists means that every Boolean algebra B has a free monadic extension A; the algebra A is uniquely determined to within a monadic isomorphism that is equal to the identity on B. The purpose of what follows is to give a "constructive" proof of this fact, i.e., a proof that exhibits A by means of certain set-theoretic constructions based on B (instead of a structurally not very informative proof via equivalence classes of strings of symbols). A by-product of the proof is a rather satisfying insight into the structure of the (Stone) dual space of the free monadic extension. The theorem can (and will) be formulated in such a way as to subsume the results of Hyman Bass [**1**] on the cardinal number of finitely generated free monadic algebras.

The idea of the construction is this. Step 1: form a free Boolean algebra generated by a copy of B. Step 2: adjoin the elements of that

Received by the editors July 24, 1958.

[1] Research supported in part by a contract with the U. S. Air Force.

free algebra to those of B. Step 3: let $\exists$ be the mapping from B into the enlarged algebra that assigns to each element of B the corresponding free generator, and reduce the enlarged algebra by the relations that hold when $\exists$ is an increasing hemimorphism (i.e., $\exists 0=0$, $p \leqq \exists p$, and $\exists(p \vee q) = \exists p \vee \exists q$). Step 4: extend $\exists$ to the entire enlarged (and then reduced) algebra so as to ensure that the elements of the Boolean algebra generated by $\exists B$ are invariant under $\exists$ (and hence that $\exists$ is indeed a quantifier). A little reflection on freedom (especially for Boolean algebras) suggests that the best approach is via duality. The desired algebra will be constructed as the dual of its dual space, and, in order to construct that dual space, the steps of the outline sketched above will be carried out in dualized form. The details go as follows.

STEP 1. Let W be the set of all 2-valued functions on B. (The symbol "2" here and throughout denotes the two-element Boolean algebra.) Endowed with the product-space topology, W is a Boolean space (i.e., a totally disconnected compact Hausdorff space).

STEP 2. Let Y be the (closed) subspace of W that consists of all 2-valued homomorphisms on B, and form the Cartesian product $Y \times W$. The space Y is the dual of the algebra B.

STEP 3. Let V be the (closed) subspace of W that consists of all those 2-valued hemimorphisms on B that map 1 onto 1, and let X be the set of all those points (y, v) of the product $Y \times V$ for which y is dominated by v. A 2-valued function v on B is a hemimorphism if

$$v0 = 0 \tag{1}$$

and

$$v(p \vee q) = vp \vee vq \tag{2}$$

whenever p and q are in B. To say that v dominates a 2-valued function y on B (notation: $y \leqq v$) means that

$$yp \leqq vp \tag{3}$$

whenever p is in B. Since yp and vp vary continuously with (y, v), the set X is a closed subset of $Y \times V$, and, consequently, X is a Boolean space on its own right.

STEP 4. Let A be the dual algebra of the space X, and write

$$(\exists p)(y, v) = \bigvee_u p(u, v) \tag{4}$$

whenever p is in A. The supremum in (4) is extended over all those u in Y for which (u, v) is in X (i.e., for which u and v satisfy (1), (2), and (3)). To say that A is the dual of X means that A is the set of all 2-valued continuous functions on X.

The notation indicates, and it is in fact true, that the preceding construction produces the desired monadic algebra A; it is, however, not yet clear how the given Boolean algebra B is to be embedded in A. This is achieved by means of the dual h of the natural projection c from X to Y. The projection c is, of course, given by

$$c(y, v) = y.$$

The dual mapping h sends each element p of B onto the element hp of A defined by

$$(hp)(y, v) = yp. \tag{5}$$

The principal result of this note can now be stated as follows.

THEOREM. *The mapping h defined by* (5) *is a monomorphism from the Boolean algebra B into the Boolean algebra A; the operator $\exists$ defined by* (4) *is a quantifier on A; the monadic algebra A with the quantifier $\exists$ is a free monadic extension of its Boolean subalgebra hB.*

Since the projection c is obviously continuous, and since h is indeed its dual (i.e., $(hp)(x) = c(x)p$ whenever x is in X and p is in A), the mapping h is a homomorphism from B into A. To prove that h is one-to-one is the same as to prove that c maps X onto Y. This, in turn, is almost obvious; if y is in Y, then the point (y, y) belongs to X, and, of course, its image by c is exactly y.

A useful tool is the projection d from X to V defined by

$$d(y, v) = v.$$

An examination of the notation shows that

$$(\exists p)(x_0) = \bigvee \{p(x): d(x) = d(x_0)\}$$

whenever x is in X and p is in A. In view of this connection between $\exists$ and d, it is an immediate consequence of §12 of [2] that $\exists$ is a quantifier on A if and only if d is a Boolean mapping on X, i.e., if and only if d is a continuous and open mapping of X onto V. The next part of the proof is devoted to showing that d does indeed have these properties. (The necessity for considering both the projections c and d emerged in the course of a conversation with F. B. Wright. The projection c is open also, but this fact is not needed below.)

The continuity of d is immediate. To prove that d is open, it is sufficient to prove that the image of each open set in a basis for X is open. A basis for X is obtained by forming a basis for $Y \times V$ and intersecting each set in that basis with X. This implies that as q varies over B and U varies over open subsets of V, the subset G of X, defined by

$$G = \{(y, v) \in X: yq = 0, v \in U\},$$

varies over a basis for X. It is therefore sufficient to prove that

$$dG = \{v \in V: vq' = 1\} \cap U$$

for each such G; the openness of the right side is a consequence of the definition of the product-space topology in V.

If v is in dG, so that in particular $y \leqq v$, then $yq' \leqq vq'$, so that $vq' = 1$ (and, of course, v is in U). Suppose now, conversely, that $vq' = 1$ and v is in U. If a 2-valued function w on B is defined by

$$wp = \begin{cases} 0 \text{ if } vp = 0 \text{ or if } p \leqq q, \\ 1 \text{ otherwise,} \end{cases}$$

then w is a 2-valued hemimorphism on B and $w \leqq v$. The hemimorphism w is not identically zero (since $wq' = 1$). Suppose now that there exists a 2-valued homomorphism y on B such that $y \leqq w$. Since $wq = 0$, this supposition implies that $yq = 0$ and hence that (y, v) is in G; it follows that v is in dG, and (except for the unsupported supposition) this completes the proof of the openness of d. If the characterization of dG is applied to the case in which $q = 0$ and $U = V$, it follows that d maps X onto V.

The supposition of the preceding paragraph is a fact. To every 2-valued hemimorphism w on B there corresponds at least one 2-valued homomorphism y in B such that $y \leqq w$. It is even true that

$$wp = \bigvee \{yp: y \leqq w\}, \tag{6}$$

where p is in B and the supremum is extended over all 2-valued homomorphisms y on B that are dominated by w. This fact is a simple special case of the duality theorem for hemimorphisms [2, Theorem 8]; the special case has an easy independent proof also. (It would be of interest to know how far (6) can be generalized beyond the 2-valued case. In other words: when is a hemimorphism the supremum of the homomorphisms below it?)

For further progress it is convenient to digress to a brief study of functions of two variables that really depend on only one variable. A function p on a product space such as $Y \times V$ is *independent* of Y if $p(y_1, v) = p(y_2, v)$ whenever y_1 and y_2 are in Y and v is in V; the definition of independence of V is similar. It is well known and easy to prove that the Boolean algebra generated by the set of all 2-valued continuous functions that are independent of either Y or V is equal to the algebra of all 2-valued continuous functions on $Y\times V$. (Every clopen set in $Y\times V$ is a finite union of clopen rectangles.)

Consider now functions on a closed subspace of $Y\times V$, such as, say, X. The definitions of independence of Y or of V still make sense; it is necessary only to insist that all the points of $Y\times V$ that are mentioned belong to the subspace X. Is it true that the algebra generated by all "one-variable" 2-valued continuous functions is equal to the algebra A of all 2-valued continuous functions on X? The answer is yes. Indeed, the process of restriction to X is a Boolean homomorphism from the algebra of all 2-valued continuous functions on $Y\times V$ *onto* the algebra of all 2-valued continuous functions on X; the asserted affirmative answer follows from the fact that the restriction of a one-variable function is a one-variable function.

One more auxiliary comment will be useful: every one-variable function on X is the restriction of some one-variable function on $Y\times V$ (2-valued and continuous understood throughout). The emphasis here (as opposed to the preceding paragraph) is that restriction, already known to be an onto map, maps a useful subset onto a useful subset. For the proof, suppose that p on X is independent of, say, Y. There exists a unique (necessarily 2-valued and necessarily continuous) function q on V such that $p(y, v) = q(v)$ whenever (y, v) is in X. If r is defined on $Y\times V$ by $r(y, v) = q(v)$, then r is independent of Y and the restriction of r to X is exactly p.

In terms of the concepts just discussed it is possible to formulate a useful fact about the way h embeds B into A; the assertion is that hB consists exactly of those functions in A that are independent of V. Inclusion one way is obvious: if p is in B, then hp is independent of V by the definition (5). If, conversely, q is in A and q is independent of V, then q is the restriction to X of some 2-valued continuous function r on $Y\times V$ that is independent of V. The elementary theory of Boolean duality implies that there exists an element p of B such that $r(y, v) = yp$ identically in y and v, and hence $q = hp$.

Knowledge about hB can be exploited to get some knowledge about $\exists hB$. The basic fact is that

$$(\exists hp)(y, v) = vp \tag{7}$$

whenever p is in B and (y, v) is in X. For the proof, note first that

$$(\exists hp)(y, v) \leqq vp.$$

The fact that this upper bound is always attained follows from (6).

The preceding paragraph shows that every function in $\exists hB$ is independent of Y. (Caution: $\exists hB$ is not in general a Boolean algebra.) In the converse direction it is true that the Boolean algebra generated by $\exists hB$ includes all the functions in A that are independent of Y. The reason is that the coordinate functions on V generate the algebra of all 2-valued continuous functions on V; by (7), the restriction of the coordinate functions to X consists exactly of the functions in $\exists hB$.

It is now clear that the monadic algebra generated by hB is A. Indeed, that generated monadic algebra includes the Boolean algebra generated by $hB \cup \exists hB$, hence it contains all one-variable functions in A, and, consequently, it coincides with A.

It remains to prove the third and crucial condition that makes A a free monadic extension of hB. Suppose, accordingly, that g is a Boolean homomorphism that maps hB into a monadic algebra C. Let Z be the dual space of C, i.e., the set of all 2-valued homomorphisms on C. There are natural (and obviously continuous) mappings a and b from Z into Y and V, respectively, defined by

$$(az)p = zghp \quad \text{and} \quad (bz)p = z\exists ghp$$

whenever p is in B. (Note that "$\exists$" on the right side of the last equation denotes quantification in C.) It is easy to verify that (az, bz) is in X for all z in Z. The desired extension f of g associates with each element q of A that element fq of C for which

$$zfq = q(az, bz)$$

for all z in Z.

The verification that f is a Boolean homomorphism from A into C is straightforward. To prove that f is an extension of g, suppose that p is in B and $q = hp$. Since

$$\begin{aligned} zfq = q(az, bz) &= (hp)(az, bz) = (az)p \\ &= zghp = zgq \end{aligned}$$

for every z in Z, it follows that fq is indeed equal to gq.

The last thing to prove is that f is a monadic homomorphism, i.e., that

$$zf\exists q = z\exists fq$$

whenever z is in Z and q is in A. Assume first that $q = hp$ for some p in B; then

$$\begin{aligned} zf\exists q &= zf\exists hp = (\exists hp)(az, bz) \text{ (by the definition of } f) \\ &= (bz)p \text{(by (7))} = z\exists ghp \text{ (by the definition of } b) \\ &= z\exists fhp = z\exists fq. \end{aligned}$$

It is now tempting to argue as follows: $f\exists$ agrees with $\exists f$ on hB and hB generates the monadic algebra A, hence $f\exists$ agrees with $\exists f$ throughout A. This argument assumes that the set of elements on which $f\exists$ agrees with $\exists f$ is closed under the Boolean operations and under quantification. The assumption is not justified. The trouble is caused by infima and complements; the mappings $f\exists$ and $\exists f$ are not homomorphisms but merely hemimorphisms.

A minor refinement of the argument proposed above works. In its unrefined form the argument applies to every set of monadic generators of A (and fails); the refinement makes use of the fact that the particular generating set at hand (namely hB) is a Boolean subalgebra of A. This fact implies (and in any case it was proved above) that the Boolean algebra generated by $hB \cup \exists hB$ is A. It follows that every element of A is a finite supremum of elements of the form

$$r = hp \wedge (\exists hq_1)^{\pm} \wedge \cdots \wedge (\exists hq_n)^{\pm},$$

where p and all the q's are in B and the ambiguous signs refer to possible complementations. Since both $f\exists$ and $\exists f$ preserve suprema, the desired end can be achieved by proving that $f\exists r$ and $\exists fr$ always coincide. The verification that this is so is a simple computation; the crucial step uses the fact that $f\exists$ and $\exists f$ do agree on hB.

The proof of the principal theorem is now complete. Since h may be regarded as an embedding of B into A, the subalgebra hB may be identified with B; the fact that A first presents itself as a free monadic extension of hB (and not of B) is merely a matter of notation.

The proof of uniqueness is an easy exercise in universal algebra. Suppose, indeed, that both A_1 and A_2 are free monadic extensions of B. If g_1 and g_2 are the natural injections of B into A_1 and A_2, respectively, then (by freedom) there exist monadic homomorphisms f_1 (from A_2 into A_1) and f_2 (from A_1 into A_2) that extend g_1 and g_2. Since f_1f_2 is a monadic endomorphism of A_1 that agrees with the identity on B, it must be equal to the identity throughout A_1, and, similarly, f_2f_1 is equal to the identity throughout A_2. This means that f_1 is a monadic isomorphism from A_2 onto A_1 with inverse f_2; the existence of such an isomorphism (equal to the identity on B) is what is meant by saying that the free monadic extension is essentially unique.

If S is an arbitrary set, the preceding results can be applied to the free Boolean algebra B generated by S. An arbitrary mapping from S into an arbitrary monadic algebra C has a (necessarily unique) extension to a Boolean homomorphism g that maps B into C. The

Boolean homomorphism g, in turn, has a (unique) extension f that maps A into C. Conclusion: the free monadic extension of a free Boolean algebra is a free monadic algebra.

If the number n of elements of S is finite, then, as is well known, B has 2^n atoms; this says exactly that the dual space Y has 2^n elements. How many elements are there in X? To find the answer, choose an arbitrary element y in Y (there are 2^n ways of doing this), and ask for the number of elements v in V that can be appended to it so as to yield an element of X. In order that a 2-valued mapping v on B have this desired property, it is necessary and sufficient that v be a 2-valued hemimorphism on B that dominates y. A hemimorphism on B is uniquely determined by its values on the atoms of B, and those values can be assigned arbitrarily; this implies that the number of 2-valued hemimorphisms on B is the number of sets of atoms, i.e., 2^{2^n}. If p is the element of B that is mapped onto 1 by y and onto 0 by all other elements of Y, then a necessary and sufficient condition that a hemimorphism v dominate y is that it send p onto 1 (recall that a hemimorphism is monotone). This observation finishes the enumeration; the number of hemimorphisms v for which $vp=1$ is just half the total number, and, therefore, the number of elements in X is $2^n \cdot 2^{2^n-1}$. Conclusion: the free monadic algebra with n generators has $2^n \cdot 2^{2^n-1}$ atoms. This result was first obtained by Bass [1].

It is of some interest to observe that such enumeration results are of a strictly monadic character; the situation becomes completely different in the presence of two quantifiers. (This observation was made orally by Alfred Tarski.) The fact is that a single generator acted upon by the Boolean operations and two (commutative) quantifiers yields an infinite free algebra. The most efficient way of proving this is to exhibit one concrete (but not necessarily free) algebra of this kind that is generated by one element but has altogether infinitely many elements.

For this purpose, let K be the set of all pairs of natural numbers, and for each subset p of K let $\exists_0 p$ [or $\exists_1 p$] be the union of all horizontal [or vertical] lines that meet p. (Lines here consist, of course, of lattice points only.) The operators $\exists_0$ and $\exists_1$ are commutative quantifiers on the algebra 2^K. Let A be the least Boolean subalgebra of 2^K that contains the "triangle"

$$p_0 = \{(x_0, x_1) \in K: x_0 \leqq x_1\}$$

and is closed under the actions of both $\exists_0$ and $\exists_1$. Assertion: A has infinitely many elements. For the proof, write $q_0 = p_0'$ and define the sequences $\{p_n\}$ and $\{q_n\}$ inductively by writing

$$p_{n+1} = p_n \wedge \exists_1 q_n \quad \text{and} \quad q_{n+1} = q_n \wedge \exists_0 p_{n+1}.$$

A casual examination of either the geometry or the logic of the situation shows that the ranges of both these sequences are infinite; in fact

$$p_n = \{(x_0, x_1) \in K: n \leqq x_0 \leqq x_1\}$$

and

$$q_n = \{(x_0, x_1) \in K: x_0 > x_1 \geqq n\}.$$

The construction of this example was inspired by a conversation with Dana Scott.

References

1. Hymann Bass, *Finite monadic algebras*, Proc. Amer. Math. Soc. vol. 9 (1958) pp. 258–268.

2. Paul R. Halmos, *Algebraic logic*, I. *Monadic Boolean algebras*, Compositio Math. vol. 12 (1955) pp. 217–249

University of Chicago and
Institute for Advanced Study

V

GENERAL THEORY

Algebraic logic (II)

Homogeneous locally finite polyadic Boolean algebras of infinite degree

Table of contents

Introduction

The purpose of this paper is to do for the lower functional calculus what the first paper of the series did for the monadic special case. To a very great extent, however, the two papers are independent of each other. The situation is similar to that in the theory of differential equations. Once the basic but relatively elementary concept of differentiation is known, it can be used to give a mathematical formulation of the problems of classical Newtonian mechanics, usually in the form of ordinary differential equations. It can also be used to give a mathematical formulation to some of the more sophisticated problems of modern physics, usually in the form of partial differential equations. Similarly, once the algebraic version of the logical operation of quantification is defined, it can be applied to the algebraic study of either classical or modern logic. The analogy is quite close. Traditional Aristotelean logic can be viewed as the theory of propositional functions of a single variable; the lower functional calculus, on the other hand, treats propositional functions of any number of variables.

The analogy between physics and logic can be used to illuminate not only the role of this paper in the series, but, for that matter, the purpose of algebraic logic as a whole. It is fashionable nowadays to give a mathematical exposition of quantum mechanics in terms of certain algebraic systems. The concepts occurring in such an exposition, the terms used to describe them, and the problems selected for special emphasis are all suggested by the facts of physics. The algebra does not pretend to solve physical problems or to give new information about the state of the universe; its primary aim is to study a mathematical subject by mathematical methods. The mathematical machinery set up for this purpose is (unlike its extramathematical source) aesthetically attractive to and intellectually fathomable by the professional mathematician. It can happen that the mathematics will eventually come to be applicable to its own parent and capable of contributing to its growth, but the mathematics is of value even if that does not happen. It is of value not only in an intrinsic mathematical sense, but also because it can be used as a map with which the professional mathematician can venture into the territory of the professional physicist. If in the preceding description of the role of algebraic treatments of quantum mechanics words such as "physics" and "physicist" are replaced by "logic" and "logician", the result is a fair description of the intended role of algebraic logic.

Not all the problems that interest a logician have their counterpart in algebraic logic. The reason for this is best seen in another analogy: roughly speaking, ordinary logic is to permutation groups as algebraic logic is to abstract (axiomatically defined) groups. Since the problems of transitivity and primitivity cannot even be expressed in the language of abstract group theory, the generality and elegance of that subject are obtained at the cost of some terminological and conceptual sacrifice. For some purposes a more illuminating ratio is this: ordinary logic is to free groups as algebraic logic is to abstract groups. This ratio is very suggestive. The logician's customary exposition of the propositional calculus is, in fact, a description of the free Boolean algebra on a countable set of generators; similarly, the customary treatments of the functional calculi can be viewed as studies in the theory of free polyadic algebras.

In group theory both the algebraic and the combinatorial methods have an important place; the reason for the existence of algebraic logic is the belief that the same is true in logic.

It is not claimed that the algebraic point of view simplifies all definitions and shortens all proofs; in many cases just the opposite is true. The pertinent analogy here is the relation between the classical theory of Hermitian matrices and the approach to the same subject via Hilbert space. The definition of the conjugate transpose of a complex matrix can be given with perfect rigor on the first page of an elementary book. The corresponding definition in a book on Hilbert space must wait its turn; many properties of inner products and linear functionals have to be established before the adjoint of an operator can make sense in an invariant, algebraic setting. An analogous situation in logic occurs, for instance, in connection with the concept of an individual constant. This simple concept is usually (and quite properly) dismissed with a few words in ordinary treatments of the subject; in the algebraic treatment (cf. § 12) it becomes a part of the most profound development of the theory.

The basic concepts of algebraic logic can be traced back to the work of Tarski and his collaborators; cf., for instance, [4], [6], and [7]. (Unfortunately the proofs of Tarski's results on cylindric algebras have not yet been published, and it is, consequently, not easy to assess the exact extent of the overlap between the theories of cylindric algebras and polyadic algebras. It is probable that the basic difficulties are the same; the concepts and the techniques are likely to be different.) It will also be obvious to the reader familiar with some of the recent literature of logic that this paper was strongly influenced by Henkin's thesis [3] and by some of the techniques introduced by Rasiowa and Sikorski [5]. It is a pleasure, finally, to express my indebtedness to A. F. Bausch and to Alfred Tarski. I had many stimulating conversations about functional calculi with Bausch and about cylindric algebras with Tarski; I am grateful for their friendly advice and encouragement.

The principal results of the paper are the functional representation (Theorem 10.9), the adjunction of variables (Theorem 11.9), the adjunction of witnessing constants (Theorem 19.6), and the representation of simple algebras (Theorem 17.3). The reader who wants to get a quick snapshot of the subject, avoiding most of the technicalities, can proceed along the following course: glance at § 1 for the notation, read § 2, omit § 3 and § 4, read the definition of local finiteness in § 5, omit § 6 and § 7, read § 8 and § 9, read the remarks and the statements of the theorems (but not the proofs) in § 10, read the definition of dilations in § 11 and the definition of constants in § 12, omit § 13 and § 14, and read the definitions and the theorems in § 15, § 16, and § 17.

§ 1. Monadic algebras

For convenience of reference this section contains a summary of those definitions and theorems from the theory of monadic algebras that will be used in the sequel. The theorems in this summary all belong to the elementary part of the subject; their proofs are easy consequences of the definitions and of the standard theory of Boolean algebras. The details and the heuristic motivation are given in [2].

We begin by establishing the notation that will be used for Boolean algebras. The symbols $\bigvee$ and $\bigwedge$ are used to indicate the supremum and the infimum, respectively, of sets in a Boolean algebra. Other Boolean symbols, which will be used without any further explanation below, are as follows: $p \vee q$ is the supremum of p and q, $p \wedge q$ is the infimum of p and q, p' is the complement of p, 0 is the zero element of the algebra, 1 is the unit element of the algebra, $p-q(=p \wedge q')$ is the difference between p and q, $p+q\big(=(p-q)\vee(q-p)\big)$ is the Boolean sum (or symmetric difference) of p and q. The symbols $\leqslant$ and $\geqslant$ will denote the natural order in a Boolean algebra, so that, for instance, $p \leqslant q$ means that $p \wedge q = p$. The identity mapping of a Boolean algebra onto itself will be denoted by e. It is understood that, by definition, a Boolean algebra contains at least two distinct elements. The symbol $\boldsymbol{O}$ will be used to denote the two-element Boolean algebra $\{0,1\}$, so that $\boldsymbol{O}$ is a Boolean subalgebra of every Boolean algebra.

A *quantifier* (or, properly speaking, an *existential quantifier*) is a normalized, increasing, and quasi-multiplicative mapping $\exists$ of a Boolean algebra into itself; in other words, it is a mapping $\exists$ such that if p and q are in the algebra, then

$$\exists 0 = 0\,,$$
$$p \leqslant \exists p\,,$$
$$\exists(p \wedge \exists q) = \exists p \wedge \exists q\,.$$

The *discrete* quantifier is the identity mapping e; the *simple* quantifier is the one for which $\exists p = 1$ whenever $p \neq 0$. A *hemimorphism* is a normalized and additive mapping f from one Boolean algebra into another; in other words, it is a mapping f such that $f0 = 0$ and such that if p and q are in its domain, then $f(p \vee q) = fp \vee fq$. Clearly every Boolean homomorphism (and, in particular, every Boolean endomorphism) is a hemimorphism.

The following simple facts are basic in the theory of quantifiers.

(1.1) *A quantifier is idempotent (i. e.,* $\exists\exists = \exists$*).*

(1.2) *A quantifier is monotone (i. e., if* $p \leqslant q$*, then* $\exists p \leqslant \exists q$*).*

(1.3) *The range of a quantifier is a Boolean algebra.*

(1.4) *A quantifier is additive* (*i. e.*, $\exists(p \vee q) = \exists p \vee \exists q$).

(1.5) *If* $\exists$ *is a quantifier and if* p *and* q *are elements of its domain, then* $\exists p + \exists q \leqslant \exists(p+q)$.

We observe that in view of (1.4) a quantifier is a hemimorphism; from this and from the elementary fact that a hemimorphism is monotone we can recapture (1.2). The reason for stating the facts in the order used above is that in the proof of (1.4) it is convenient to make use of (1.2).

A *monadic algebra* is a Boolean algebra $\boldsymbol{A}$ together with a quantifier $\exists$ on $\boldsymbol{A}$. A subset $\boldsymbol{B}$ of a monadic algebra $\boldsymbol{A}$ is a *monadic subalgebra* of $\boldsymbol{A}$ (or a *monadic ideal* in $\boldsymbol{A}$) if and only if $\boldsymbol{B}$ is a Boolean subalgebra of $\boldsymbol{A}$ (or a Boolean ideal in $\boldsymbol{A}$) such that $\exists p$ belongs to $\boldsymbol{B}$ whenever p belongs to $\boldsymbol{B}$. A mapping f between monadic algebras is a *monadic homomorphism* if it is a Boolean homomorphism such that $f\exists p = \exists f p$ for every p in its domain. (The adjective "monadic" will be used with "subalgebra", "ideal", "homomorphism", etc., whenever it is advisable to emphasize the distinction from other kinds of subalgebras, ideals, homomorphisms, etc. – *e. g.*, from the plain Boolean kind. Usually, however, the adjective will be omitted and the context will unambigously indicate what is meant. A similar remark applies to the polyadic concepts that will be introduced later.) A monadic algebra is *simple* if $\{0\}$ is the only proper ideal in it. A monadic ideal is *maximal* if it is not a proper subset of any proper ideal. A monadic algebra is *semisimple* if the intersection of all maximal ideals in it is $\{0\}$.

(1.6) *A monadic algebra is simple if and only if its quantifier is simple.*

(1.7) *Every monadic algebra is semisimple.*

If X is a non-empty set and if $\boldsymbol{B}$ is a Boolean algebra, the set of all functions from X into $\boldsymbol{B}$ is a Boolean algebra with respect to the pointwise operations. A $\boldsymbol{B}$-valued *functional monadic algebra* with domain X is a Boolean subalgebra $\boldsymbol{A}$ of the algebra of all functions from X into $\boldsymbol{B}$ such that, for each p in $\boldsymbol{A}$, the range of p has a supremum in $\boldsymbol{B}$, and such that the (constant) function, whose value at every point of X is that supremum, belongs to $\boldsymbol{A}$. If that constant function is denoted by $\exists p$, then $\exists$ is a quantifier on $\boldsymbol{A}$, so that a functional monadic algebra is a monadic algebra.

(1.8) *An* $\boldsymbol{O}$*-valued functional monadic algebra is simple.*

A *constant* of a monadic algebra $\boldsymbol{A}$ (with quantifier $\exists$) is a Boolean endomorphism c on $\boldsymbol{A}$ such that $c\exists = \exists$ and $\exists c = c$.

(1.9) *A constant is idempotent* (*i. e.*, $cc = c$).

(1.10) *If* c *is a constant of a monadic algebra* $\boldsymbol{A}$ *with quantifier* $\exists$, *then* $cp \leqslant \exists p$ *for every* p *in* $\boldsymbol{A}$.

§ 2. Functional polyadic algebras

If "proposition" is taken to mean an element of a Boolean algebra $\boldsymbol{B}$, then it is natural to interpret "propositional function" as a function from some set into $\boldsymbol{B}$. A propositional function of several variables becomes in this scheme a function from a Cartesian product of several sets into $\boldsymbol{B}$. In order to keep the notation from becoming cumbersome, we shall restrict our attention to Cartesian powers, *i. e.*, to sets of the form X^I, where I is an arbitrary index set. (Cf. also the discussion at the end of § 4.) An element of X^I is, by definition, a function from I into X; the value of such a function x at an index i will be denoted by x_i.

The set of all functions from a non-empty set into a Boolean algebra is a Boolean algebra with respect to the pointwise operations. According to the usual conventions the set X^I is empty if and only if X is empty but I is not. In order to ensure the non-emptiness of X^I we shall always assume that the set X is not empty. Since the elements of the set X constitute the universe of discourse with which the propositions and propositional functions of the theory are concerned, this assumption is a reasonable one.

The main thing that distinguishes the abstractly given Boolean algebra $\boldsymbol{B}$ from the Boolean algebra of all $\boldsymbol{B}$-valued functions on X^I is that the concrete representation of the latter automatically endows it with a lot of new structure. The additional structure comes from the presence of variables. Substituting some variables for others, or holding some variables fixed while forming a supremum over the others, we obtain new propositional functions out of old ones.

To make these indications more specific, we consider the set of all transformations on I, *i. e.*, the set of all transformations whose domain is the set I and whose range is included in I. Since the transformations we consider are not necessarily one-to-one and not necessarily onto, the set of all such transformations does not form a group. Since, however, the product (*i. e.*, composite) of two transformations on I is again a transformation on I, and since transformation multiplication is associative, the set of all transformations on I does form a semigroup. Since the identity mapping from I onto itself (which we shall always denote by δ) is a transformation on I, the semigroup contains a (necessarily unique) unit.

The idea of "substituting some variables for others" is to apply a transformation on I to the indices of the arguments of a function from X^I into $\boldsymbol{B}$. The desideratum is to associate with each transformation τ on I, in a natural manner, a transformation $\boldsymbol{S}(\tau)$ that sends propositional functions (*i. e.*, functions from X^I into $\boldsymbol{B}$) into propositional functions.

The definition of $S(\tau)$ is most easily given in terms of an auxiliary transformation τ_* on X^I. The transformation τ_* is defined by

$$(2.1) \qquad (\tau_* x)_i = x_{\tau i}$$

for all i in I and for all x in X^I. The functional transformation $S(\tau)$ associated with a transformation τ on I is defined by

$$(2.2) \qquad S(\tau)p(x) = p(\tau_* x)$$

for all x in X^I and for all functions p from X^I into B. (The symbol $S(\tau)p(x)$ is an abbreviation for $(S(\tau)p)(x)$. Such abbreviations will be used frequently without any further explanation.) The precise version of "substituting some variables for others" is the application of a transformation such as $S(\tau)$.

The formation of partial suprema is a slightly more delicate matter. If J is a subset of I (not necessarily proper), then there is associated with J in a natural manner a transformation $\exists(J)$ that sends some (but not necessarily all) propositional functions into others. The definition of $\exists(J)$ is most easily given in terms of an auxiliary binary relation J_* in X^I. The relation J_* is, by definition, such that

$$(2.3) \qquad xJ_*y \quad \textit{if and only if} \quad x_i = y_i \quad \textit{whenever} \quad i \notin J$$

for all x and y in X^I. (The symbols $\in$ and $\notin$ are used to indicate the relations of belonging and non-belonging, respectively.) The functional transformation $\exists(J)$ is defined by

$$(2.4) \qquad \exists(J)p(x) = \bigvee \{p(y) \colon xJ_*y\},$$

provided that the indicated supremum exists for every x in X^I. The idea is to form the supremum of the values of the function p over the points obtained from x by holding fixed the variables outside J and varying the variables in J arbitrarily.

Since $\exists(J)p$ need not exist for every function p from X^I into B, we cannot continue to insist on studying the Boolean algebra of all functions from X^I into B. We must and we shall be content with considering a Boolean subalgebra of that algebra, say A. By requiring that $\exists(J)p$ exist and belong to A for every subset J of I and for every element p of A we can circumvent the difficulty caused by the possible non-existence of suprema. Now, however, another difficulty arises. If τ is a transformation on I, then $S(\tau)p$ is a function from X^I into B whenever p is such, but if we assume that $p \in A$, that does not guarantee that $S(\tau)p \in A$. This difficulty can be circumvented also, simply by requiring that A be closed under the operation $S(\tau)$ for every transformation τ on I.

We can summarize the preceding considerations as follows. Suppose that $\boldsymbol{B}$ is a Boolean algebra and that X and I are sets, with X non-empty. The elements of I will play for us the role of what are usually called *variables* (although they do not, in any sense of the word, vary), and the set X will play the role of the *domain* of the variables (although, in general, the variables do not belong to the domain). The elements of the Boolean algebra $\boldsymbol{B}$ will play the role of the *propositions* that can occur as the values of the propositional functions to be considered. A *functional polyadic (Boolean) algebra* is a Boolean subalgebra $\boldsymbol{A}$ of the algebra of all functions from X^I into $\boldsymbol{B}$, such that $\boldsymbol{S}(\tau)p \in \boldsymbol{A}$ whenever $p \in \boldsymbol{A}$ and τ is a transformation on I, and such that $\exists(J)p$ exists and belongs to $\boldsymbol{A}$ whenever $p \in \boldsymbol{A}$ and J is a subset of I. Whenever it becomes desirable to indicate the constituents used in the definition, $\boldsymbol{A}$ will be called a $\boldsymbol{B}$-valued functional polyadic algebra with domain X and variables I, or, more simply, a $\boldsymbol{B}$-valued I-algebra over X. Functional polyadic algebras are the concrete objects in terms of which propositional functions can be studied; our main purpose is to investigate the abstract algebraic objects of which functional polyadic algebras are the typical representations.

§ 3. Substitutions

In order to find out what are the algebraic properties of the operators $S(\tau)$ and $\exists(J)$ on a functional polyadic algebra, we shall make use of a concept in terms of which we can treat both kinds of operators at the same time. The concept (to be called *substitution*) is not an intuitive one; the only reason for introducing it here is that it is efficient for the purpose at hand. By a substitution in a set I we shall mean a function σ whose domain (dom σ) and range (ran σ) are subsets of I. The fact that the domain of a substitution is not necessarily equal to I will be important for us. In what follows we shall reserve the use of the word *transformation* for a substitution τ such that $\operatorname{dom}\tau = I$; this is in agreement with the use of that word in the preceding section.

If σ and τ are substitutions, their *product* $\sigma\tau$ is defined to be the substitution whose domain is the set of all those elements i of I for which $i \in \operatorname{dom}\tau$ and $\tau i \in \operatorname{dom}\sigma$ and whose value for each i in its domain is defined by $(\sigma\tau)i = \sigma(\tau i)$. It is easy to verify that substitution multiplication is associative, so that, with respect to multiplication as the law of composition, the set of all substitutions in I is a semigroup. This semigroup has a unit, namely the identity transformation δ.

There is another and more familiar semigroup associated with the set I, namely the set of all subsets of I, with respect to the formation of unions as the law of composition. This semigroup also has a unit, namely the empty set $\emptyset$. There is a close relation between (all) subsets of I and (some) substitutions in I. In order to study this relation we make correspond to each subset J the identity mapping, to be denoted by $\chi(J)$, on $I-J$. More explicitly, $\chi(J)$ is the substitution such that $\operatorname{dom}\chi(J)=I-J$ and $\chi(J)i=i$ for all i in $I-J$.

(3.1) LEMMA. *The correspondence χ is a one-to-one correspondence from the set of subsets of I into the set of substitutions in I such that $\chi(\emptyset)=\delta$ and $\chi(J\cup K)=\chi(J)\chi(K)$.*

Proof. Since $\operatorname{dom}\chi(J)\chi(K)=(I-J)\cap(I-K)=I-(J\cup K)$, it follows that $\chi(J)\chi(K)=\chi(J\cup K)$. The proof is completed by the observation that $\chi(J)=\delta$ if and only if $J=\emptyset$.

We note that, in algebraic terms, the result of Lemma 3.1 is simply that the correspondence χ is a semigroup monomorphism.

There is a possible source of minor confusion in connection with the equation $\chi(\emptyset)=\delta$. The empty set $\emptyset$ is also a substitution, namely the unique substitution with empty domain and empty range. It is notationally convenient to make a distinction between the empty set as a subset of I and the empty set as a substitution in I; we shall denote the latter by θ. It is clear that the substitution θ belongs to the range of the correspondence χ; in fact, $\theta=\chi(I)$.

Transformations on I and subsets of I are more or less natural objects, whereas substitutions are apparently artificial constructs. In fact, however, substitutions arise quite naturally from transformations and subsets, via the correspondence χ.

(3.2) LEMMA. *Every substitution σ can be written in the form $\tau\chi(J)$, where τ is a transformation on I and J is a subset of I. The set J in this representation is uniquely determined as $I-\operatorname{dom}\sigma$; the transformation τ must agree with σ outside J and is arbitrary in J.*

Proof. If $\sigma=\tau\chi(J)$, then $\operatorname{dom}\sigma=\operatorname{dom}\chi(J)=I-J$, and, if $i\in I-J$, then $\tau i=\tau\chi(J)i=\sigma i$. This proves the uniqueness assertions. To prove the rest, let τ be an arbitrary transformation that agrees with σ in $\operatorname{dom}\sigma$; an immediate verification shows that if $J=I-\operatorname{dom}\sigma$, then $\sigma=\tau\chi(J)$.

What is the connection between the representation established in Lemma 3.2 and substitution multiplication? The next two results provide the answer to this question.

(3.3) LEMMA. *If τ is a transformation and $J\subset I$, then $\chi(J)\tau=\tau\chi(\tau^{-1}J)$.*

Proof. An element i of I belongs to $\operatorname{dom}\chi(J)\tau$ if and only if $\tau i\in\operatorname{dom}\chi(J)=I-J$, *i. e.*, if and only if $i\in\tau^{-1}(I-J)=I-\tau^{-1}J$. An element i of I belongs to $\operatorname{dom}\tau\chi(\tau^{-1}J)$ if and only if $i\in\operatorname{dom}\chi(\tau^{-1}J)=I-\tau^{-1}J$.

In other words, $\chi(J)\tau$ and $\tau\chi(\tau^{-1}J)$ have the same domain; if i belongs to this common domain, then $\chi(J)\tau i = \tau i$ and $\tau\chi(\tau^{-1}J)i = \tau i$.

(3.4) LEMMA. *If $\sigma_1 = \tau_1\chi(J_1)$ and $\sigma_2 = \tau_2\chi(J_2)$, where τ_1 and τ_2 are transformations, then $\sigma_1\sigma_2 = \tau_1\tau_2\chi(\tau_2^{-1}J_1 \cup J_2)$.*

Proof. Observe that $\sigma_1\sigma_2 = \tau_1\chi(J_1)\tau_2\chi(J_2)$, and apply Lemma 3.3 to the product $\chi(J_1)\tau_2$.

We proceed now to introduce, without any immediately apparent motivation, a concept that will presently turn out to be useful. For substitutions σ and τ we shall say that σ *follows* τ, and we shall write $\sigma \succcurlyeq \tau$, if τ never maps two distinct elements onto the same element of $I - \operatorname{dom}\sigma$. (The notation is not intended to suggest, and it is not in fact true, that the relation thereby defined is a partial order.) The definition can be rephrased as follows: $\sigma \succcurlyeq \tau$ if and only if the conditions $i \neq j$ and $\tau i = \tau j = k$ imply that $k \in \operatorname{dom}\sigma$. Equivalently: $\sigma \succcurlyeq \tau$ if and only if τ is one-to-one on the set $\tau^{-1}(I - \operatorname{dom}\sigma)$. While none of these conditions has much intuitive appeal, the point in considering them is that in addition to being equivalent to each other they also turn out to be equivalent to an algebraically elegant and intuitively natural condition (cf. Theorem 3.8).

To prepare the ground for later work, and also as an aid in understanding this somewhat peculiar concept, we explicitly mention some useful special cases.

(3.5) LEMMA. *If σ is a transformation, then $\sigma \succcurlyeq \tau$ for all τ. If τ is one-to-one, and, in particular, if $\tau = \chi(J)$ for some J, then $\sigma \succcurlyeq \tau$ for all σ.*

Proof. The first assertion follows from the fact that if σ is a transformation, then $I - \operatorname{dom}\sigma$ is empty. The second assertion is obvious.

(3.6) LEMMA. *A necessary and sufficient condition that $\chi(J) \succcurlyeq \tau$ is that τ be one-to-one on $\tau^{-1}J$.*

Proof. The condition means that the restriction of the function τ to the set $\tau^{-1}J$ is one-to-one, and this, in turn, means exactly that τ never maps two distinct elements onto the same element of J. Since $J = I - \operatorname{dom}\chi(J)$, the proof is complete.

Now we consider again a Cartesian power X^I, where X is a non-empty set; we shall discuss the way in which the action of the substitutions discussed above is reflected in the set X^I. This is most easily done in terms of an auxiliary binary relation σ_* in X^I associated with each substitution σ in I. The relation σ_* (called the *dual* of σ) is, by definition, such that

(3.7) *$x\sigma_* y$ if and only if $x_{\sigma i} = y_i$ whenever $i \in \operatorname{dom}\sigma$*

for all x and y in X^I.

The duals of transformations and the duals of substitutions of the form $\chi(J)$ are particularly simple. If τ is a transformation, then $x\tau_* y$ means that $x_{\tau i}=y_i$ for all i. Hence, in this case, y is uniquely determined by x, so that the relation τ_* is a function; the value of the function τ_* at a point x of X is given by $(\tau_* x)_i=x_{\tau i}$. (We have thus recaptured (2.1) as a special case of (3.7).) Note in particular that δ_* is the identity mapping on X^I. If $\sigma=\chi(J)$, then $x\sigma_* y$ means that $x_i=y_i$ whenever $i \notin J$. (We have thus recaptured (2.3) as a special case of (3.7).) In this case σ_* is obviously an equivalence relation; it is exactly the relation that we denoted by J_* in the preceding section. Since $\chi(\emptyset)=\delta$, the fact that δ_* is the relation of equality in X^I is the relation version of the characterization of δ_* stated just above in the language of mappings. We note that, at the opposite extreme, $\theta_*=I_*$ is the trivial equivalence relation that places all the points of X^I into the same equivalence class. Another worth while observation is that if $J\subset K$, then $J_*\subset K_*$. The inclusion sign here has its usual meaning if it is recalled that a relation is a set of ordered pairs; explicitly $J_*\subset K_*$ means that xJ_*y implies that xK_*y whenever x and y are in X^I.

The terminology and the notation are designed to indicate that the mapping $\sigma\to\sigma_*$ behaves like duality mappings behave in many other parts of mathematics. Experience with such mappings makes it reasonable to conjecture that $(\sigma\tau)_*=\tau_*\sigma_*$ (in the sense of relation product). Unfortunately, however, this equation does not always hold. It is at this point that our concept of one substitution "following" another becomes useful; that apparently *ad hoc* condition is, in all non-trivial cases, necessary and sufficient for the validity of the desired equation.

(3.8) THEOREM. *If* $\sigma \succeq \tau$, *then* $(\sigma\tau)_*=\tau_*\sigma_*$. *If, conversely, X consists of more than one point, and* $(\sigma\tau)_*=\tau_*\sigma_*$, *then* $\sigma\succeq\tau$.

Remark. The relation product $\tau_*\sigma_*$ is defined so that $x(\tau_*\sigma_*)z$ holds if and only if there exists a point y with $y\tau^*z$ and $x\sigma_*y$. This order of events is in accordance with the standard functional notation. Indeed, if both τ_* and σ_* are single-valued, then $\tau_*\sigma_*$ is single-valued and $x(\tau_*\sigma_*)z$ means that $z=\tau_*\sigma_*x$. If, in this case, $y=\sigma_*x$, then $z=\tau_*y$. These considerations have at least a mnemonic value even in the general case; the proper notational set-up for any relation product can be instantaneously rederived by pretending that the factors are functions.

If X consists of just one point, then the same is true of X^I. In this case σ_* is the identity mapping of X^I onto itself for every σ, and, consequently, $(\sigma\tau)_*=\tau_*\sigma_*$ is universally valid.

Proof. Assume first that $x(\tau_*\sigma_*)z$ and let y be a witness to this connection; assume, in other words, that $y\tau_*z$ and $x\sigma_*y$. Suppose now that $i \in \operatorname{dom} \sigma\tau$, so that $i \in \operatorname{dom} \tau$ and $\tau i \in \operatorname{dom} \sigma$. Since $y \tau_* z$ and $i \in \operatorname{dom} \tau$, it follows that $y_{\tau i} = z_i$. Since $x \sigma_* y$ and $\tau i \in \operatorname{dom} \sigma$, it follows that $x_{\sigma(\tau i)} = y_{\tau i}$. Conclusion: $x_{(\sigma\tau)i} = z_i$ whenever $i \in \operatorname{dom} \sigma\tau$, and therefore $x(\sigma\tau)_* z$. We have proved so far that $\tau_* \sigma_* \subset (\sigma\tau)_*$; note that the assumption $\sigma \succ \tau$ was not used yet.

Assume next that $x(\sigma\tau)_* z$; this means that $x_{(\sigma\tau)i} = z_i$ whenever $i \in \operatorname{dom} \tau$ and $\tau i \in \operatorname{dom} \sigma$. To prove that $x(\tau_* \sigma_*) z$, a witness y to this connection will be constructed in three stages. (i) If $j \in \operatorname{dom} \sigma$, put $y_j = x_{\sigma j}$. (ii) If $j \in \operatorname{ran} \tau - \operatorname{dom} \sigma$, then $j \in I - \operatorname{dom} \sigma$ and therefore (here is where the assumption $\sigma \succ \tau$ comes in) there is at most one i such that $j = \tau i$; since also $j \in \operatorname{ran} \tau$, one such i (and therefore exactly one such i) does exist. In this case we may without ambiguity put $y_j = z_i$. (iii) If $j \in I - (\operatorname{ran} \tau \cup \operatorname{dom} \sigma)$, define y_j arbitrarily; say, put $y_j = x_j$. This three-stage construction defines an element y_j of X^I. The fact that $x \sigma_* y$ follows from (i) and from the definition of σ_*. To prove that $y \tau_* z$, consider an element i in $\operatorname{dom} \tau$. If $\tau i \in \operatorname{dom} \sigma$, then $x_{(\sigma\tau)i} = z_i$ (because of the assumption $x(\sigma\tau)_* z$), and $y_{\tau i} = x_{\sigma(\tau i)}$ (by (i)). If, on the other hand, $\tau i \in I - \operatorname{dom} \sigma$, then $\tau i \in \operatorname{ran} \tau - \operatorname{dom} \sigma$ and therefore $y_{\tau i} = z_i$ (by (ii)). We have proved thus that $x(\tau_* \sigma_*) z$, and hence that $(\sigma\tau)_* \subset \tau_* \sigma_*$; this completes the proof of the first assertion of the theorem.

To prove the second assertion, we assume that X consists of more than one point and that σ and τ are such that $\sigma \succ \tau$ is false. The latter assumption means that there exist two distinct elements j and k in $\operatorname{dom} \tau$ such that $\tau j = \tau k \in I - \operatorname{dom} \sigma$. Consider now an arbitrary point x in X^I and use it to define a point z as follows. If $i \in \operatorname{dom} \sigma\tau$, put $z_i = x_{\sigma\tau i}$; define z_j and z_k arbitrarily, subject only to the proviso that $z_j \neq z_k$; for all other indices i, define z_i completely arbitrarily. It follows from this definition that $x(\sigma\tau)_* z$. On the other hand, it is false that $x(\tau_* \sigma_*) z$; in fact, $y \tau_* z$ is false for all y. The reason is that $\tau j = \tau k$ implies that $y_{\tau j} = y_{\tau k}$ for all y; since $z_j \neq z_k$, this makes it impossible that $y_{\tau i}$ should be equal to z_i for all i in $\operatorname{dom} \tau$. The proof of the theorem is complete.

§ 4. Polyadic algebras

Throughout this section (and, in fact, in most of the paper) we shall continue to use the notation established above; in particular the symbols $\boldsymbol{B}$, X, and I will always retain their meanings.

If p is an arbitrary function from X^I into $\boldsymbol{B}$, if $x \in X^I$, and if σ is a substitution in I, then $\{p(y)\colon x \sigma_* y\}$ is a subset of $\boldsymbol{B}$, and, as such, it may or may not have a supremum in $\boldsymbol{B}$. If p and σ are such that the supremum exists for all x, and if the value of the supremum for each x in X^I is denoted by $q(x)$, then, of course, q is a function from X^I into $\boldsymbol{B}$.

In this situation we shall write $q=S(\sigma)p$, or, equivalently, we shall say that $S(\sigma)p$ exists and has the value q. Explicitly

$$S(\sigma)p(x)=\bigvee\{p(y):\ x\sigma_* y\} \tag{4.1}$$

whenever the indicated supremum exists.

The point in considering substitutions at all is that they enable us to treat simultaneously the two kinds of operators that enter into the definition of a functional polyadic algebra. Indeed, if τ is a transformation on I, then the set $\{p(y):\ x\tau_* y\}$ consists of the single element $p(\tau_* x)$; this shows that the use of the symbol S in (4.1) is consistent with its earlier use in (2.2). If J is a subset of I, and if $\sigma=\chi(J)$, then, as we have already observed, the relation σ_* that occurs in (4.1) is identical with the relation J_* in (2.4), so that $S(\chi(J))=\exists(J)$.

(4.2) Lemma. *If p is a function from X^I into B and if σ and τ are substitutions in I with $\sigma \succ \tau$, then $S(\sigma\tau)p=S(\sigma)S(\tau)p$, in the following sense: if both $S(\tau)p$ and $S(\sigma)S(\tau)p$ exist, then $S(\sigma\tau)p$ exists and the equality holds, and conversely, if τ is a transformation on I such that $S(\sigma\tau)p$ exists, then $S(\sigma)S(\tau)p$ exists and the equality holds again.*

Proof. Assume first that both $S(\tau)p$ and $S(\sigma)S(\tau)p$ exist. If $q=S(\tau)p$, then $q(y)=\bigvee\{p(z):\ y\tau_* z\}$ for all y in X^I. It follows that

$$\begin{aligned} S(\sigma)S(\tau)p(x)&=S(\sigma)q(x)=\bigvee\{q(y):\ x\sigma_* y\}\\ &=\bigvee\big\{\bigvee\{p(z): y\tau_* z\}:\ x\sigma_* y\big\}=\bigvee\{p(z):\ x(\tau_*\sigma_*)z\}. \end{aligned}$$

The proof in this case is completed by an application of Theorem 3.8 and of the definition of $S(\sigma\tau)p$. Assume next that τ is a transformation such that $S(\sigma\tau)p$ exists. Since the definition of the product of two relations implies that $\{p(z):\ x(\tau_*\sigma_*)z\}=\{p(\tau_* y):\ x\sigma_* y\}$, it follows that

$$S(\sigma\tau)p(x)=\bigvee\{S(\tau)p(y):\ x\sigma_* y\},$$

and the proof is completed by an application of the definition of $S(\sigma)S(\tau)p$.

One consequence of the preceding result is that a functional polyadic algebra admits more operations than its definition demands.

(4.3) Lemma. *A necessary and sufficient condition that a Boolean algebra A of functions from X^I into B be a functional polyadic algebra is that $S(\sigma)p$ exist and belong to A whenever $p\in A$ and σ is a substitution in I.*

Proof. The sufficiency of the condition is obvious: if the condition is satisfied for every substitution, then it is satisfied, in particular, for every transformation and for every substitution of the form $\chi(J)$ where J is a subset of I. To prove necessity, we recall that, by Lemma 3.2, every substitution can be written in the form $\tau\chi(J)$, where τ is a transformation on I and J is a subset of I, and that, by Lemma 3.5, $\tau \succcurlyeq \chi(J)$. The desired result now follows from Lemma 4.2.

If $\boldsymbol{A}$ is a functional polyadic algebra (more explicitly, a $\boldsymbol{B}$-valued I-algebra over X), then, for each substitution σ in I, $\boldsymbol{S}(\sigma)$ is an operator on $\boldsymbol{A}$, *i. e.*, a mapping of $\boldsymbol{A}$ into itself. In the remainder of this section we shall derive the basic properties of the operators $\boldsymbol{S}(\sigma)$.

(4.4) LEMMA. *If τ is a transformation on I, then $\boldsymbol{S}(\tau)$ is a Boolean endomorphism of the algebra of all functions from X^I into $\boldsymbol{B}$ (and hence of any functional I-algebra); the endomorphism $\boldsymbol{S}(\delta)$ is the identity mapping e.*

Proof. The assertion is an immediate consequence of (2.2) and of the fact that the Boolean operations in an algebra of functions are defined pointwise. (Recall also that δ_* is the identity mapping on X^I.)

(4.5) LEMMA. *If $J \subset I$ and if x and y are elements of X^I such that xJ_*y, then $\exists(J)p(x) = \exists(J)p(y)$ for any function p from X^I into $\boldsymbol{B}$, in the sense that if either term of the equation exists, then the other one exists and the two are equal.*

Proof. Since J_* is an equivalence relation in X^I, the assumption xJ_*y implies that either one of the conditions xJ_*z and yJ_*z is necessary and sufficient for the other. It follows that the sets $\{p(z)\colon xJ_*z\}$ and $\{p(z)\colon yJ_*z\}$ are the same and hence that if either one has a supremum, the other one has the same supremum.

(4.6) LEMMA. *If $J \subset I$, then $\exists(J)$ is a quantifier on A.*

Proof. It is clear that $\exists(J)$ is normalized, *i. e.*, that $\exists(J)0 = 0$. Since the equivalence relation J_* is reflexive, *i. e.*, xJ_*x for all x, it follows that

$$\exists(J)p(x) = \bigvee\{p(y)\colon xJ_*y\} \geqslant p(x),$$

so that $\exists(J)$ is increasing. The fact that $\exists(J)$ is quasi-multiplicative is a consequence, via Lemma 4.5, of the following computation:

$$\begin{aligned}\exists(J)\bigl(p \wedge \exists(J)q\bigr)(x) &= \bigvee\{p(y) \wedge \exists(J)q(y)\colon xJ_*y\}\\ &= \bigvee\{p(y)\colon xJ_*y\} \wedge \exists(J)q(x) = \exists(J)p(x) \wedge \exists(J)q(x).\end{aligned}$$

It follows from Lemmas 3.5 and 4.2 that if σ and τ are transformations on I, then $\boldsymbol{S}(\sigma\tau) = \boldsymbol{S}(\sigma)\boldsymbol{S}(\tau)$. Since $\chi(\emptyset) = \delta$, the fact that $\boldsymbol{S}(\delta) = e$ can also be expressed by saying that $\exists(\emptyset)$ is the discrete quantifier on $\boldsymbol{A}$. From Lemma 3.1 we infer (using Lemmas 3.5 and 4.2 as before) that $\exists(J \cup K) = \exists(J)\exists(K)$ and hence, in particular, that $\exists(J)\exists(K) = \exists(K)\exists(J)$ whenever J and K are subsets of I. All these facts are rather near the surface; the next two results, though easy, are less obvious.

(4.7) Lemma. *If σ and τ are transformations on I, if $J \subset I$, and if $\sigma = \tau$ outside J, then $\boldsymbol{S}(\sigma)\exists(J)p = \boldsymbol{S}(\tau)\exists(J)p$ whenever p is a function from X^I into $\boldsymbol{B}$ such that $\exists(J)p$ exists.*

Proof. The last assumption means that $\sigma i = \tau i$ whenever $i \in I - J$ and hence that $\sigma\chi(J) = \tau\chi(J)$. The conclusion follows from Lemmas 3.5 and 4.2.

(4.8) Lemma. *If p is a function from X^I into $\boldsymbol{B}$, if τ is a transformation on I and J is a subset of I such that τ is one-to-one on $\tau^{-1}J$, and if $\exists(\tau^{-1}J)p$ exists, then $\exists(J)\boldsymbol{S}(\tau)p$ exists and $\exists(J)\boldsymbol{S}(\tau)p = \exists(\tau)\exists(\tau^{-1}J)p$.*

Proof. By Lemma 3.3, $\chi(J)\tau = \tau\chi(\tau^{-1}J)$. By Lemma 3.6, the assumption on τ means that $\chi(J) \succ \tau$. By Lemma 3.5, $\tau \succ \chi(\tau^{-1}J)$. By Lemma 4.2, $\boldsymbol{S}(\tau)\exists(\tau^{-1}J)p = \boldsymbol{S}(\tau\chi(\tau^{-1}J))p$, so that $\boldsymbol{S}(\chi(J)\tau)p$ exists; the conclusion follows from another application of the same lemma.

The superficially complicated conclusions of Lemmas 4.7 and 4.8 are merely a condensed summary of the usual intuitively obvious relations between quantification and substitution. A couple of examples will make them clearer. Suppose that i and j are distinct elements of I and let τ be the transformation that maps i onto j and maps everything else (including j) onto itself. Since τ agrees with δ outside i, it follows from Lemma 4.7 that $\boldsymbol{S}(\tau)\exists(i) = \exists(i)$. (If J is a singleton, $J = \{i\}$, we write $\exists(i)$ instead of $\exists(J)$.) This equation corresponds to the familiar fact that once a variable has been quantified, the replacement of that variable by another one has no further effect. To get another example, note, for the same τ, that $\tau^{-1}i$ is empty. It follows from Lemma 4.8 that $\exists(i)\boldsymbol{S}(\tau) = \boldsymbol{S}(\tau)$. This equation corresponds to the familiar fact that once a variable has been replaced by another one, a quantification on the replaced variable has no further effect.

We are now ready to define the central concept of this paper. Abstracting from the functional case, we shall say that a *polyadic (Boolean) algebra* is a quadruple $(\boldsymbol{A}, I, \boldsymbol{S}, \exists)$, where $\boldsymbol{A}$ is a Boolean algebra, I is a set, $\boldsymbol{S}$ is a mapping from transformations on I to Boolean endomorphisms of $\boldsymbol{A}$, and $\exists$ is a mapping from subsets of I to quantifiers on $\boldsymbol{A}$ such that

(P_1) *$\boldsymbol{S}(\delta)$ is the identity mapping on $\boldsymbol{A}$ (i. e., $\boldsymbol{S}(\delta) = e$),*

(P_2) *$\boldsymbol{S}(\sigma\tau) = \boldsymbol{S}(\sigma)\boldsymbol{S}(\tau)$ whenever σ and τ are transformations on I,*

(P_3) *$\exists(\emptyset)$ is the discrete quantifier on $\boldsymbol{A}$ (i. e., $\exists(\emptyset) = e$),*

(P_4) *$\exists(J \cup K) = \exists(J)\exists(K)$ whenever J and K are subsets of I,*

(P_5) *if σ and τ are transformations on I, if $J \subset I$, and if $\sigma = \tau$ outside J, then $\boldsymbol{S}(\sigma)\exists(J) = \boldsymbol{S}(\tau)\exists(J)$,*

(P$_6$) *if τ is a transformation on I, if $J \subset I$, and if τ is one-to-one on $\tau^{-1}J$, then $\exists(J)S(\tau) = S(\tau)\exists(\tau^{-1}J)$.*

From this definition and from our preceding results it follows immediately that a functional polyadic algebra is a polyadic algebra.

Most of the time we shall use the same symbols (S and $\exists$) for the endomorphism and quantifier mappings of every polyadic algebra; only rarely will we find it necessary to use a more detailed notation in order to avoid confusion. We shall also commit the common simplifying solecism of identifying the Boolean algebra A with the polyadic algebra $(A, I, S, \exists)$. We shall, accordingly, use expressions such as "the polyadic algebra A", or "the polyadic algebra A with variables I", or simply "the I-algebra A". The cardinal number of I will be called the *degree* of the algebra and an algebra of degree n will be called an *n-adic* algebra.

If I is empty, then there is only one transformation, namely δ, and there is only one subset, namely $\emptyset$; in this extreme case $\theta = \delta$. Consequently, if A is an arbitrary Boolean algebra, if I is the empty set $\emptyset$, and if both $S(\delta)$ and $\exists(\emptyset)$ are defined to be e, then A becomes a polyadic algebra of degree 0. We see thus that the classical theory Boolean algebras is subsumed under the 0-adic case of the theory of polyadic algebras.

If I consists of a single element, then there is only one transformation, namely δ, and there are two distinct subsets, namely $\emptyset$ and I. Consequently if A is an arbitrary monadic algebra with quantifier $\exists$, if I is a singleton, and if we write, by definition, $S(\delta) = \exists(\emptyset) = e$ and $\exists(I) = \exists$, then A becomes a polyadic algebra of degree 1. We see thus that the theory of monadic algebras is subsumed under the 1-adic case of the theory of polyadic algebras.

In a certain sense the definition of polyadic algebras above (and, correspondingly, the definition of functional polyadic algebras in § 2) is not sufficiently general. It often happens (*e. g.*, in axiomatizations of Euclidean geometry) that the propositional functions to be considerep are functions of several different kinds of variables (*e. g.*, points, lines, and planes). The appropriate way of allowing for this phenomenon from the point of view of functional polyadic algebras is to consider not one domain and its Cartesian powers but several possibly distinct sets and their Cartesian product. More explicitly, suppose that to each element i of I there corresponds a non-empty set X_i and let X_I be the Cartesian product of this family of sets. The sets X_i may overlap or coincide among themselves quite arbitrarily. Functions from X_I to B form a Boolean algebra as before, and the theory of the operators $\exists(J)$ carries over to the generalized situation with no significant change. If, however, τ is a transformation on I, then the definition of $\tau_* x$ is not always meaningful (since $X_{\tau i}$ and X_i may have no points in common). The simplest way out of the difficulty is to restrict attention to such transformations τ ("special" transformations) for which $X_{\tau i}=X_i$ for all i in I. The set of special transformations is a subsemigroup (containing the unit) of the set of all transformations on I. If we consider only the transformations in this subsemigroup (and, correspondingly, consider only those substitutions σ on I for which $X_{\sigma i}=X_i$ for all i in domσ), then the results of § 2 and § 3 go through without any difficulties. The definition of a functional polyadic algebra will, naturally, become slightly different; in place of closure under $\boldsymbol{S}(\tau)$ for all transformations τ, the modified definition requires only closure under $\boldsymbol{S}(\tau)$ when τ is a special transformation.

The considerations of the preceding paragraph suggest a modification in the definition of abstract polyadic algebras also. The conditions referring to $\exists$ alone remain unmodified, but the conditions referring to $\boldsymbol{S}$ are required to hold only for a suitable subsemigroup of the semigroup of all transformations on I. The resulting concept is a generalization of the one defined above; the special case is obtained from the generalization by the consideration of the improper subsemigroup of all transformations. Polyadic algebras in the special sense will be called *homogeneous* algebras; they are the only ones with which we shall be concerned in this paper. The simplest and at the same time the most interesting polyadic algebras are homogeneous; most of the techniques that apply to homogeneous algebras apply with only minor modifications to the general case; and, finally, there are some known techniques (cf. [8]) for generalizing results about homogeneous polyadic algebras to arbitrary polyadic algebras. For these reasons, the details of the study of non-homogeneous polyadic algebras may safely be postponed to a later occasion.

§ 5. Locally finite algebras

In order to see the direction in which the theory of abstract polyadic algebras should develop, we take another look at propositional functions, *i. e.*, at functions from a Cartesian power X^I into a Boolean algebra $\boldsymbol{B}$. While, for logical purposes, it is frequently essential that the set of variables be infinite, each particular propositional function usually depends on a finite number of variables only. The precise definition of a function depending on a certain set of variables is best approached indirectly. A function p from X^I into $\boldsymbol{B}$ is *independent* of a subset J of I if $p(x)=p(y)$ whenever xJ_*y. In other words, p is independent of J if and only if the replacement of an argument x of p by a point whose coordinates differ from those of x only when their index is in J leaves the value of p unchanged. Roughly speaking, p is independent of J when coordinates in J can be changed arbitrarily without changing p.

It turns out to be easy to express the concept of independence in algebraic terms.

(5.1) Lemma. *If p is an element of a functional I-algebra and if $J \subset I$, then a necessary and sufficient condition that p be independent of J is that $\exists(J)p=p$.*

Proof. If p is independent of J, then

$$\exists(J)p(x)=\bigvee\{p(y):\ xJ_*y\}=p(x)\,.$$

If, conversely, $\exists(J)p=p$, then, by Lemma 4.5, xJ_*y implies that

$$p(x)=\exists(J)p(x)=\exists(J)p(y)=p(y)\,.$$

In view of Lemma 5.1 it is natural to define independence in an arbitrary polyadic algebra as follows: an element p of an I-algebra A is *independent* of a subset J of I if and only if $\exists(J)p=p$. We note that the concept of independence is the algebraic substitute for familiar logical concepts usually described in terms of "free" and "bound" variables.

Dependence could now be defined in terms of independence in the obvious way. It is grammatically more convenient, however, to use a different term. We shall say that a subset J of I is a *support* of p (equivalent expression: *J supports p*) if and only if p is independent of $I-J$. The element p of A will be called *finite-dimensional*, or simply *finite*, if it has a finite support, or, equivalently, if there exists a cofinite subset J of I such that p is independent of J. (A *cofinite* subset of I is one whose complement is finite.) Roughly speaking, p is finite if and only if it is independent of almost all (*i. e.*, all but a finite number) of the variables. The algebra A is *locally finite-dimensional*, or simply *locally finite*, if every one of its elements is finite. Throughout the rest of this paper we shall deal with locally finite polyadic algebras only. *Unless in a special context we explicitly say the opposite, we shall automatically (and often tacitly) assume that every polyadic algebra we refer to is locally finite.* The main reason for this procedure is that almost nothing is known as yet about polyadic algebras that are not locally finite. The "infinite logics" that give rise to such algebras might repay study; for the time being, however, we concentrate our attention on the algebraic systems whose logical counterparts are of already proven value.

The first thing we can do for locally finite algebras is to prove that, just as the functional algebras that suggested them, they admit more operations than their definition demands; the additional operations, moreover, have the same desirable multiplicative properties that they have in functional algebras.

(5.2) THEOREM. *If A is a locally finite I-algebra, then there exists a unique mapping $\tilde{S}$ from substitutions in I to operators on A such that* (i) $\tilde{S}(\tau)=S(\tau)$ *whenever τ is a transformation on I,* (ii) $\tilde{S}(\chi(J))=\exists(J)$ *whenever J is a subset of I, and* (iii) $S(\sigma\tau)=\tilde{S}(\sigma)\tilde{S}(\tau)$ *whenever σ and τ are substitutions in I such that $\sigma\succcurlyeq\tau$.*

Remark. A kind of converse of Theorem 5.2 is true and easy to prove (even in the absence of local finiteness.) Suppose in fact that A is a Boolean algebra and that $\tilde{S}$ is a mapping from substitutions in a certain set I to operators on A satisfying (iii) above and satisfying also the following three conditions. (a) $\tilde{S}(\delta)$ is the identity mapping on A. (b) If τ is a transformation on I, then $\tilde{S}(\tau)$ is a Boolean endomorphism of A. (c) If $J\subset I$, then $\tilde{S}(\chi(J))$ is a quantifier on A. It follows that if S and $\exists$ are defined by $S(\tau)=\tilde{S}(\tau)$ (whenever τ is a transformation on I) and $\exists(J)=\tilde{S}(\chi(J))$ (whenever J is a subset of I), then A becomes an I-algebra. Indeed (P_1) and (P_3) follow from (a), and (P_2), (P_4), (P_5), and (P_6) follow from (b) and (c) via (iii) and Lemma 3.5. In other words, the conditions (a), (b), (c), and (iii) yield an alternative definition of polyadic algebras.

Once Theorem 5.2 is proved, there is no more necessity for maintaining a careful distinction between S and its extension $\tilde{S}$; in the future (after the proof of Theorem 5.2 is over) we shall use the same symbol $S(\tau)$ whether τ is a transformation on I or an arbitrary substitution in I.

Proof. If σ is a substitution in I, then, by Lemma 3.2, σ can be written in the form $\tau\chi(J)$, where τ is a transformation on I and J is a subset of I. Using (i), (ii), (iii), and Lemma 3.5, we conclude that $\tilde{S}(\sigma)=S(\tau)\exists(J)$. This proves the uniqueness assertion, and shows, incidentally, that $\tilde{S}(\sigma)$ is always a hemimorphism from A into itself.

To prove the existence assertion, we first note that if $\tau_1\chi(J_1)=\tau_2\chi(J_2)$, where τ_1 and τ_2 are transformations, then Lemma 3.2 and (P_5) imply that $S(\tau_1)\exists(J_1)=S(\tau_2)\exists(J_2)$. This proves that whenever $\sigma=\tau\chi(J)$ is a substitution, where τ is a transformation and $J\subset I$, then the equation $\tilde{S}(\sigma)=S(\tau)\exists(J)$ unambigously defines a hemimorphism $\tilde{S}(\sigma)$ on A. The assertions (i) and (ii) are now obvious; it remains only to prove that $\tilde{S}$ satisfies (iii).

Suppose therefore that τ_1 and τ_2 are transformations and J_1 and J_2 are sets such that if $\sigma_1=\tau_1\chi(J_1)$ and $\sigma_2=\tau_2\chi(J_2)$, then $\sigma_1\succcurlyeq\sigma_2$. It is to be proved that if $p\in A$, then $\tilde{S}(\sigma_1\sigma_2)p=\tilde{S}(\sigma_1)\tilde{S}(\sigma_2)p$. Let J be a cofinite set such that $\exists(J)p=p$ and write $\tilde{J}_2=J_2\cup J$. Since $I-\tilde{J}_2$ is finite, $I-\tau_2(I-\tilde{J}_2)$ is cofinite and has a cardinal number greater than or equal to that of $\tilde{J}_2$. It follows that there exists a transformation $\tilde{\tau}_2$ that agrees with τ_2 outside $\tilde{J}_2$ and that maps $\tilde{J}_2$ into $I-\tau_2(I-\tilde{J}_2)$ in a one-to-one manner. Since $\tilde{\tau}_2$ maps $\tilde{J}_2$ and $I-\tilde{J}_2$ into disjoint sets, and since $\tilde{\tau}_2$ is one-to-one on $\tilde{J}_2$, the only time that $\tilde{\tau}_2$ can send two distinct elements i and j onto the same element k is when i and j belong to $I-\tilde{J}_2$ and $\tau_2 i = \tau_2 j = k$. Since $I-\tilde{J}_2 \subset I-J_2$, it follows that both i and j belong to $\operatorname{dom}\sigma_2$ and $\sigma_2 i = \sigma_2 j = k$. The assumption $\sigma_1 \succsim \sigma_2$ implies therefore that $k \in \operatorname{dom}\sigma_1 = \operatorname{dom}\chi(J_1)$. This proves that $\chi(J_1) \succsim \tilde{\tau}_2$; equivalently, by Lemma 3.6, $\tilde{\tau}_2$ is one-to-one on $\tilde{\tau}_2^{-1}J_1$. The proof of Theorem 5.2 can now be completed by the following computation:

$$\begin{aligned}
\widetilde{\mathbf{S}}(\sigma_1\sigma_2)p &= \widetilde{\mathbf{S}}\big(\tau_1\chi(J_1)\tau_2\chi(J_2)\big)p = \widetilde{\mathbf{S}}\big(\tau_1\tau_2\chi(\tau_2^{-1}J_1 \cup J_2)\big)p \\
&= \mathbf{S}(\tau_1\tau_2)\boldsymbol{\exists}(\tau_2^{-1}J_1 \cup J_2)\boldsymbol{\exists}(J)p = \mathbf{S}(\tau_1\tau_2)\boldsymbol{\exists}(\tau_2^{-1}J_1 \cup \tilde{J}_2)p \\
&= \widetilde{\mathbf{S}}\big(\tau_1\tau_2\chi(\tau_2^{-1}J_1 \cup \tilde{J}_2)\big)p = \widetilde{\mathbf{S}}\big(\tau_1\chi(J_1)\tau_2\chi(\tilde{J}_2)\big)p \\
&= \widetilde{\mathbf{S}}\big(\tau_1\chi(J_1)\tilde{\tau}_2\chi(\tilde{J}_2)\big)p = \widetilde{\mathbf{S}}\big(\tau_1\tilde{\tau}_2\chi(\tilde{\tau}_2^{-1}J_1 \cup \tilde{J}_2)\big)p \\
&= \mathbf{S}(\tau_1\tilde{\tau}_2)\boldsymbol{\exists}(\tilde{\tau}_2^{-1}J_1 \cup \tilde{J}_2)p = \mathbf{S}(\tau_1)\mathbf{S}(\tilde{\tau}_2)\boldsymbol{\exists}(\tilde{\tau}_2^{-1}J_1)\boldsymbol{\exists}(\tilde{J}_2)p \\
&= \mathbf{S}(\tau_1)\boldsymbol{\exists}(J_1)\mathbf{S}(\tilde{\tau}_2)\boldsymbol{\exists}(\tilde{J}_2)p = \widetilde{\mathbf{S}}\big(\tau_1\chi(J_1)\big)\widetilde{\mathbf{S}}\big(\tilde{\tau}_2\chi(\tilde{J}_2)\big)p \\
&= \widetilde{\mathbf{S}}(\sigma_1)\widetilde{\mathbf{S}}\big(\tau_2\chi(\tilde{J}_2)\big)p = \widetilde{\mathbf{S}}(\sigma_1)\mathbf{S}(\tau_2)\boldsymbol{\exists}(\tilde{J}_2)p \\
&= \widetilde{\mathbf{S}}(\sigma_1)\mathbf{S}(\tau_2)\boldsymbol{\exists}(J_2)\boldsymbol{\exists}(J)p = \widetilde{\mathbf{S}}(\sigma_1)\widetilde{\mathbf{S}}(\sigma_2)p.
\end{aligned}$$

§ 6. Supports and independence

It is intuitively clear that if a propositional function depends on a finite number of variables only, then most of the action of a substitution on such a propositional function is wasted; all that matters is the behavior of the substitution on the finite set of variables in question. In order to make these considerations precise, we now make a digression into the appropriate part of the theory of substitutions.

We shall say that a subset J of a set I is a *support* of a substitution σ in I (equivalent expression: *J supports σ*) if σ agrees with the identity transformation outside the set J, or, more explicitly, if $I-J \subset \operatorname{dom}\sigma$ and $\sigma i = i$ whenever $i \in I-J$.

(6.1) LEMMA. *The set of all supports of a substitution σ in I is a Boolean filter (dual ideal) in the Boolean algebra of all subsets of I.*

Proof. It follows immediately from the definition that J supports σ if and only if

$$J_0 = \operatorname{dom}\sigma \cap \{i\colon\ \sigma i \neq i\} \subset J\,.$$

This implies that I always supports σ, and hence that the set of supports of σ is never empty. It implies also that if J supports σ and $J \subset K$, then K supports σ, and that if both J and K support σ, then so does $J \cap K$.

Remark. The proof shows more than the lemma asserts. The filter of supports of σ is, in fact, the principal filter generated by J_0, and, consequently, J_0 is the minimal support of σ. Despite this fact the language of supports is useful; it is often more convenient to use "a support" of σ than to insist on using "the support".

The preceding result characterizes all supports of a fixed substitution; the following result goes in the other direction.

(6.2) LEMMA. *The set of all substitutions with support J is a subsemigroup, containing the unit, in the semigroup of all substitutions.*

Proof. Since δ agrees with δ outside J, it is trivial that J supports δ. If J supports both σ and τ, and if $i \in I-J$, then $i \in \operatorname{dom}\sigma \cap \operatorname{dom}\tau$ and $\sigma i = \tau i = i$; it follows that $i \in \operatorname{dom}\sigma\tau$ and $(\sigma\tau)i = \sigma(\tau i) = \sigma i = i$.

(6.3) LEMMA. *If J supports σ and K supports τ, then $J \cup K$ supports $\sigma\tau$.*

Proof. By Lemma 6.1, $J \cup K$ supports both σ and τ; by Lemma 6.2, $J \cup K$ supports $\sigma\tau$.

A substitution σ is *finite* if it leaves fixed almost all the elements of I. More precisely, σ is finite if $\operatorname{dom}\sigma$ includes a cofinite subset of I on which σ agrees with the identity transformation δ. In the language of supports, σ is finite if and only if it has a finite support.

(6.4) LEMMA. *The set of all finite substitutions is a subsemigroup, containing the unit, in the semigroup of all substitutions.*

Proof. Since there exists a finite set J (*e. g.*, $\emptyset$) such that δ agrees with δ outside J, it is trivial that δ is finite. The fact that the set of all finite substitutions is closed under multiplication is an immediate consequence of Lemma 6.3.

(6.5) LEMMA. *A necessary and sufficient condition that $\chi(J)$ be a finite substitution is that J be a finite set.*

Proof. If J is finite, then, since $\chi(J) = \delta$ outside J, it follows that $\chi(J)$ is finite. If, conversely, $\chi(J) = \delta$ outside some finite set K, then it follows from the definition of $\chi(J)$ that $J \subset K$ and hence that J is finite.

(6.6) LEMMA. *Every finite substitution σ can be written in the form $\tau\chi(J)$, where τ is a finite transformation on I and J is a finite subset of I.*

Proof. By Lemma 3.2, $\sigma = \tau\chi(J)$ with $J = I - \operatorname{dom}\sigma$ and $\sigma = \tau$ outside J. The finiteness of σ implies that J is finite. The finiteness of σ and J, together with the fact that $\sigma = \tau$ outside J, implies that $\tau = \sigma$ outside a finite set, *i. e.*, that τ is finite.

We proceed now to investigate the concepts of independence and support in an I-algebra $\boldsymbol{A}$ and the relation between these concepts and the algebraic structure of $\boldsymbol{A}$. It is convenient to be able to phrase the results in the language of either independence or supports; because, however, of the intimate relation between these two concepts, it is generally sufficient to give the proof for only one of the two cases.

The first five of the results that follow are in a curious situation. They are valid for algebraic systems that are somewhat more general than polyadic algebras, and, in the next section, their generalized versions are needed. In order not to complicate the presentation, we shall state and prove them here for polyadic algebras only, and, in the next section, we shall point out the minor modifications that suffice to reach the generalization. Subsequently we shall refer to the generalized results by the same number as is borne by the special case; to avoid any possible confusion, systematic cross-references are provided at the appropriate places.

(6.7) LEMMA. *If $p \in A$, then the set of all sets J such that p is independent of J is a Boolean ideal, and the set of all supports of p is a Boolean filter, in the Boolean algebra of all subsets of I.*

Proof (cf. 7.1)). Since $\exists(\emptyset)p = p$, it is trivial that p is independent of $\emptyset$. If $\exists(J)p = p$ and $K \subset J$, then

$$\exists(K)p = \exists(K)\exists(J)p = \exists(K \cup J)p = \exists(J)p = p .$$

If $\exists(J)p = \exists(K)p = p$, then

$$\exists(J \cup K)p = \exists(J)\exists(K)p = \exists(J)p = p .$$

(6.8) LEMMA. *If $J \subset I$, then the set of all elements p in A such that p is independent of J is a Boolean subalgebra of A, and the set of all elements p in A such that J supports p is a Boolean subalgebra of A.*

Proof (cf. (7.2)). The set of all those elements p for which $\exists(J)p = p$ is simply the range of the quantifier $\exists(J)$. The desired result follows from the fact (1.3) that the range of a quantifier is always a Boolean algebra.

(6.9) LEMMA. *If p is independent of J, then $\exists(K)p = \exists(K-J)p$ for every K; if J supports p, then $\exists(K)p = \exists(K \cap J)p$ for every K.*

Proof (cf. (7.3)). For independence: $\exists(K)p = \exists(K-J)\exists(K \cap J)p$ $= \exists(K-J)p$ (by Lemma 6.7). For supports: apply the result for independence to $I-J$.

(6.10) LEMMA. *If p is independent of J (if J supports p), then $\exists(K)p$ is independent of $J \cup K$ ($J-K$ supports $\exists(K)p$).*

Proof (cf. (7.4)). $\exists(J \cup K)\exists(K)p = \exists(J \cup K)p = \exists(K)\exists(J)p = \exists(K)p$.

(6.11) LEMMA. *If p is independent of J and if σ and τ are transformations that agree outside J, then $S(\sigma)p = S(\tau)p$.*

Proof (cf. (7.5)). $S(\sigma)p = S(\sigma)\exists(J)p = S(\tau)\exists(J)p$ (by (P_5)) $= S(\tau)p$.

For the proof of the main non-trivial relation between independence and transformations (Lemma 6.14) the following two technical lemmas are needed.

(6.12) LEMMA. *If τ is a transformation, if $J \subset I$, and if $K = I - \tau(I-J)$, then $\tau^{-1}K \subset J$.*

Proof. This purely set-theoretic lemma is based on the fact that $J \subset \tau^{-1}\tau J$. Applying this inclusion to $I-J$ in place of J and then forming the complement of both terms, we obtain

$$I - \tau^{-1}\tau(I-J) \subset J.$$

Since $\tau^{-1}K = \tau^{-1}\big(I - \tau(I-J)\big) = I - \tau^{-1}\tau(I-J)$, the proof is complete.

(6.13) LEMMA. *If τ is a transformation, if $J \subset I$, and if $K = I - \tau(I-J)$, then $\chi(K) \succ \tau\chi(J)$.*

Proof. If $\tau\chi(J)i = j$, then $i \in I-J$ and therefore $j \in \tau(I-J)$, so that $j \in I-K$. In other words, the entire range of $\tau\chi(J)$ is included in $I-K$. Under these circumstances the substitution $\tau\chi(J)$ obviously never maps two distinct elements onto the same element of $I - \operatorname{dom}\chi(K) = K$, because, in fact, it never maps anything at all into K.

(6.14) LEMMA. *If τ is a transformation, if p is independent of J, and if $K = I - \tau(I-J)$, then $\boldsymbol{S}(\tau)p$ is independent of K. If J supports p, then τJ supports $\boldsymbol{S}(\tau)p$.*

Remark. The consideration of simple examples, and, in particular, of δ in the role of τ, shows that the result is in general best possible.

Proof. $\exists(K)\boldsymbol{S}(\tau)p = \exists(K)\boldsymbol{S}(\tau)\exists(J)p$ (by the assumed independence) $= \boldsymbol{S}\big(\chi(K)\big)\boldsymbol{S}\big(\tau\chi(J)\big)p$ (by Theorem 5.2) $= \boldsymbol{S}\big(\chi(K)\tau\chi(J)\big)p$ (by Lemma 6.13 and Theorem 5.2) $= \boldsymbol{S}\big(\tau\chi(\tau^{-1}K)\chi(J)\big)p$ (by Lemma 3.3) $= \boldsymbol{S}\big(\tau\chi(J)\big)p$ (by Lemma 6.12) $= \boldsymbol{S}(\tau)\exists(J)p$ (by Theorem 5.2) $= \boldsymbol{S}(\tau)p$ (by the assumed independence).

It is to be noted that up to now the results of this section (together with their proofs) are valid even in the absence of local finiteness. In the following two results the assumption of local finiteness is essential.

(6.15) LEMMA. *A necessary and sufficient condition that p be independent of J (that J support p) is that $\exists(K)p = p$ whenever K is a finite subset of J (of $I-J$).*

Proof. The necessity of the condition is trivial from Lemma 6.7. To prove sufficiency, apply the definition of local finiteness to find a cofinite set J_0 such that p is independent of J_0 and note that $\exists(J)p = \exists(J-J_0)p$ (by Lemma 6.9) $= p$ (by assumption, since $J-J_0$ is finite).

(6.16) LEMMA. *If* $p \in A$, *then* (i) *to every subset* K *of* I *there corresponds a finite subset* K_0 *of* I *such that* $\exists(K)p = \exists(K_0)p$, (ii) *to every transformation* τ *on* I *there corresponds a finite transformation* τ_0 *on* I *such that* $S(\tau)p = S(\tau_0)p$, *and* (iii) *to every substitution* σ *in* I *there corresponds a finite substitution* σ_0 *in* I *such that* $S(\sigma)p = S(\sigma_0)p$.

Proof. By the definition of local finiteness, p has a finite support J. The conclusion (i) follows from Lemma 6.9. To prove (ii), write $\tau_0 i = \tau i$ when $i \in J$ and $\tau_0 i = i$ otherwise. Since J is finite, τ_0 is finite; (ii) follows from Lemma 6.11. The conclusion (iii) follows from (i) and (ii), together with Lemma 3.2, Theorem 5.2, and Lemma 6.4.

§ 7. Quasi-polyadic algebras

We began the preceding section with a vaguely formulated conjecture to the effect that in the study of (locally finite) polyadic algebras there is no loss of generality in restricting attention to finite substitutions only. Lemma 6.16 may be viewed as a precise formulation (and proof) of that vague conjecture. In this section we shall prove a much stronger result of the same type; for convenience in formulating that strengthened result we now introduce an auxiliary concept. The concept of a *quasi-polyadic algebra* is more general than that of a (locally finite) polyadic algebra; the difference between the two is that in the definition of quasi-polyadic algebras the mappings S and $\exists$ are assumed to be defined only for finite transformations on I and finite subsets of I. To avoid all possible misunderstanding, we formulate the pertinent definition explicitly as follows. A quasi-polyadic algebra is a quadruple $(A, I, S, \exists)$, where A is a Boolean algebra, I is a set, S is a mapping from finite transformations on I into Boolean endomorphisms of A, and $\exists$ is a mapping from finite subsets of I to quantifiers on A, such that

(Q_1) $S(\delta)$ *is the identity mapping on* A,

(Q_2) $S(\sigma\tau) = S(\sigma)S(\tau)$ *whenever* σ *and* τ *are finite transformations on* I,

(Q_3) $\exists(\emptyset)$ *is the discrete quantifier on* A,

(Q_4) $\exists(J \cup K) = \exists(J)\exists(K)$ *whenever* J *and* K *are finite subsets of* I,

(Q_5) *if* σ *and* τ *are finite transformations on* I, *if* J *is a finite subset of* I, *and if* $\sigma = \tau$ *outside* J, *then* $S(\sigma)\exists(J) = S(\tau)\exists(J)$,

(Q_6) *if* τ *is a finite transformation on* I, *if* J *is a finite subset of* I, *and if* τ *is one-to-one on* $\tau^{-1}J$, *then* $\exists(J)S(\tau) = S(\tau)\exists(\tau^{-1}J)$,

(Q_7) *if* $p \in A$, *then there exists a cofinite subset* J *of* I *such that* $\exists(K)p = p$ *whenever* K *is a finite subset of* J.

A quasi-polyadic algebra is exactly what was called a polyadic algebra in [1]. The principal result of this section asserts essentially that the two definitions are equivalent, *i. e.*, that every quasi-polyadic algebra is a (locally finite) polyadic algebra (or, rather, can be converted into one in a natural manner).

For a quasi-polyadic algebra the expression $\exists(J)$ makes sense only if J is finite, and, consequently, the concept of independence cannot be defined in such an algebra the same way as in a polyadic algebra. On the basis of (Q_7) (cf. also Lemma 6.15) the remedy is clear; we shall say that p is *independent* of J if $\exists(K)p=p$ whenever K is a finite subset of J. The concept of *support* is defined in terms of independence, just as before.

It will follow from Theorem 7.6 that independence and support behave the same way in quasi-polyadic algebras as in ordinary polyadic algebras. In the course of the proof of Theorem 7.6 it is convenient to be able to make use of some of the elementary facts about independence and support. These facts (namely Lemmas 6.7-6.11) are almost as easy to prove for quasi-polyadic algebras as for polyadic algebras (and sometimes easier); in order to justify their use in the proof of Theorem 7.6 we proceed to indicate, very briefly, how the proofs go in the generalized situation.

(7.1) Proof of (6.7). If $K \subset J$, then finite subsets of K are finite subsets of J; this proves that if p is independent of J, then p is independent of K. The proof that independence is additive uses the additivity of $\exists$ not for the given sets but for their finite subsets.

(7.2) Proof of (6.8). For each finite set K, the set of all p with $\exists(K)p=p$ is a Boolean algebra; the desired result follows from the consideration of the intersection of all such Boolean algebras over all finite subsets K of J.

(7.3) Proof of (6.9). The assertion makes sense for finite sets K only; for them the proof is essentially the same as before.

(7.4) Proof of (6.10). Again K must be finite; in the proof replace J by an arbitrary finite subset of J.

(7.5) Proof of (6.11). This time σ and τ must be finite. If (cf. Lemma 6.1) K is a finite set that supports them both, then $\sigma=\tau$ outside $J \cap K$; using this comment, we modify the proof by using $J \cap K$ in place of J.

We are now ready to fulfill our promise; we shall show that the concept of a polyadic algebra as defined in [1] is essentially equivalent to the concept of a locally finite polyadic algebra as defined in this paper.

(7.6) THEOREM. *If* $(A, I, S, \exists)$ *is a quasi-polyadic algebra, then* (i) *there exists a mapping* $\tilde{S}$ *from transformations on* I *to Boolean endomorphisms of* A *such that* $\tilde{S}(\tau)=S(\tau)$ *whenever* τ *is a finite transformation,* (ii) *there exists a mapping* $\tilde{\exists}$ *from subsets of* I *to quantifiers on* A *such that* $\tilde{\exists}(J) =\exists(J)$ *whenever* J *is a finite set,* (iii) *the quadruple* $(A, I, \tilde{S}, \tilde{\exists})$ *is a locally finite polyadic algebra, and* (iv) *the mappings* $\tilde{S}$ *and* $\tilde{\exists}$ *are uniquely determined by* (i), (ii), *and* (iii).

Proof. (iv) Uniqueness is an immediate consequence of Lemma 6.16.

(i) If $p \in A$ and if J_0 is a finite support of p, it is natural to try to define $\tilde{S}(\tau)p$, for each transformation τ on I, by finding an appropriate finite transformation τ_0 on I and writing $\tilde{S}(\tau)p=S(\tau_0)p$. The idea is easy: we write $\tau_0 i=\tau i$ when $i \in J_0$ and $\tau_0 i=i$ otherwise. The difficulty with this procedure is that $\tilde{S}(\tau)p$ appears to depend on the choice of the support J_0. To show that it does not in fact do so, we suppose that both J_1 and J_2 are finite supports of p and that τ_1 and τ_2 are defined in terms of J_1 and J_2, respectively, just as τ_0 was defined in terms of J_0; we must prove that $S(\tau_1)p=S(\tau_2)p$. By Lemma 6.7, $J_1 \cap J_2$ supports p, and, by definition, τ_1 and τ_2 agree in $J_1 \cap J_2$; the desired equality follows from Lemma 6.11.

If τ itself happens to be finite, then the finite support J_0 of p can be chosen so as to be a support of τ at the same time. If that is done, then $\tau_0=\tau$, and it follows that $\tilde{S}$ is indeed an extension of S.

If p_1 and p_2 are in A, then a finite set J_0 can be found that supports them both, and, consequently, supports their supremum as well. It follows that

$$\tilde{S}(\tau)(p_1 \vee p_2)=S(\tau_0)(p_1 \vee p_2)=S(\tau_0)p_1 \vee S(\tau_0)p_2=\tilde{S}(\tau)p_1 \vee \tilde{S}(\tau)p_2 .$$

The proof that $\tilde{S}(\tau)$ preserves complementation is similar, but even easier. In other words, each $\tilde{S}(\tau)$ is a Boolean endomorphism of A, and the proof of (i) is complete.

(ii) If $p \in A$ and if J_0 is a finite support of p, it is natural to try to define $\tilde{\exists}(J)p$, for each subset J of I, as $\exists(J \cap J_0)p$. In order to justify this, however, we must show first that $\exists(J \cap J_0)p$ is the same no matter which support J_0 of p is used. In other words, we must prove that if both J_1 and J_2 are finite supports of p, then $\exists(J \cap J_1)p=\exists(J \cap J_2)p$. Since, by Lemma 6.7, $J_1 \cap J_2$ supports p, the desired result follows, via Lemma 6.9, from the fact that both $\exists(J \cap J_1)p$ and $\exists(J \cap J_2)p$ are equal to $\exists(J \cap J_1 \cap J_2)p$.

In J happens to be finite, then the finite support J_0 of p can be chosen so that $J \subset J_0$. If that is done, then $J \cap J_0=J$, and it follows that $\tilde{\exists}$ is indeed an extension of $\exists$.

Since $\emptyset$ supports 0, it follows that $\tilde{\exists}(J)0=\exists(J\cap\emptyset)0=\exists(\emptyset)0=0$, so that $\tilde{\exists}(J)$ is normalized. To prove that $\tilde{\exists}(J)$ is increasing, we merely note that $p\leqslant\exists(J\cap J_0)p$. The proof of quasi-multiplicativity is, naturally, the hardest. Given p and q, we first find a finite set J_0 that supports them both. Next we note that J_0 supports $\tilde{\exists}(J)q$ (recall that $\tilde{\exists}(J)q=\mathrm{E}(J\cap J_0)q$ and apply Lemma 6.10 and Lemma 6.7), and that J_0 supports $p\wedge\tilde{\exists}(J)q$ (by Lemma 6.8). Quasi-multiplicativity now follows from the computation:

$$\begin{aligned}\tilde{\exists}(J)\big(p\wedge\tilde{\exists}(J)q\big)&=\exists(J\cap J_0)\big(p\wedge\exists(J\cap J_0)q\big)\\&=\exists(J\cap J_0)p\wedge\exists(J\cap J_0)q=\tilde{\exists}(J)p\wedge\tilde{\exists}(J)q.\end{aligned}$$

In other words, each $\tilde{\exists}(J)$ is a quantifier on $\boldsymbol{A}$, and the proof of (ii) is complete.

(iii) We must now prove that $\boldsymbol{A}$ together with $\tilde{\boldsymbol{S}}$ and $\tilde{\exists}$ is a locally finite I-algebra. Local finiteness is immediate from (Q_7) and Lemma 6.15; it remains to prove that $\tilde{\boldsymbol{S}}$ and $\tilde{\exists}$ satisfy (P_1)-(P_6).

(P_1) Immediate from (Q_1) and (i).

(P_2) Suppose that σ and τ are transformations on I, that $p\in\boldsymbol{A}$, and that J_0 and K_0 are finite supports of p and of $\tilde{\boldsymbol{S}}(\tau)p$, respectively. (If we knew that Lemma 6.14 is valid for quasi-polyadic algebras, we could take $K_0=\tau J_0$.) If, as before, $\tau_0 i=\tau i$ when $i\in J_0$ and $\tau_0 i=i$ otherwise, and if $\sigma_0 j=\sigma j$ when $j\in K_0\cup\tau J_0$ and $\sigma_0 j=j$ otherwise, then $\tilde{\boldsymbol{S}}(\tau)p=\boldsymbol{S}(\tau_0)p$ and $\tilde{\boldsymbol{S}}(\sigma)\tilde{\boldsymbol{S}}(\tau)p=\boldsymbol{S}(\sigma_0)\tilde{\boldsymbol{S}}(\tau)p=\boldsymbol{S}(\sigma_0)\boldsymbol{S}(\tau_0)p=\boldsymbol{S}(\sigma_0\tau_0)p$. Since $\sigma_0\tau_0$ is a finite transformation, and since $\sigma_0\tau_0 i=\sigma\tau i$ when $i\in J_0$, it follows easily that $\tilde{\boldsymbol{S}}(\sigma\tau)p=\boldsymbol{S}(\sigma_0\tau_0)p$.

(P_3) Immediate from (Q_3) and (ii).

(P_4) If J and K are subsets of I, if $p\in\boldsymbol{A}$, and if J_0 is a finite support of p, then J_0 supports $\tilde{\exists}(K)p$ (by Lemma 6.10 and Lemma 6.7, since $\tilde{\exists}(K)p=\exists(K\cap J_0)p$). It follows that

$$\tilde{\exists}(J)\tilde{\exists}(K)p=\exists(J\cap J_0)\exists(K\cap J_0)p=\exists\big((J\cup K)\cap J_0\big)p=\tilde{\exists}(J\cup K)p.$$

(P_5) If σ and τ are transformations on I, if $J\subset I$, and if $\sigma=\tau$ outside J, it is to be proved that $\tilde{\boldsymbol{S}}(\sigma)\tilde{\exists}(J)p=\tilde{\boldsymbol{S}}(\tau)\tilde{\exists}(J)p$ for each p in $\boldsymbol{A}$. If J_0 is a finite support of p, and if σ_0 and τ_0 are defined as usual ($\sigma_0=\sigma$ and $\tau_0=\tau$ in J_0, and $\sigma_0=\tau_0=\delta$ outside J_0), then $\sigma_0=\tau_0$ outside $J\cap J_0$ and therefore

$$\tilde{\boldsymbol{S}}(\sigma)\tilde{\exists}(J)p=\boldsymbol{S}(\sigma_0)\exists(J\cap J_0)p=\boldsymbol{S}(\tau_0)\exists(J\cap J_0)p=\tilde{\boldsymbol{S}}(\tau)\tilde{\exists}(J)p.$$

(P_6) This is the combinatorially most difficult part of the proof; it will be presented in three steps. We are given a transformation τ on I and a subset J of I, such that τ is one-to-one on $\tau^{-1}J$. We are to prove that $\tilde{\mathfrak{I}}(J)\tilde{S}(\tau)=\tilde{S}(\tau)\tilde{\mathfrak{I}}(\tau^{-1}J)$.

(a) Assume first that J is finite. If $p \in A$, write $q=\tilde{\mathfrak{I}}(\tau^{-1}J)p$; note that since τ is one-to-one on $\tau^{-1}J$, it follows that $\tau^{-1}J$ is finite. Let J_0 be a finite support of p; then J_0 supports q. If σ is the transformation that agrees with τ on $J_0 \cup J \cup \tau^{-1}J$ and with δ outside that set, then σ is a finite transformation on I. The transformation σ is such that $\sigma^{-1}J=\tau^{-1}J$. Indeed, if $i \in \tau^{-1}J$, then $\sigma i=\tau i \in J$, so that $i \in \sigma^{-1}J$; this proves that $\tau^{-1}J \subset \sigma^{-1}J$. If $j \in \sigma^{-1}J$, then (since σ always agrees with either τ or δ) either $\sigma j=j \in J$ or $\sigma j=\tau j \in J$. In the first case $\tau j=j$ (since σ and τ agree on J), so that, in either case, $\tau j=\sigma j$, and therefore $j \in \tau^{-1}J$; this proves that $\sigma^{-1}J \subset \tau^{-1}J$.

Since $\sigma^{-1}J=\tau^{-1}J$ and since $\sigma=\tau$ on $\tau^{-1}J$, it follows that σ is one-to-one on $\sigma^{-1}J$. Consequently

$$\begin{aligned}\tilde{\mathfrak{I}}(J)\tilde{S}(\tau)p&=\mathfrak{I}(J)S(\sigma)p=S(\sigma)\mathfrak{I}(\sigma^{-1}J)p=S(\sigma)\mathfrak{I}(\tau^{-1}J)p\\&=S(\sigma)\tilde{\mathfrak{I}}(\tau^{-1}J)p=S(\sigma)q=\tilde{S}(\tau)q=\tilde{S}(\tau)\tilde{\mathfrak{I}}(\tau^{-1}J)p\,.\end{aligned}$$

(b) Assume next that $p \in A$, $\tilde{\mathfrak{I}}(J)p=p$, and $\tilde{\mathfrak{I}}(J)\tilde{S}(\tau)p=\tilde{S}(\tau)p$; let J_0 be a simultaneous finite support of both p and $\tilde{S}(\tau)p$. It follows that J_0-J is a simultaneous support of p and $\tilde{S}(\tau)p$, and we may therefore assume, without any loss of generality, that J_0 is disjoint from J.

If $K=\tau^{-1}J \cap J_0$, then $\tilde{S}(\tau)\tilde{\mathfrak{I}}(\tau^{-1}J)p=\tilde{S}(\tau)\mathfrak{I}(K)p$. If $H=\tau K$, then, clearly, $H \subset J$ and $K=\tau^{-1}H$ (here is where the assumed one-to-one property of τ is used). Since, moreover, J_0 is finite, it follows that H and K are finite. Hence

$$\begin{aligned}\tilde{S}(\tau)\tilde{\mathfrak{I}}(\tau^{-1}J)p&=\tilde{S}(\tau)\tilde{\mathfrak{I}}(\tau^{-1}H)p=\mathfrak{I}(H)\tilde{S}(\tau)p \qquad \text{(by (a) above)}\\&=\tilde{S}(\tau)p \qquad (\text{since } H \subset J \text{ and } \tilde{\mathfrak{I}}(J)\tilde{S}(\tau)p=\tilde{S}(\tau)p)\,.\end{aligned}$$

Since we have also assumed that $\tilde{\mathfrak{I}}(J)\tilde{S}(\tau)p=\tilde{S}(\tau)p$, it follows that, in this case,

$$\tilde{S}(\tau)\tilde{\mathfrak{I}}(\tau^{-1}J)p=\tilde{\mathfrak{I}}(J)\tilde{S}(\tau)p,$$

as desired.

(c) If $p \in A$, let J_0 be a simultaneous finite support of p and $\tilde{S}(\tau)p$. If $J_1=J \cap J_0$ and $J_2=J-J_0$, then $J=J_1 \cup J_2$, the set J_1 is finite, and the set J_2 is such that $\tilde{\mathfrak{I}}(J_2)p=p$ and $\mathfrak{I}(J_2)\tilde{S}(\tau)p=\tilde{S}(\tau)p$ (since J_2 is disjoint from J_0). Moreover, since J_1 and J_2 are subsets of J, the transformation τ is one-to-one on $\tau^{-1}J_1$ and on $\tau^{-1}J_2$. It follows that

$$\begin{aligned}\tilde{\exists}(J)\tilde{S}(\tau)p &= \tilde{\exists}(J_1)\tilde{\exists}(J_2)\tilde{S}(\tau)p \\ &= \tilde{\exists}(J_1)\tilde{S}(\tau)\tilde{\exists}(\tau^{-1}J_2)p \qquad \text{(by (b) above)} \\ &= \tilde{S}(\tau)\tilde{\exists}(\tau^{-1}J_1)\tilde{\exists}(\tau^{-1}J_2)p \qquad \text{(by (a) above)} \\ &= \tilde{S}(\tau)\tilde{\exists}(\tau^{-1}J)p.\end{aligned}$$

The proof of Theorem 7.6 is complete.

§ 8. Algebraic theory

The elementary algebraic theory of polyadic algebras is an easy part of universal algebra. The basic concepts are, as always, those of subalgebra, homomorphism, and ideal.

A subset $\boldsymbol{B}$ of a polyadic algebra $\boldsymbol{A}$ is a *(polyadic) subalgebra* of $\boldsymbol{A}$ if it is a Boolean subalgebra of $\boldsymbol{A}$ and if it is a polyadic algebra with respect to the operator mappings of $\boldsymbol{A}$. More precisely, if $\boldsymbol{A}$ is an I-algebra and if $\boldsymbol{B}$ is a Boolean subalgebra of $\boldsymbol{A}$ such that, for all p in $\boldsymbol{B}$, $\boldsymbol{S}(\tau)p \in \boldsymbol{B}$ whenever τ is a transformation on I, and $\exists(J)p \in \boldsymbol{B}$ whenever J is a subset of I, then the I-algebra $\boldsymbol{B}$ is called a polyadic subalgebra of $\boldsymbol{A}$. In the locally finite case this definition can be improved in two distinct directions. It follows from Theorem 5.2 that a Boolean subalgebra $\boldsymbol{B}$ of an I-algebra $\boldsymbol{A}$ is a polyadic subalgebra of $\boldsymbol{A}$ if and only if $\boldsymbol{S}(\sigma)p \in \boldsymbol{B}$ whenever $p \in \boldsymbol{B}$ and σ is a substitution in I. It follows from Lemma 6.16 that a Boolean subalgebra $\boldsymbol{B}$ of an I-algebra $\boldsymbol{A}$ is a polyadic subalgebra of $\boldsymbol{A}$ if and only if, for all p in $\boldsymbol{B}$, $\boldsymbol{S}(\tau)p \in \boldsymbol{B}$ whenever τ is a finite transformation on I and $\exists(J)p \in \boldsymbol{B}$ whenever J is a finite subset of I. We observe that every polyadic subalgebra of an I-algebra is itself an I-algebra. If $\boldsymbol{A}$ is an I-algebra and if I_0 is a subset of I, then, by appropriately restricting the operator mappings of $\boldsymbol{A}$ to I_0 we obtain an I_0-algebra, but that I_0-algebra is not a polyadic subalgebra of $\boldsymbol{A}$. We observe also that every polyadic subalgebra of a locally finite algebra is locally finite.

If $\boldsymbol{A}$ and $\boldsymbol{B}$ are polyadic algebras (with the same set I of variables), a *(polyadic) homomorphism* is a mapping f from $\boldsymbol{A}$ into $\boldsymbol{B}$ such that f is a Boolean homomorphism and such that, for all p in $\boldsymbol{A}$, $f\boldsymbol{S}(\tau)p = \boldsymbol{S}(\tau)fp$ whenever τ is a transformation on I and $f\exists(J)p = \exists(J)fp$ whenever J is a subset of I. In the locally finite case this definition can be improved, just as for subalgebras, in two distinct directions. A Boolean homomorphism f from $\boldsymbol{A}$ into $\boldsymbol{B}$ is a polyadic homomorphism if and only if $f\boldsymbol{S}(\sigma)p = \boldsymbol{S}(\sigma)fp$ whenever $p \in \boldsymbol{A}$ and σ is a substitution in I. Alternatively, a Boolean homomorphism f from $\boldsymbol{A}$ into $\boldsymbol{B}$ is a polyadic homomorphism if and only if, for all p in $\boldsymbol{A}$, $f\boldsymbol{S}(\tau)p = \boldsymbol{S}(\tau)fp$ whenever

τ is a finite transformation on I and $f\exists(J)p=\exists(J)fp$ whenever J is a finite subset of I. We observe that the concept of a polyadic homomorphism is defined only between algebras with the same set of variables. A polyadic algebra is, after all, a Boolean algebra with operators, and it is quite natural to insist that a homomorphism preserve the operator structure. The situation is analogous to the one encountered in the theory of groups with operators. We observe also that a homomorphism preserves supports and hence that every homomorphic image of a locally finite algebra is locally finite.

If $\boldsymbol{A}$ is an I-algebra, a *(polyadic) ideal* in $\boldsymbol{A}$ is a Boolean ideal $\boldsymbol{M}$ in $\boldsymbol{A}$ such that, for all p in $\boldsymbol{M}$, $\boldsymbol{S}(\tau)p \in \boldsymbol{M}$ whenever τ is a transformation on I, and $\exists(J)p \in \boldsymbol{M}$ whenever J is a subset of I. In the locally finite case a Boolean ideal $\boldsymbol{M}$ is a polyadic ideal if and only if $\boldsymbol{S}(\sigma)p \in \boldsymbol{M}$ whenever $p \in \boldsymbol{M}$ and σ is a substitution in I. Alternatively, a Boolean ideal $\boldsymbol{M}$ is a polyadic ideal if and only if, for all p in $\boldsymbol{M}$, $\boldsymbol{S}(\tau)p \in \boldsymbol{M}$ whenever τ is a finite transformation on I, and $\exists(J)p \in \boldsymbol{M}$ whenever J is a finite subset of I. For polyadic ideals (but not for subalgebras and homomorphisms) there is an even simpler characterization that is frequently very useful.

(8.1) LEMMA. *A Boolean ideal* $\boldsymbol{M}$ *in* $\boldsymbol{A}$ *is a polyadic ideal if and only if* $\exists(I)p \in \boldsymbol{M}$ *whenever* $p \in \boldsymbol{M}$.

Proof. Half of the assertion is trivial; if $\boldsymbol{M}$ is a polyadic ideal, then $\boldsymbol{M}$ is invariant under all the quantifiers of $\boldsymbol{A}$, and, in particular, under $\exists(I)$. Suppose, to prove the converse, that $\boldsymbol{M}$ is a Boolean ideal invariant under $\exists(I)$. If $p \in \boldsymbol{M}$ and if τ is a transformation on I, then the inequality $p \leqslant \exists(I)p$, together with the fact that $\boldsymbol{S}(\tau)$ is a Boolean endomorphism of $\boldsymbol{A}$, implies that $\boldsymbol{S}(\tau)p \leqslant \boldsymbol{S}(\tau)\exists(I)p$. Since, trivially, $\tau=\delta$ outside I, it follows from (P_5) and (P_1) that $\boldsymbol{S}(\tau)\exists(I)=\boldsymbol{S}(\delta)\exists(I)=\exists(I)$, and hence that $\boldsymbol{S}(\tau)p \leqslant \exists(I)p$. Since $\exists(I)p \in \boldsymbol{M}$ and since $\boldsymbol{M}$ is a Boolean ideal, we conclude that $\boldsymbol{S}(\tau)p \in \boldsymbol{M}$. Similarly, if $p \in \boldsymbol{M}$ and $J \subset I$, then $\exists(J)p \leqslant \exists(I)\exists(J)p=\exists(I)p$, and therefore $\exists(J)p \in \boldsymbol{M}$; the proof of the lemma is complete.

If $p \in \boldsymbol{A}$ and if J is a finite subset of I that supports p, then $\exists(K)\exists(J)p=\exists(J)p$ for all K. (If all the variables that p depends on are quantified out, the result is invariant under all further quantification.) The quantifier-invariant element thus associated with p will be called the *logical closure* of p; according to an equivalent definition the logical closure of p is the element $\exists(I)p$. In accordance with this definition, we shall speak also of the *closed* elements of $\boldsymbol{A}$, meaning thereby the elements invariant under $\exists(I)$, or, equivalently, the range of the quantifier $\exists(I)$. The concept of logical closure as here defined is not quite the same as the concept usually covered by the same term. The usual closure is essentially the dual of the present one; it is obtained by prefixing an adequate supply of universal quantifiers (instead of existential ones). The present terminology seems preferable because of its harmony with the theory of closure algebras: an existential quantifier is, after all, a closure operation, whereas its dual, a universal quantifier, is not.

If $\boldsymbol{A}$ is an I-algebra and if J is a subset of I, then the Boolean algebra $\boldsymbol{A}$, with the quantifier $\exists(J)$, is a monadic algebra that we shall denote by $\boldsymbol{A}(J)$. With this notation the result of Lemma 8.1 can be stated in the following form: a Boolean ideal $\boldsymbol{M}$ in $\boldsymbol{A}$ is a polyadic ideal in $\boldsymbol{A}$ if and only if it is a monadic ideal in $\boldsymbol{A}(I)$. In these terms we can also give a useful characterization of simple polyadic algebras. By definition, a polyadic algebra $\boldsymbol{A}$ is *simple* if $\{0\}$ is the only proper polyadic ideal in $\boldsymbol{A}$.

(8.2) LEMMA. *An I-algebra $\boldsymbol{A}$ is simple if and only if the operation of forming logical closure is a simple quantifier (i. e., $\exists(I)p=1$ whenever $p\neq 0$), or, equivalently, if and only if the monadic algebra $\boldsymbol{A}(I)$ is simple.*

Proof. The equivalence of the simplicity of the polyadic algebra $\boldsymbol{A}$ with that of the monadic algebra $\boldsymbol{A}(I)$ follows immediately from Lemma 8.1; the characterization in terms of $\exists(I)$ is a consequence of the corresponding monadic result (1.6).

If f is a polyadic homomorphism with domain $\boldsymbol{A}$, then its *kernel*, *i. e.*, the set $\boldsymbol{M}=\{p\colon fp=0\}$, is a polyadic ideal. Since f is in particular a Boolean homomorphism, so that $f1=1$, it follows that this polyadic ideal is proper (*i. e.*, $\boldsymbol{M}\neq\boldsymbol{A}$). The homomorphism theorem for polyadic algebras asserts that, conversely, every proper ideal is a kernel. Suppose, indeed, that $\boldsymbol{A}$ is an I-algebra and that $\boldsymbol{M}$ is a proper polyadic ideal in $\boldsymbol{A}$. Let $\boldsymbol{B}$ be the Boolean quotient algebra $\boldsymbol{A}/\boldsymbol{M}$ and let f be the corresponding natural Boolean homomorphism from $\boldsymbol{A}$ onto $\boldsymbol{B}$. We proceed to show that there is a unique, natural way of converting $\boldsymbol{B}$ into an I-algebra so that f becomes a polyadic homomorphism (with kernel $\boldsymbol{M}$). For this purpose it is convenient to make use of the remark following Theorem 5.2. To define $\boldsymbol{S}(\sigma)q$, where $q\in\boldsymbol{B}$ and σ is a substitution in I, find p in $\boldsymbol{A}$ so that $fp=q$ and write $\boldsymbol{S}(\sigma)q=f\boldsymbol{S}(\sigma)p$. It is to be proved, of course, that this definition is unambiguous, *i. e.*, that if $fp_1=fp_2$, then $f\boldsymbol{S}(\sigma)p_1=f\boldsymbol{S}(\sigma)p_2$. Equivalently, it is to be proved that if $p_1+p_2\in\boldsymbol{M}$, then $\boldsymbol{S}(\sigma)p_1+\boldsymbol{S}(\sigma)p_2\in\boldsymbol{M}$. To prove this, write $\sigma=\tau\chi(J)$, where τ is a transformation on I and J is a subset of I. It follows that

$$\boldsymbol{S}(\sigma)p_1+\boldsymbol{S}(\sigma)p_2=\boldsymbol{S}(\tau)\exists(J)p_1+\boldsymbol{S}(\tau)\exists(J)p_2=\boldsymbol{S}(\tau)\big(\exists(J)p_1+\exists(J)p_2\big)$$

(since $\boldsymbol{S}(\tau)$ is a Boolean endomorphism of $\boldsymbol{A}$) and hence that

$$\boldsymbol{S}(\sigma)p_1+\boldsymbol{S}(\sigma)p_2\leqslant\boldsymbol{S}(\tau)\exists(J)(p_1+p_2)$$

(by (1.5), together with the fact that a Boolean endomorphism preserves order). Since $S(\tau)\exists(J)(p_1+p_2)$ belongs to M along with p_1+p_2, and since M contains all elements smaller than any of its elements, it follows, as desired, that M contains $S(\sigma)p_1+S(\sigma)p_2$. A routine verification shows that the mapping S from substitutions in I into operators on B satisfies the conditions described in the remark following Theorem 5.2, so that B is an I-algebra, the *quotient algebra* of A modulo the ideal M.

Caution: the homomorphism theorem, together with Lemma 8.1, has the effect of making a false assertion seem plausible. If f is a polyadic homomorphism from A into B with kernel M, then, of course, f is in particular a monadic homomorphism from $A(I)$ into $B(I)$ with kernel M. If, conversely, f is a monadic homomorphism from $A(I)$ into $B(I)$ with kernel M, then, in view of Lemma 8.1, M is a polyadic ideal in A, and it might seem plausible to conjecture that f must be a polyadic homomorphism. The conjecture is false; finite, dyadic counter examples are easily constructed.

A polyadic ideal is *maximal* if it is a proper ideal that is not a proper subset of any other proper ideal. The connection between maximal ideals and simple algebras is an elementary part of universal algebra: the kernel of a homomorphism is a maximal ideal if and only if its range is a simple algebra. A polyadic algebra is *semisimple* if the intersection of all maximal ideals in it is $\{0\}$.

§ 9. Logics

The concepts of simplicity and semisimplicity play a central role in the theories of Boolean algebras and monadic algebras. Their relevance in the theory of polyadic algebras can be established by almost exactly the same argument as was used in sections 5, 6, and 7 of [2]; cf. also sections 4 and 5 of [1]. For the sake of completeness we proceed to review the most important points of that argument.

A *polyadic logic* is a pair (A, M), where A is a polyadic algebra and M is a polyadic ideal in A. If, heuristically, the elements of A are thought of as propositional functions, then the elements of M are to be thought of as the refutable ones among them. A moment's thought shows that, from the point of view of logic, it is reasonable that the set of refutable elements form a polyadic ideal. Indeed, just as in the propositional calculus (*i. e.*, the theory of Boolean logics) that set should form a Boolean ideal. That it should also form a polyadic ideal comes from the intuitively obvious requirement that if a propositional function p is refutable, then so also is every propositional function obtained from p by existential quantifications and substitutions of variables.

Within the theory of polyadic logics almost all logical terms are translatable into algebraic language. Thus, for example, a polyadic logic $(\boldsymbol{A}, \boldsymbol{M})$ is called *simply consistent*, or, simply, *consistent*, if $\boldsymbol{M}$ is a proper ideal in $\boldsymbol{A}$ (*i. e.*, if not everything is refutable), and a polyadic logic $(\boldsymbol{A}, \boldsymbol{M})$ is called *simply complete*, or, simply, *complete* if either $\boldsymbol{M}=\boldsymbol{A}$ or else $\boldsymbol{M}$ is a maximal ideal in $\boldsymbol{A}$. We note that $(\boldsymbol{A}, \boldsymbol{M})$ is consistent if and only if for no element p of $\boldsymbol{A}$ do both p and p' belong to $\boldsymbol{M}$. (Indeed, if p and p' are both in $\boldsymbol{M}$, then so also is 1, since $1=p \vee p'$, and therefore $\boldsymbol{M}=\boldsymbol{A}$; if, on the other hand, $\boldsymbol{M}=\boldsymbol{A}$, then $1 \in \boldsymbol{M}$ and therefore there is at least one p, namely 0, such that both p and p' are in $\boldsymbol{M}$.) This formulation of the definition of consistency is probably the one that accords most closely with our intuitive ideas on the subject.

The algebraic properties of a consistent polyadic logic $(\boldsymbol{A}, \boldsymbol{M})$ are most conveniently studied in terms of the associated quotient algebra $\boldsymbol{A}/\boldsymbol{M}$. The concept of simple completeness, in particular, has an easy formulation in these terms.

(9.1) LEMMA. *A consistent polyadic logic* $(\boldsymbol{A}, \boldsymbol{M})$ *is simply complete if and only if the polyadic algebra* $\boldsymbol{A}/\boldsymbol{M}$ *is simple.*

The proof of this lemma is an immediate consequence of the relation between simple algebras and maximal ideals. With the aid of this lemma the concept of simple completeness can be reformulated in still other ways that bring it closer to our intuitive notions. We note, first of all, that the set $\boldsymbol{B}$ of all closed elements in an I-algebra $\boldsymbol{A}$ is a Boolean subalgebra of $\boldsymbol{A}$; this follows from the fact that $\boldsymbol{B}$ is the range of the quantifier $\exists(I)$. If, moreover, $\boldsymbol{M}$ is a polyadic ideal in $\boldsymbol{A}$, then $\boldsymbol{N}=\boldsymbol{B} \cap \boldsymbol{M}$ is a Boolean ideal in $\boldsymbol{B}$. We assert now that a consistent polyadic logic $(\boldsymbol{A}, \boldsymbol{M})$ is simply complete if and only if $\boldsymbol{N}$ is a maximal ideal in $\boldsymbol{B}$. Indeed, let f be the natural homomorphism from $\boldsymbol{A}$ onto $\boldsymbol{A}/\boldsymbol{M}$. If $\boldsymbol{A}/\boldsymbol{M}$ is simple, and if $p \in \boldsymbol{B}$, then $p=\exists(I)p$ and therefore $fp=f\exists(I)p=\exists(I)fp$. Since (cf. Lemma 8.2) $\exists(I)$ is a simple quantifier on $\boldsymbol{A}/\boldsymbol{M}$, it follows that either $fp=0$ or $fp=1$. If $fp=0$, then $p \in \boldsymbol{M}$, and if $fp=1$, then $p' \in \boldsymbol{M}$. In other words, if p is an arbitrary element of $\boldsymbol{B}$, then either $p \in \boldsymbol{N}$ or $p' \in \boldsymbol{N}$; this proves that $\boldsymbol{N}$ is a maximal ideal in $\boldsymbol{B}$. If, conversely, $\boldsymbol{N}$ is a maximal ideal in $\boldsymbol{B}$, and if $p \in \boldsymbol{A}$, then $\exists(I)p \in \boldsymbol{B}$ and therefore either $\exists(I)p \in \boldsymbol{N}$ or $(\exists(I)p)' \in \boldsymbol{N}$. Since $f\exists(I)p=\exists(I)fp$, it follows that either $\exists(I)fp=0$ or $\exists(I)fp=1$, *i. e.*, that $\exists(I)$ is a simple quantifier on $\boldsymbol{A}/\boldsymbol{M}$, and hence that $\boldsymbol{A}/\boldsymbol{M}$ is simple.

It follows from the result (or, more directly, from the argument) of the preceding paragraph that a polyadic logic $(\boldsymbol{A}, \boldsymbol{M})$ is simply complete if and only if, for each closed p in $\boldsymbol{A}$, either $p \in \boldsymbol{M}$ or $p' \in \boldsymbol{M}$. In logical terms, therefore, simple completeness means that, for every (closed) proposition p, either p or its negation is refutable (and therefore either p or its negation is provable). We see thus that the celebrated Gödel incompleteness theorem asserts that certain important polyadic logics are either (simply) incomplete or else (simply) inconsistent. In other words, if $(\boldsymbol{A}, \boldsymbol{M})$ is one of those logics, then either the ideal $\boldsymbol{M}$ is very large ($\boldsymbol{M}=\boldsymbol{A}$) or else it is rather small (non-maximal). In a later paper of this series we shall study the exact class of logics to which this assertion applies; most of the remainder of this paper is devoted to the problems centering around the concept of semisimplicity.

Just as simplicity turns out to be the algebraic counterpart of simple completeness, semisimplicity is the algebraic counterpart of another logical concept, called semantic completeness. The proof of this assertion is, however, quite deep; a significant part of the structure theory of polyadic algebras has to be developed first.

A *model* is an $\boldsymbol{O}$-valued functional polyadic algebra. This definition is the same as the one given in [1]; it differs, in the monadic case, from the definition given in [2], where a model was defined not as an algebra but as a logic. Since the reduction of consistent logics to algebras is an easy step, the present, simpler definition seems to be more appropriate. If $\boldsymbol{A}$ is an $\boldsymbol{O}$-valued functional algebra with domain X, then usually the set X itself is called a model. These differences in terminology are not significant; each author's choice among them is dictated merely by which aspect of the concept he wishes to emphasize.

Once the concept of a model is defined, the other so-called "semantic" concepts of logic are easily translated into algebraic form. Thus, an *interpretation* of a polyadic algebra $\boldsymbol{A}$ is a homomorphism from $\boldsymbol{A}$ into a model; since the range of such a homomorphism is a subalgebra of the model and since a subalgebra of a model is itself a model, we could also have defined an interpretation as a homomorphism onto a model. An element p of $\boldsymbol{A}$ is *false* in an interpretation f if $fp=0$, *i. e.*, if p belongs to the kernel of f; an element p is *universally invalid* (or *contravalid*) if it is false in every interpretation. The algebra $\boldsymbol{A}$ is *semantically complete* if 0 is the only universally invalid element. The justification of this terminology is the same for polyadic algebras as for monadic algebras; cf. section 6 of [2].

If we knew that (a) every model is simple, and (b) every simple algebra is (isomorphic to) a model, then the class of all kernels of interpretations would coincide with the class of all maximal ideals, and we could conclude that a polyadic algebra is semantically complete if and only if it is semisimple. In fact, (a) is true, and (b) is true in all important cases, but the clarification and proof of the latter assertion will have to be postponed for a while. We shall conclude our discussion of these generalities by proving (a) and by proving a result that, in the presence of (b), is the algebraic version of Gödel's completeness theorem.

(9.2) THEOREM. *Every model is simple.*

Proof. If $\boldsymbol{A}$ is an $\boldsymbol{O}$-valued functional polyadic algebra, then

$$\exists(I)p(x) = \bigvee\{p(y)\colon xI_*y\}$$

whenever $p \in \boldsymbol{A}$ and $x \in X^I$. Since xI_*y holds for every y, it follows that the value of $\exists(I)p$ at each x in X^I is the supremum of the range of p. In other words, the monadic algebra $\boldsymbol{A}(I)$ is an $\boldsymbol{O}$-valued functional monadic algebra and therefore (1.8) $\boldsymbol{A}(I)$ is simple. The desired result follows from Lemma 8.2.

(9.3) THEOREM. *Every polyadic algebra is semisimple.*

Proof. Suppose that $\boldsymbol{A}$ is an I-algebra and that p is a non-zero element of $\boldsymbol{A}$. By the semisimplicity of monadic algebras (1.7) there exists a maximal monadic ideal $\boldsymbol{M}$ in $\boldsymbol{A}(I)$ not containing p. By Lemma 8.1, $\boldsymbol{M}$ is a polyadic ideal in $\boldsymbol{A}$, and, moreover, a maximal one, for any polyadic ideal including $\boldsymbol{M}$ would also be a monadic ideal (in $\boldsymbol{A}(I)$) including the maximal monadic ideal $\boldsymbol{M}$. This completes the proof.

§ 10. Functional representation

Since the only concrete examples of polyadic algebras at our disposal are the functional polyadic algebras, it is natural to conjecture that they are the only ones that exist. The conjecture says, in other words, that if $\boldsymbol{A}$ is an arbitrary I-algebra, then it is possible to construct a set X and a Boolean algebra $\boldsymbol{B}$ so that $\boldsymbol{A}$ is isomorphic to a functional polyadic algebra of functions from X^I into $\boldsymbol{B}$. The main purpose of this section is to prove that the conjecture is true at least for the polyadic algebras of greatest interest in logic.

We begin by introducing some terminology and notation to be used in connection with transformations on I. If a transformation τ maps I one-to-one onto itself, we shall say that τ is a *permutation*; if, on the other hand, $\tau^2 = \tau$, we shall say that τ is a *retraction*. If i and j are in I, the transformation τ for which $\tau i = j$, $\tau j = i$, and $\tau k = k$ whenever k is distinct from both i and j, will be denoted by (i, j); any such transformation will be called a *transposition*. If $\tau = (i,j)$ is a transposition, the corresponding endomorphism $\boldsymbol{S}(\tau)$ will be denoted by $\boldsymbol{S}(i,j)$. If i and j are in I, the transformation σ for which $\sigma i = j$, $\sigma j = j$, and $\sigma k = k$ whenever k is distinct from both i and j will be denoted by (i/j); any such transformation will be called a *replacement*. If $\tau = (i/j)$ is a replacement, the corresponding endomorphism $\boldsymbol{S}(\tau)$ will be denoted by $\boldsymbol{S}(i/j)$. Clearly every transposition is a finite permutation and every replacement is a finite retraction.

With every subset J of I there is associated a binary relation J_* between transformations on I; the relation J_* is, by definition, such that

$$\sigma J_* \tau \quad \textit{if and only if} \quad \sigma i = \tau i \quad \textit{whenever} \quad i \notin J.$$

This concept is, as the notation indicates, a special case of the one defined by (2.3); the specialization is obtained by choosing the set X to be the same as I.

(10.1) THEOREM. *If A is a locally finite I-algebra of infinite degree (i. e., if I is infinite), then*

$$S(\tau)\exists(J)p = \bigvee\{S(\sigma)p\colon\ \sigma J_*\tau\} \tag{10.2}$$

whenever $p \in A$, $J \subset I$, and τ is a transformation on I; the supremum indicated in (10.2) is extended over all those transformations σ on I for which $\sigma J_* \tau$.

Remark. (a) It is, of course, not being assumed that A is a complete Boolean algebra, and, consequently, the supremum in (10.2) does not automatically exist. It is intended to be a part of the conclusion of the theorem that the supremum does exist.

(b) If $\tau = \delta$, (10.2) becomes

$$\exists(J)p = \bigvee\{S(\sigma)p\colon\ \sigma J_*\delta\}\,. \tag{10.3}$$

Warning: (10.2) cannot be recaptured from (10.3) simply by applying $S(\tau)$. The trouble is that even if the algebraic hurdle were conquered, *i. e.*, even if we had already shown that $\bigvee\{S(\tau)S(\sigma)p\colon\ \sigma J_*\delta\} = \bigvee\{S(\sigma)p\colon\ \sigma J_*\tau\}$, we would still have to prove that that is the right thing to show. There is no reason to assume that $S(\tau)$, which, being an endomorphism of A, preserves the finite Boolean operations, necessarily preserves the infinite Boolean operations also.

(c) If the set J is a singleton, say $J = \{i\}$, then (10.3) becomes

$$\exists(i)p = \bigvee\{S(\sigma)p\colon\ \sigma i_*\delta\}\,. \tag{10.4}$$

(We recall that $\exists(i)$ stands for $\exists(\{i\})$; similarly i_* stands for $\{i\}_*$.) Now if $\sigma i_* \delta$, then σ must have the form (i/j) for some variable j; consequently (10.4) may be rewritten in the form

$$\exists(i)p = \bigvee\{S(i/j)p\colon\ j \in I\}\,. \tag{10.5}$$

In this special form the result is due to Rasiowa and Sikorski [5]. Equation (10.5) acquires intuitive significance if we (illegally) verbalize it by equating "p holds for some i" with "the result of replacing i by j in p holds for some j".

(d) Finally we observe that the assumption that I be infinite cannot be omitted. Suppose, in fact, that $I=\{0,1\}$, and that the algebra $\boldsymbol{A}$ consists of the four elements $\{0,p,p',1\}$. The actions of the transformations on $\boldsymbol{A}$ are uniquely determined by the equations $\boldsymbol{S}(0/1)p=\boldsymbol{S}(1/0)p=0$ and $\boldsymbol{S}(0,1)p=p$. If the quantifiers $\exists(0)$, $\exists(1)$, and $\exists(0,1)$ $(=\exists(\{0,1\}))$ are all required to coincide with the simple quantifier on $\boldsymbol{A}$, then, with respect to this system of operators, $\boldsymbol{A}$ becomes an I-algebra. The relation (10.5), which in this case reduces to $\exists(1)p=p\vee\boldsymbol{S}(0/1)p$, does not hold; indeed $\exists(1)p=1$ and $p\vee\boldsymbol{S}(0/1)p=p$.

Proof. If $\sigma J_*\tau$, then $\boldsymbol{S}(\sigma)p\leqslant\boldsymbol{S}(\sigma)\exists(J)p=\boldsymbol{S}(\tau)\exists(J)p$ (by (P_5)). This observation proves half the theorem. What must now be proved is the assertion

(10.6) *if* $\boldsymbol{S}(\sigma)p\leqslant q$ *whenever* $\sigma J_*\tau$, *then* $\boldsymbol{S}(\tau)\exists(J)p\leqslant q$.

Whenever the implication (10.6) holds for some triple (p,J,τ), then the equation (10.2) holds for that triple. Relying on this comment, we proceed to prove (10.6) in three successive stages of generality.

(I) *τ and J are both finite.* Let K be a set having the same number of elements as J, say $J=\{j_1,\dots,j_n\}$, $K=\{k_1,\dots,k_n\}$, selected so that it is far from the scene of action as far as p, q, τ, and J are concerned. Precisely speaking, we require that $\exists(K)p=p$, $\exists(K)q=q$, $\tau=\delta$ on K and $\tau^{-1}K=K$, and K is disjoint from J. Since I is infinite, a subset K of I can be found satisfying all these conditions.

We introduce the auxiliary transformations π, ϱ and σ defined by

$$\pi=(j_1,k_1)\dots(j_n,k_n),\qquad \varrho=(j_1/k_1)\dots(j_n/k_n),\qquad \sigma=\tau\varrho.$$

We note that if $j\notin J$, then $\sigma j=\tau j$; in other words $\sigma J_*\tau$. Since $\exists(K)p=p$ and $\pi=\varrho$ outside K, it follows from Lemma 6.11 that

$$\boldsymbol{S}(\pi)p=\boldsymbol{S}(\varrho)p\,; \tag{10.7}$$

similarly, since $\exists(K)\exists(J)p=\exists(J)p$ and $\pi=\delta$ outside $J\cup K$, we have

$$\boldsymbol{S}(\pi)\exists(J)p=\exists(J)p\,. \tag{10.8}$$

The proof of (I) is now completed as follows:

$$\begin{aligned}
q&=\exists(K)q && \text{(by assumption)}\\
&\geqslant\exists(K)\boldsymbol{S}(\sigma)p && (\text{since } \sigma J_*\tau)\\
&=\exists(K)\boldsymbol{S}(\tau)\boldsymbol{S}(\varrho)p && (\text{by the definition of } \sigma)\\
&=\exists(K)\boldsymbol{S}(\tau)\boldsymbol{S}(\pi)p && \text{(by (10.7))}\\
&=\boldsymbol{S}(\tau)\exists(K)\boldsymbol{S}(\pi)p && \text{(by assumption)}\\
&=\boldsymbol{S}(\tau)\boldsymbol{S}(\pi)\exists(J)p && (\text{since } \pi^{-1}K=J)\\
&=\boldsymbol{S}(\tau)\exists(J)p && \text{(by (10.8))}.
\end{aligned}$$

(II) τ *is finite.* Let K be a cofinite set such that $\exists(K)p=p$; it follows that $\exists(J)p=\exists(J-K)p$. Since $J-K$ is finite, it follows from (I) that

$$S(\tau)\exists(J)p=S(\tau)\exists(J-K)p=\bigvee\{S(\sigma)p\colon\ \sigma(J-K)_*\tau\}.$$

If $\sigma(J-K)_*\tau$, then, *a fortiori*, $\sigma J_*\tau$, and therefore $S(\sigma)p\leqslant q$. It follows that, in this case also, $S(\tau)\exists(J)p\leqslant q$, as desired.

(III) The general case will be reduced to (II) pretty much the same way as (II) was reduced to (I). First we find a finite set K that supports p (and therefore also supports $\exists(J)p$, cf. Lemma 6.10), and we define the (finite) transformation τ_0 to be τ in K and δ outside K. It follows (cf. Lemma 6.11) that $S(\tau)p=S(\tau_0)p$, and, more relevantly, $S(\tau)\exists(J)p=S(\tau_0)\exists(J)p$. Hence, by (II), $S(\tau)\exists(J)p=\bigvee\{S(\sigma)p\colon\ \sigma J_*\tau_0\}$.

Now we maintain that $S(\sigma_0)p\leqslant q$ whenever $\sigma_0 J_*\tau_0$. If this is admitted for a moment, then it follows from the preceding paragraph that $S(\tau)\exists(J)p\leqslant q$, and, therefore, the proof is complete. To prove the assertion we define a transformation σ (not necessarily finite) as follows: $\sigma=\sigma_0$ in K and $\sigma=\tau$ outside K. It follows that $S(\sigma)p=S(\sigma_0)p$. If $i\in I-J$, then either $i\in K$, in which case $\sigma i=\sigma_0 i$ (by the definition of σ) $=\tau_0 i$ (by the assumption on σ_0) $=\tau i$ (by the definition of τ_0), or $i\notin K$, in which case $\sigma i=\tau i$ directly from the definition of σ. In other words $\sigma J_*\tau$, and therefore (by the assumption on q) $S(\sigma)p\leqslant q$. The last two results imply that $S(\sigma_0)p\leqslant q$, and, therefore, complete the proof of the theorem.

We come now to the main theorem of this section. The result is powerful, and it looks even more powerful than it is. On first glance it looks as if it could single-handedly solve all the interesting problems about polyadic algebras. It cannot, but it is good to know anyway.

(10.9) THEOREM. *Every locally finite polyadic algebra of infinite degree is isomorphic to a functional polyadic algebra whose domain has the same cardinal number as the set of variables.*

Proof. Let A be an I-algebra with an infinite I; we are to manufacture a Boolean algebra B, a set X, and a one-to-one mapping f so that f maps A onto an appropriate class of functions from X^I into B in such a way that the polyadic operations are preserved. The proper choice of B and X looks deceptively simple; we write $B=A$ and $X=I$. The choice is deceptively simple because the very fact that no complicated constructions were used to manufacture B and X will make for notational collisions. The major collision occurs in the consideration of X^I. An element of X^I is a function from I into X. Hence, in the present instance, an element of X^I is a function from I into itself, *i. e.*, a transformation on I. The transformation point of view will be relevant in what follows; we shall call an object a point of X^I or a transformation on I, depending on which of its two roles needs to be emphasized at the moment.

If σ is a transformation on I and if $\tau \in X^I$, what does $\sigma_* \tau$ mean? It means the element of X^I whose value at each i in I is $\tau(\sigma i)$, and hence the element $\tau\sigma$; in other words $\sigma_* \tau = \tau\sigma$.

We are now ready to define the mapping f (which will turn out to be the desired polyadic isomorphism) that associates a function $\tilde{p}$ from X^I into $\boldsymbol{B}$ (*i. e.*, into $\boldsymbol{A}$) with each element p of $\boldsymbol{A}$. If $\tau \in X^I$, we define $\tilde{p}(\tau)$ to be $\boldsymbol{S}(\tau)p$; we proceed to investigate the properties of the mapping $p \to \tilde{p} = fp$. The easiest thing to prove is that f is a Boolean isomorphism. If $p \in \boldsymbol{A}$, $\tilde{p} = fp$, and $\tau \in X^I$, then $fp'(\tau) = \boldsymbol{S}(\tau)p' = (\boldsymbol{S}(\tau)p)'$ (since $\boldsymbol{S}(\tau)$ is a Boolean endomorphism) $= (\tilde{p}(\tau))' = \tilde{p}'(\tau)$ (by the definition of complementation in functional algebras), so that $fp' = (fp)'$. If also $q \in \boldsymbol{A}$ and $\tilde{q} = fq$, then $f(p \vee q)(\tau) = \boldsymbol{S}(\tau)(p \vee q) = \boldsymbol{S}(\tau)p \vee \boldsymbol{S}(\tau)q = \tilde{p}(\tau) \vee \tilde{q}(\tau) = (\tilde{p} \vee \tilde{q})(\tau)$, so that $f(p \vee q) = fp \vee fq$. If, finally, $\tilde{p} = 0$, then, in particular, $\tilde{p}(\delta) = 0$; it follows that $p = \boldsymbol{S}(\delta)p = 0$, so that f is indeed a Boolean isomorphism.

It must now be proved that (i) $f(\boldsymbol{A})$ is a functional polyadic algebra and that (ii) f is a polyadic isomorphism between $\boldsymbol{A}$ and $f(\boldsymbol{A})$. The first of these assertions means that (i)′ if $\tilde{p} \in f(\boldsymbol{A})$ and if σ is a transformation on I, then $\boldsymbol{S}(\sigma)\tilde{p} \in f(\boldsymbol{A})$ (cf. (2.2)), and (i)″ if $\tilde{p} \in f(\boldsymbol{A})$ and J is a subset of I, then $\exists(J)\tilde{p}$ exists and belongs to $f(\boldsymbol{A})$ (cf. (2.4)). The second assertion means, of course, that, under the same assumptions on $\tilde{p}$, σ, and J, (ii)′ $f\boldsymbol{S}(\sigma)p = \boldsymbol{S}(\sigma)fp$ and (ii)″ $f\exists(J)p = \exists(J)fp$. It turns out that the proofs of these two assertions (or, rather, two pairs of assertions) need not be considered separately. The fact that $\boldsymbol{S}(\sigma)fp \in f(\boldsymbol{A})$ is most easily proved by showing that $\boldsymbol{S}(\sigma)fp = f\boldsymbol{S}(\sigma)p$, and, similarly, both the existence and location of $\exists(J)fp$ are settled by proving that the pertinent supremum is equal to $f\exists(J)p$.

Write $q = \boldsymbol{S}(\sigma)p$, $\tilde{q} = fq$, and $\tilde{p} = fp$; let τ be an arbitrary element of X^I and write $\pi = \tau\sigma$. It is to be proved that $\tilde{q}(\tau) = \boldsymbol{S}(\sigma)\tilde{p}(\tau)$. The proof consists of the following computation:

$$\begin{aligned}\tilde{q}(\tau) &= \boldsymbol{S}(\tau)q = \boldsymbol{S}(\tau)\boldsymbol{S}(\sigma)p = \boldsymbol{S}(\pi)p \\ &= \tilde{p}(\pi) = \tilde{p}(\tau\sigma) = \tilde{p}(\sigma_*\tau) = \boldsymbol{S}(\sigma)\tilde{p}(\tau).\end{aligned}$$

The result concerning quantification is obtained similarly. Write $q = \exists(J)p$, $\tilde{q} = fq$, and $\tilde{p} = fp$. Then $\tilde{q}(\tau) = \boldsymbol{S}(\tau)q = \boldsymbol{S}(\tau)\exists(J)p$. Using Theorem 10.1 (here is where the assumption of infinite degree comes in), we conclude that

$$\tilde{q}(\tau) = \bigvee\{\boldsymbol{S}(\sigma)p \colon \sigma J_* \tau\} = \exists(J)\tilde{p}(\tau).$$

The proof of Theorem 10.9 is complete. Methods similar to the ones used in the proof can be found in the work of Henkin [3] and, more explicitly, in the work of Rasiowa and Sikorski [5].

§ 11. Dilations

If I and I^+ are sets such that $I \subset I^+$, then there is a natural way of associating a transformation τ^+ on I^+ with every transformation τ on I; the transformation τ^+ is defined by

$$\tau^+ i = \tau i \quad \textit{when} \quad i \in I, \qquad \tau^+ i = i \quad \textit{when} \quad i \in I^+ - I. \tag{11.1}$$

It follows immediately from this definition that the mapping $\tau \to \tau^+$ is a homomorphism from the semigroup of all transformations on I into the semigroup of all transformations on I^+, *i. e.*, that

$$(\sigma\tau)^+ = \sigma^+\tau^+, \tag{11.2}$$

whenever σ and τ are transformations on I. We observe explicitly that the image δ^+ of the identity transformation δ on I is the identity transformation on I^+, and we note also that if a subset J of I supports τ, then J supports τ^+ also. It follows in particular that if τ is finite, then the same is true of τ^+.

Suppose now that $(A^+, I^+, S^+, \exists^+)$ is a polyadic algebra and write A for the range of the quantifier $\exists^+(I^+ - I)$. (Equivalently, A consists of all the elements of A^+ independent of $I^+ - I$.) Clearly A is a Boolean subalgebra of A^+. More than this is true: the Boolean algebra A can be converted into an I-algebra in a simple and canonical manner. If τ is a transformation on I, we write

$$S(\tau)p = S^+(\tau^+)p, \tag{11.3}$$

and if J is a subset of I, we write

$$\exists(J)p = \exists^+(J)p \tag{11.4}$$

for every element p of A. Since $(\tau^+)^{-1}(I^+ - I) = I^+ - I$, and since τ^+ is one-to-one on $I^+ - I$, and since $\exists^+(I^+ - I)p = p$, it follows that

$$\exists^+(I^+ - I)\, S^+(\tau^+)p = S^+(\tau^+)\exists^+(I^+ - I)p = S^+(\tau^+)p.$$

Since, even easier,

$$\exists^+(I^+ - I)\exists^+(J)p = \exists^+(J)\exists^+(I^+ - I)p = \exists^+(J)p,$$

we may conclude that the algebra A is closed under the operations $S(\tau)$ and $\exists(J)$ defined in (11.3) and (11.4). It is easy to verify that $(A, I, S, \exists)$ is a polyadic algebra. Indeed, (P_1) follows from the fact that δ^+ is the identity transformation on I^+, and (P_2) follows from the multiplicativity relation (11.2). Since in the passage from $\exists(J)$ to $\exists^+(J)$ nothing changes, (P_3) and (P_4) are automatic. The condition (P_5) follows from the fact that if $\sigma=\tau$ outside J, then $\sigma^+=\tau^+$ outside J, and, similarly, (P_6) follows from the fact that if τ is one-to-one on $\tau^{-1}J$, then τ^+ is one-to-one on $(\tau^+)^{-1}J$ and $(\tau^+)^{-1}J=\tau^{-1}J$.

We shall describe the relation between the I-algebra A and the I^+-algebra A^+ discussed in the preceding paragraph by saying that A is a *compression* of A^+ and A^+ is a *dilation* of A. More explicitly, if I and I^+ are sets such that $I\subset I^+$, then to say that an I^+-algebra A^+ is a dilation of an I-algebra A means that A consists of all the elements of A^+ independent of I^+-I and that the operator mappings S and $\exists$ of A are related to the operator mappings S^+ and $\exists^+$ of A^+ by the equations (11.3) and (11.4). It follows from this definition that if A^+ is locally finite, then the same is true of A. Indeed, if $p\in A$, then, since $p\in A^+$ and A^+ is locally finite, there exists a cofinite subset J^+ of I^+ such that $\exists^+(J^+)p=p$. If we write $J=J^+\cap I$, then J is a cofinite subset of I and $\exists(J)p=\exists^+(J)p=\exists^+(J)\exists^+(J^+-I)p$ (since $p\in A$)$=\exists^+(J^+)p=p$; this proves, as desired, that every element of A has finite support.

The process of going from an algebra to one of its dilations might be thought of as the adjunction of new variables. It is not at all obvious that new variables can always be adjoined to an arbitrary polyadic algebra; this section is devoted to the discussion of the most important special cases of the problem. We shall discuss, in particular, certain functional algebras with value algebra B, say, and domain X. In this connection, and, in fact, in most discussions of the theory of dilations, it is useful to make a distinction between the operator mappings on I-algebras and those on I^+-algebras; we shall always denote the former by S and $\exists$, as before, and the latter by S^+ and $\exists^+$. For transformations and for subsets we shall continue to use familiar symbols (such as σ, τ, J, and K); a symbol such as τ^+ will be used only in its special meaning defined by (11.1).

(11.5) Lemma. *If A is a B-valued I-algebra over X and if $I\subset I^+$, then there exists a mapping f from A onto a Boolean algebra of functions from X^{I^+} into B such that* (i) *f is a Boolean isomorphism between A and $f(A)$*, (ii) *fp is independent of I^+-I for all p in A*, (iii) *$fS(\tau)p=S^+(\tau^+)fp$ whenever τ is a transformation on I and $p\in A$, and* (iv) *$f\exists(J)p=\exists^+(J)fp$ whenever J is a subset of I and $p\in A$.*

Remark. The concept of independence referred to in (ii) is to be interpreted in the sense of the functional independence defined at the beginning of § 5. The operators $S^+(\tau^+)$ in (iii) and $\exists^+(J)$ in (iv) are to be interpreted in the functional sense defined in (2.2) and (2.4). The assertion of the equation in (iv) is intended to imply that its right term exists for every subset J of I and every element p of A.

Proof. Let φ be the natural (projection) mapping from X^{I^+} onto X^I, *i. e.*, if $x^+ \in X^{I^+}$, then $(\varphi x^+)_i = x_i^+$ for all i in I. We define the mapping f by writing $(fp)(x^+) = p(\varphi x^+)$ for all x^+ in X^{I^+}. The fact that f is a Boloean homomorphism follows immediately from the fact that the Boolean operations in algebras of functions are defined pointwise. Since, moreover, φ is onto, it follows that the kernel of f is trivial and hence that f is one-to-one. This proves (i). Suppose next that x^+ and y^+ are points of X^{I^+} such that $x^+(I^+ - I)_* y^+$, *i. e.*, such that $x_i^+ = y_i^+$ whenever $i \in I$. It follows that $\varphi x^+ = \varphi y^+$ and hence that $fp(x^+) = fp(y^+)$ for all p in A; this proves (ii). If τ is a transformation on I, then $(\tau_* \varphi x^+)_i = (\varphi x^+)_{\tau i} = x_{\tau i}^+$ and $(\varphi \tau_*^+ x^+)_i = (\tau_*^+ x^+)_i = x_{\tau^+ i}^+$ whenever $x^+ \in X^{I^+}$ and $i \in I$. It follows that $\tau_* \varphi = \varphi \tau_*^+$ and hence that

$$
\begin{aligned}
fS(\tau)p(x^+) &= S(\tau)p(\varphi x^+) = p(\tau_* \varphi x^+) \\
&= p(\varphi \tau_*^+ x^+) = fp(\tau_*^+ x^+) = S^+(\tau^+) fp(x^+)
\end{aligned}
$$

for all p in A; this proves (iii). We observe finally that, because of (ii),

$$
\begin{aligned}
\{fp(y^+) \colon x^+ J_* y^+\} &= \{fp(y^+) \colon x^+ \big(J \cup (I^+ - I)\big)_* y^+\} \\
&= \{p(\varphi y^+) \colon \varphi x^+ J_* \varphi y^+\} = \{p(y) \colon \varphi x^+ J_* y\}
\end{aligned}
$$

whenever $x^+ \in X^{I^+}$ and $J \subset I$. This implies that $\exists^+(J) fp(x^+)$ exists and is equal to $\exists(J) p(\varphi x^+) = f\exists(J) p(x^+)$. Thus (iv) is proved and the proof of the lemma is complete.

(11.6) Lemma. *If A is a locally finite, B-valued, functional I-algebra of infinite degree over X, and if I^+ is a set including I and containing exactly one new element, $I^+ = I \cup \{i^+\}$, then there exists a locally finite B-valued, functional I^+-algebra A^+ over X such that A is isomorphic to a compression of A^+.*

Remark. The lemma is true for every set I^+ including I and the idea of the proof of the more general assertion is the same. Since, however, the notation and the details in the general case are monstrously complicated, it is preferable to state and prove the lemma in the special case, and then, if necessary in the applications, apply it many times. This process of repeated application leads in general to transfinite induction, whereas an application of the general form of the lemma does not. Thus the facts are that the transfinite inductions that will be occasionally mentioned below are avoidable, but the price for avoiding them is a significant loss in perspicuity.

Proof. By Lemma 11.5, identifying $\boldsymbol{A}$ with its image under the mapping there described, we may and do assume that the elements of $\boldsymbol{A}$ are functions (independent of i^+) from X^{I^+} into $\boldsymbol{B}$. Lemma 11.5 guarantees that if $\boldsymbol{S}^+$ and $\boldsymbol{\exists}^+$ denote the functional operations defined for functions from X^{I^+} into $\boldsymbol{B}$ by (2.2) and (2.4), and if $\boldsymbol{S}$ and $\boldsymbol{\exists}$ are defined for $\boldsymbol{A}$ by (11.3) and (11.4), then $(\boldsymbol{A}, I, \boldsymbol{S}, \boldsymbol{\exists})$ is a polyadic algebra isomorphic to the originally given functional algebra. (The elements of $\boldsymbol{A}$ are now not functions from X^I into $\boldsymbol{B}$, but functions from X^{I^+} into $\boldsymbol{B}$, and, consequently, we are not allowed to say that in its present form $\boldsymbol{A}$ is a functional I-algebra.)

We are now ready to define the algebra $\boldsymbol{A}^+$ that will turn out to be the desired dilation of $\boldsymbol{A}$. The elements of $\boldsymbol{A}^+$ are, by definition, all functions of the form $\boldsymbol{S}^+(i/i^+)p$, where $i \in I$ and $p \in \boldsymbol{A}$. It is clear that $\boldsymbol{A} \subset \boldsymbol{A}^+$ (choose i so that p is independent of it). What we must show is that (i) $\boldsymbol{A}^+$ is a Boolean algebra, (ii) if $p^+ \in \boldsymbol{A}^+$ and if τ is a finite transformation on I^+, then $\boldsymbol{S}^+(\tau)p^+ \in \boldsymbol{A}^+$, (iii) if $p^+ \in \boldsymbol{A}^+$ and if J is a finite subset of I^+, then $\boldsymbol{\exists}^+(J)p^+$ exists and belongs to $\boldsymbol{A}^+$, (iv) if $p^+ \in \boldsymbol{A}^+$, then there exists a cofinite subset J of I^+ such that p^+ is independent of J, and (v) if $p^+ \in \boldsymbol{A}^+$ and p^+ is independent of i^+, then $p^+ \in \boldsymbol{A}$. Since every element of $\boldsymbol{A}$ is independent of i^+, (v) implies that $\boldsymbol{A}$ consists exactly of all the elements of $\boldsymbol{A}^+$ independent of i^+. Hence, as soon as we know that $\boldsymbol{A}^+$ is an I^+-algebra, we can deduce from (v) that $\boldsymbol{A}$ is a compression of $\boldsymbol{A}^+$. The proof that $\boldsymbol{A}^+$ is a locally finite functional I^+-algebra is an easy consequence of (i)-(iv) (cf. also Lemma 6.16 and Theorem 7.6). We take this easy argument for granted and proceed to complete the proof of the lemma by establishing (i)-(v).

We begin with an auxiliary comment. The representation of an element of $\boldsymbol{A}^+$ in the form $\boldsymbol{S}^+(i/i^+)p$ (as described in the preceding paragraph) is far from unique. Indeed, if $j \in I$, if p is independent of j, and if $q = \boldsymbol{S}(i/j)p$, then

$$(11.7) \quad \boldsymbol{S}^+(j/i^+)q = \boldsymbol{S}^+(j/i^+)\boldsymbol{S}^+(i/j)p = \boldsymbol{S}^+(i/i^+)\boldsymbol{S}(j/i^+)p = \boldsymbol{S}^+(i/i^+)p.$$

(Warning: since we do not as yet know that $\boldsymbol{A}^+$ is a polyadic algebra, we are not allowed to apply to $\boldsymbol{A}^+$ our results about polyadic algebras. Thus the last step in (11.7), *i. e.*, the conclusion that since p is independent of j therefore $\boldsymbol{S}^+(j/i^+)p = p$, is a consequence not of our algebraic results but of the elementary geometric facts about Cartesian products that were the motivating background of our algebraic considerations all along. Similar warnings apply on a few other occasions in the course of this proof.)

(i) If $p^+ \epsilon A^+$, say $p^+ = S^+(i/i^+)p$ with $i \epsilon I$ and $p \epsilon A$, then $(p^+)' = S^+(i/i^+)p'$; this proves that A^+ is closed under complementation. Suppose next that p^+ and q^+ are in A^+, so that $p^+ = S^+(i/i^+)p$ and $q^+ = S^+(j/i^+)q$, with i and j in I, and p and q in A. Let k be an element of I such that both p and q are independent of k. (The existence of k follows from the facts that A is locally finite and I is infinite.) By the preceding paragraph we may assume that both i and j are equal to k. Since now $p^+ = S^+(k/i^+)p$ and $q^+ = S^+(k/i^+)q$ (with not necessarily the same p and q as before), it follows that $p^+ \vee q^+ = S^+(k/i^+)(p \vee q)$; this proves that A^+ is closed under the formation of suprema.

(ii) To prove that A^+ is closed under the action of $S^+(\tau)$ for all finite transformations τ on I^+, we first show that if τ is a finite transformation on I^+, then $S^+(\tau)p \epsilon A^+$ for every p in A. In the proof of this assertion we may assume that $\tau i^+ = i^+$, for (since p is independent of i^+) we can always change τ so as to force it to have this property without changing $S^+(\tau)p$. Let i be an element of I such that p is independent of i and such that $\tau^{-1}\{i\} = \{i\}$. If σ is the transformation on I (not I^+) such that $\sigma j = i$ when $\tau j = i^+$ and $\sigma j = \tau j$ otherwise, then $\tau = (i/i^+)\sigma$ outside i; it follows that

$$S^+(\tau)p = S^+(i/i^+)\, S^+(\sigma^+)p = S^+(i/i^+)\, S(\sigma)p \,\epsilon\, A^+.$$

The general case is easy now: if τ is a finite transformation on I^+, then $S^+(\tau)S^+(i/i^+)p \epsilon A^+$ whenever $i \epsilon I$ and $p \epsilon A$ simply because $\tau(i/i^+)$ is a finite transformation on I^+.

(iii) Suppose first that J is a finite subset of I and that $p^+ \epsilon A^+$, so that $p^+ = S^+(i/i^+)p$ with $i \epsilon I$ and $p \epsilon A$. Since $(i/i^+)^{-1}J = J - \{i\}$, it follows that (i/i^+) is one-to-one (in fact, equal to the identity) on $(i/i^+)^{-1}J$. Since $\exists^+(J - \{i\})p = \exists(J - \{i\})p$ exists and belongs to A, it follows from Lemma 4.8 that $\exists^+(J)\, S^+(i/i^+)p$ exists, is equal to $S^+(i/i^+)\exists(J - \{i\})p$, and, consequently, belongs to A^+. Suppose now that $p^+ \epsilon A^+$, so that, as usual $p^+ = S^+(i/i^+)p$ with $i \epsilon I$ and $p \epsilon A$. Since $(i/i^+) = (i, i^+)$ outside i^+ and since p is independent of i^+, it follows that $p^+ = S^+(i, i^+)p$ Since, moreover, $(i, i^+)^{-1}\{i^+\} = \{i\}$, and since $\exists^+(i)p = \exists(i)p$ exists and belongs to A, it follows as before that $\exists^+(i^+)p^+$ exists and belongs to A^+. If J is an arbitrary finite subset of I^+, write $J_0 = J \cap I$ and $J_1 = J - I$. The fact that $\exists^+(J)p^+$ exists and belongs to A^+ whenever $p^+ \epsilon A^+$ follows from the preceding two special cases of this assertion together with Lemma 4.2.

(iv) If $p^+ \epsilon A^+$, say $p^+ = S^+(i/i^+)p$ with $i \epsilon I$ and $p \epsilon A$, and if J is a finite support of p, then a direct geometric verification shows (cf. also

Lemma 6.14) that $(i/i^+)J$ supports p^+ and hence that p^+ does indeed have a finite support.

(v) Suppose now that $p^+ \in \boldsymbol{A}^+$ and that p^+ is independent of i^+. We represent p^+ in the form $\boldsymbol{S}^+(i,i^+)p$, as in the proof of (iii) above. Since by now we know that $\boldsymbol{A}^+$ is an I^+-algebra, in the remainder of this argument we can make free use of the technique of polyadic algebras. We have

$$(11.8)\quad \boldsymbol{S}^+(i,i^+)p = p^+ = \exists^+(i^+)p^+ = \exists^+(i^+)\boldsymbol{S}^+(i,i^+)p = \boldsymbol{S}^+(i,i^+)\exists^+(i)p.$$

Since $\boldsymbol{S}^+(i,i^+)$ is a Boolean automorphism (recall that $(i,i^+)^2 = \delta^+$), it follows from (11.8) that $p = \exists(i)p$ and hence that

$$p^+ = \boldsymbol{S}^+(i,i^+)p = \boldsymbol{S}^+(i/i^+)p = \boldsymbol{S}^+(i/i^+)\exists(i)p = \exists(i)p = p.$$

This concludes the proof of Lemma 11.6.

The main theorem of this section is a consequence of the preceding lemma and of Theorem 10.9.

(11.9) THEOREM. *If I is an infinite set and if $I \subset I^+$, then every locally finite I-algebra is the compression of a locally finite I^+-algebra; the dilation is uniquely determined to within an isomorphism that acts as the identity on the compression.*

Proof. The theorem as stated follows from the special case in which $I^+ = I \cup \{i^+\}$ by repeated applications of that special case (*i. e.*, by transfinite induction). In the special case the only thing we still need to establish is uniqueness. Suppose that $(\boldsymbol{A}, I, \boldsymbol{S}, \exists)$ is the given algebra and that $(\boldsymbol{A}^+, I^+, \boldsymbol{S}^+, \exists^+)$ is a dilation. If $p^+ \in \boldsymbol{A}^+$ and if i is an element of I such that p^+ is independent of i, then $p^+ = \boldsymbol{S}^+(i/i^+)\boldsymbol{S}^+(i^+/i)p^+$; this shows that every element of $\boldsymbol{A}^+$ has the form $\boldsymbol{S}^+(i/i^+)p$ where $i \in I$ and $p \in \boldsymbol{A}$. If both $(\boldsymbol{A}_1^+, I^+, \boldsymbol{S}_1^+, \exists_1^+)$ and $(\boldsymbol{A}_2^+, I^+, \boldsymbol{S}_2^+, \exists_2^+)$ are dilations of $\boldsymbol{A}$, and if a mapping f from $\boldsymbol{A}_1^+$ to $\boldsymbol{A}_2^+$ is defined by $f\boldsymbol{S}_1^+(i/i^+)p = \boldsymbol{S}_2^+(i/i^+)p$, then f is a polyadic isomorphism from $\boldsymbol{A}_1^+$ onto $\boldsymbol{A}_2^+$. To prove this it must be shown that if $p^+ \in \boldsymbol{A}^+$, then fp^+ is unambiguously defined (independently of the representation of p^+ in the form $\boldsymbol{S}_1^+(i/i^+)p$), that f is a Boolean isomorphism, and that f transforms $\boldsymbol{S}_1^+$ and $\exists_1^+$ into $\boldsymbol{S}_2^+$ and $\exists_2^+$ respectively. The verifications of all these assertions are routine applications of frequently used techniques (cf., in particular, the proof of Lemma 11.6); we omit the details.

§ 12. Constants

The main thing that makes Theorem 10.9 much less useful than it appears to be is that it does not even come close to answering the question of which simple algebras are models. If Theorem 10.9 is applied to a simple algebra, the representation it yields is not by a model; if, in fact, Theorem 10.9 is applied to a model, the representation it yields is not, as it ought to be, the identity representation. In the presence of a very strong cardinality restriction (namely, that A and I be countably infinite) Rasiowa and Sikorski [5] were able to obtain the desired result, but their proof is inseparably tied to that cardinality restriction. To get a better grip on the problem we shall introduce below a new concept whose role in the theory of polyadic algebras is even more important than that of the well-known concepts of universal algebra (such as ideal and homomorphism).

The main problem to be kept in mind is that of functional representation. Given an I-algebra $\boldsymbol{A}$, we must try to construct a Boolean algebra $\boldsymbol{B}$ and a domain X so that $\boldsymbol{A}$ is isomorphic to an algebra of functions from X^I into $\boldsymbol{B}$, and, moreover, the isomorphism must be a natural one in a very strong sense of the word. We might try to formulate the desideratum precisely as follows: if $\boldsymbol{A}$ is already a functional algebra, then the general method of constructing X and $\boldsymbol{B}$ should recapture the domain and the value-algebra in terms of which $\boldsymbol{A}$ is defined. A moment's thought about the (analogous) representation theory of ordinary Boolean algebras shows that this desideratum is really more than is needed; by aiming in its general direction, however, we shall be able to get all the information we want about the problem of functional representation for polyadic algebras.

We can hope to get a clue to the solution of our problem by assuming that $\boldsymbol{A}$ is a functional I-algebra, with domain X and value-algebra $\boldsymbol{B}$, and trying to find a purely algebraic characterization of the elements of X. In order to avoid extraneous technical difficulties, we shall temporarily over-simplify the situation by assuming that $\boldsymbol{B}=\boldsymbol{O}$ and that $\boldsymbol{A}$ is the set of all functions from X into $\boldsymbol{O}$. The same intuitive considerations that motivate calling the elements of I variables suggest that the elements of X be thought of as the *constants* that constitute the subject matter of the values of the propositional functions in $\boldsymbol{A}$. We shall adopt this terminology; accordingly we may say that we are now seeking an algebraic characterization of the constants of a polyadic algebra.

If $\bar{x}$ is a constant, *i. e.*, $\bar{x} \in X$, the only way we can connect $\bar{x}$ with $\boldsymbol{A}$ is to consider an element p of $\boldsymbol{A}$ and to replace some or all of its arguments by $\bar{x}$. Suppose, to be more precise, that J is a subset of I, and associate with each point x in X^I the point x^J (this notation is only temporary) defined by $x_i^J=\bar{x}$ when $i \in J$ and $x_i^J=x_i$ otherwise. If we write $q(x)=p(x^J)$ for all x, then we may say that the function q was obtained by replacing some of the arguments of p (the ones with index in J) by $\bar{x}$.

The passage from p to q is a transformation, depending on the set J, from $\boldsymbol{A}$ into itself. We shall denote this transformation by $c(J)$, so that $c(J)p=q$. (We do not indicate the dependence of this transformation on $\bar{x}$.)

It is not hard to see, directly from the definition, how $c(J)$ depends on J and what are the relations between $c(J)$ and the algebraic structure of $\boldsymbol{A}$. Instead of listing the facts, we proceed forthwith to use them as the basis for a general definition. Abandoning functional polyadic algebras and our special assumptions, we define a *constant* of an arbitrary I-algebra $\boldsymbol{A}$ as a mapping c from subsets J of I into Boolean endomorphisms $c(J)$ of $\boldsymbol{A}$, such that

(C_1) $c(\emptyset)=e$,

(C_2) $c(J\cup K)=c(J)c(K)$,

(C_3) $c(J)\exists(K)=\exists(K)c(J-K)$,

(C_4) $\exists(J)c(K)=c(K)\exists(J-K)$,

(C_5) $c(J)\boldsymbol{S}(\tau)=\boldsymbol{S}(\tau)c(\tau^{-1}J)$,

whenever J and K are subsets of I and τ is a transformation on I. (It is easy to verify that in the special case of monadic algebras this definition reduces to the one given in § 1.) In case the set J happens to be a singleton, $J=\{i\}$, we shall write $c(i)$ instead of $c(J)$.

In the theory of constants, as throughout the theory of (locally finite) polyadic algebras, it is sufficient to restrict attention to finite sets and finite transformations. The proofs that establish this fact are similar to but not immediately derivable from the corresponding proofs for quantifiers; cf. Theorem 7.6.

(12.1) LEMMA. *If c is a constant of an I-algebra $\boldsymbol{A}$ and if $p\in\boldsymbol{A}$, then to every subset K of I there corresponds a finite subset K_0 of I such that $c(K)p=c(K_0)p$.*

Proof. Let J be a finite support of p. If $K_0=K\cap J$, then

$$\begin{aligned} c(K)p=c(K)\exists(I-J)p&=\exists(I-J)c(K\cap J)p \quad &&(\text{by } (C_3))\\ &=c(K\cap J)\exists(I-J)p \quad &&(\text{by } (C_4))\\ &=c(K_0)p. \end{aligned}$$

(12.2) LEMMA. *If c is a constant of an I-algebra $\boldsymbol{A}$, if p is an element of $\boldsymbol{A}$ with support J, and if K is a subset of I, then J supports $c(K)p$.*

Proof. Since $J\cup K$ supports p, it follows that

$$\exists(I-J)c(K)p=c(K)\exists\big(I-(J\cup K)\big)p=c(K)p.$$

Remark. On one or two occasions below we shall refer to Lemma 12.2 even when c is not quite known to be a constant of $\boldsymbol{A}$; as long as c satisfies the conditions used in the proof (*i. e.*, (C_3) and (C_4)), such references are justified.

(12.3) THEOREM. *If c is a mapping from all finite subsets of I into Boolean endomorphisms of A satisfying* (C_1)-(C_5) *whenever J and K are finite subsets of I and τ is a finite transformation on I, then there exists a unique constant $\tilde{c}$ of A such that $\tilde{c}(J)=c(J)$ for all finite subsets J of I.*

Proof. If $J \subset I$ and $p \in A$, we find a finite support J_0 of p and we write $\tilde{c}(J)p=c(J \cap J_0)p$. In order to justify this definition of $\tilde{c}(J)$, we must show first that $c(J \cap J_0)p$ is the same no matter which support J_0 of p is used. In other words, we must prove that if both J_1 and J_2 are finite supports of p, then $c(J \cap J_1)p=c(J \cap J_2)p$. Since, by Lemma 6.7, $J_1 \cap J_2$ supports p, the desired result follows from the fact that both $c(J \cap J_1)p$ and $c(J \cap J_2)p$ are equal to $c(J \cap J_1 \cap J_2)p$. The proof of this fact, in turn, is reduced to the computation

$$\begin{aligned} c(J \cap J_1)p &= c(J \cap J_1)\exists(I-J_2)p \\ &= \exists(I-J_2)c(J \cap J_1 \cap J_2)p \qquad \text{(by } (C_3)) \\ &= c(J \cap J_1 \cap J_2)\exists(I-J_2)p \qquad \text{(by } (C_4)) \\ &= c(J \cap J_1 \cap J_2)p, \end{aligned}$$

together with the one obtained by interchanging the roles of J_1 and J_2.

It J happens to be finite, then the finite support J_0 of p can be chosen so that $J \subset J_0$. If that is done, then $J \cap J_0 = J$, and it follows that $\tilde{c}$ is indeed an extension of c.

It remains to prove that $\tilde{c}$ is a constant of A. Since $\tilde{c}(\emptyset)=c(\emptyset)$, it follows that $\tilde{c}(\emptyset)$ is the identity mapping on A. For the remainder of the proof we fix an element p of A with finite support J_0.

We observe that J_0 supports $c(K \cap J_0)p$ (cf. Lemma 12.2) and hence that

$$\begin{aligned} \tilde{c}(J \cup K)p &= c\big((J \cup K) \cap J_0\big)p = c\big((J \cap J_0) \cup (K \cap J_0)\big)p \\ &= c(J \cap J_0)c(K \cap J_0)p = \tilde{c}(J)c(K \cap J_0)p = \tilde{c}(J)\tilde{c}(K)p. \end{aligned}$$

Since

$$\begin{aligned} \tilde{c}(J)\exists(K)p &= c(J \cap J_0)\exists(K)p \qquad \text{(cf. Lemma 6.10)} \\ &= \exists(K)c\big((J \cap J_0)-K\big)p = \exists(K)c\big((J-K) \cap J_0\big)p \\ &= \exists(K)\tilde{c}(J-K)p, \end{aligned}$$

and since

$$\begin{aligned} \exists(J)\tilde{c}(K)p &= \exists(J)c(K \cap J_0)p = c(K \cap J_0)\exists\big(J-(K \cap J_0)\big)p \\ &= c(K \cap J_0)\exists\big((J-K) \cup (J-J_0)\big)p \\ &= c(K \cap J_0)\exists(J-K)\exists(J-J_0)p \\ &= c(K \cap J_0)\exists(J-K)p = \tilde{c}((K)\exists(J-K)p, \end{aligned}$$

it follows that (C_3) and (C_4) hold for $\tilde{c}$.

To prove (C_5) we assume first that the transformation τ is finite. In this case

$$\begin{aligned}\tilde{c}(J)\,S(\tau)p &= c(J \cap \tau J_0)\,S(\tau)p \qquad \text{(by Lemma 6.14)}\\ &= S(\tau)c(\tau^{-1}J \cap \tau^{-1}\tau J_0)p = S(\tau)\tilde{c}(\tau^{-1}J)p,\end{aligned}$$

since $J_0 \subset \tau^{-1}\tau J_0$ and therefore $\tau^{-1}\tau J_0$ supports p. In the general case, *i. e.*, if τ is not necessarily finite, let τ_0 be a finite transformation that agrees with τ in J_0. Since J_0 supports both p and $\tilde{c}(\tau^{-1}J)p$ (cf. Lemma 12.2), it follows from Lemma 6.11 that $S(\tau)p = S(\tau_0)p$ and $S(\tau)\tilde{c}(\tau^{-1}J)p = S(\tau_0)\tilde{c}(\tau^{-1}J)p$. Observe also that the agreement of τ and τ_0 in J_0 implies that $\tau^{-1}J \cap J_0 = \tau_0^{-1}J \cap J_0$; from this it follows that

$$\tilde{c}(\tau^{-1}J)p = c(\tau^{-1}J \cap J_0)p = c(\tau_0^{-1}J \cap J_0)p = \tilde{c}(\tau_0^{-1}J)p.$$

The existence part of the proof is now completed by the observation that

$$\begin{aligned}\tilde{c}(J)\,S(\tau)p &= \tilde{c}(J)\,S(\tau_0)p = S(\tau_0)\tilde{c}(\tau_0^{-1}J)p\\ &= S(\tau_0)\tilde{c}(\tau^{-1}J)p = S(\tau)\tilde{c}(\tau^{-1}J)p.\end{aligned}$$

Since the uniqueness of $\tilde{c}$ is an immediate consequence of Lemma 12.1, the proof of Theorem 12.3 is complete.

§ 13. Factorization and commutativity

We interrupt our discussion of constants in order to insert two auxiliary results, unrelated to each other and (except indirectly) to polyadic algebras. These results (Theorem 13.3 and Lemma 13.4) are of a certain amount of interest in themselves (especially Theorem 13.3); the main reason for presenting them, however, is their use in the proof of Theorem 14.1.

Our first purpose is to generalize to arbitrary finite transformations (on a set I, say) the classical result on the expression of finite permutations by means of transpositions.

(13.1) LEMMA. *A finite transformation τ is a retraction if and only if there exists a finite class $\{J_1, \dots, J_n\}$ of finite subsets of I such that if $i \in J_k$ and $j \in J_k$, then $\tau i = \tau j \in J_k$, and if $i \notin J_1 \cup \dots \cup J_n$, then $\tau i = i$.*

Proof. If τ is a finite retraction, then there exists a cofinite set J such that $\tau i = i$ whenever $i \in J$ and such that $\tau(I-J) \subset I-J$. If $\tau(I-J) = \{j_1, \dots, j_n\}$, write $J_k = \tau^{-1}\{j_k\}$, $k = 1, \dots, n$. Clearly $J_1 \cup \dots \cup J_n = I-J$, so that, indeed, $\tau i = i$ when $i \notin J_1 \cup \dots \cup J_n$. If $i \in J_k$, then $\tau i = j_k$, and, since τ is a retraction, $\tau j_k = \tau i = j_k$, so that $j_k \in J_k$ and therefore $\tau i \in J_k$. This settles the "only if". If, conversely, the condition is satisfied, and if $i \notin J_1 \cup \dots \cup J_n$, then $\tau i = i$ and therefore $\tau^2 i = \tau i$. If $i \in J_k$, then, by assumption, $\tau i \in J_k$, and therefore, by another part of the assumption, $\tau i = \tau\tau i = \tau^2 i$, so that τ is a retraction.

(13.2) LEMMA. *Every finite transformation is the product of a finite permutation and a finite retraction.*

Proof. If τ is a finite transformation, then there exists a cofinite set J such that $\tau i = i$ whenever $i \in J$ and such that $\tau(I-J) \subset I-J$. If $\tau(I-J) = \{j_1, \ldots, j_n\}$, write $J_k = \tau^{-1}\{j_k\}$, and let i_k be an arbitrary element of J_k, $k = 1, \ldots, n$. Let π be a permutation such that $\pi i = i$ whenever $i \in J$ and such that $\pi i_k = j_k$, $k = 1, \ldots, n$; obviously π is finite. If $\sigma = \pi^{-1}\tau$, then, clearly, σ is a finite transformation; it remains to prove that σ is a retraction. Clearly $\sigma i = i$ whenever $i \notin J_1 \cup \ldots \cup J_n$. If $i \in J_k$, then $\sigma i = \pi^{-1}\tau i = \pi^{-1} j_k = i_k \in J_k$; the desired result follows from Lemma 13.1.

(13.3) THEOREM. *Every finite permutation is a finite product of transpositions; every finite retraction is a finite product of replacements; every finite transformation is a finite product of transpositions and replacements.*

Proof. The first assertion is well-known, and the last assertion follows from the first two together with Lemma 13.2. To prove the second assertion, suppose that τ is a finite retraction, and let $J_1, \ldots, J_n$ be sets with the properties described in Lemma 13.1. If $\tau i = j_k$ when $i \in J_k$, then, clearly,

$$\tau = \prod_{k=1}^{n} \prod_{i \in J_k} (i/j_k),$$

and the proof is complete. (The first proof of Theorem 13.3 was obtained in the course of a conversation with G. P. Hochschild.)

Theorem 13.3 can be used to explain why general transformations (and, in particular, permutations) are almost never used in the traditional presentation of the functional calculi. Usually the only transformations that are visible are replacements. The reason is that in the usual discussion it is always assumed that the set I is infinite, and, under that assumption, it turns out that the action of any finite transformation on any element of a (locally finite) polyadic algebra can be achieved by replacements only. To prove this assertion it is sufficient, in view of Theorem 13.3, to prove it for transpositions only. It is to be proved, in other words, that if p is an element of an I-algebra (locally finite, of course) and if i and j are in I, then there exists a transformation τ that is a finite product of replacements and for which $S(i,j)p = S(\tau)p$. To prove this, let J be a cofinite set such that $\exists(J)p = p$ and let k be an element of J such that $k \neq i$ and $k \neq j$ (here is where we need an infinite I). If $\tau = (k/j)(j/i)(i/k)$, then $(i,j) = \tau$ outside J and therefore

$$S(i,j)p = S(i,j)\exists(J)p = S(\tau)\exists(J)p = S(\tau)p.$$

It is easy to see that if the set I is finite, then the method just used breaks down, and, in fact, the result itself becomes false.

Our second auxiliary result gives a sufficient condition for the commutativity of two Boolean endomorphisms under some rather special circumstances.

(13.4) LEMMA. *If A is a monadic algebra with quantifier $\exists$, if a is a constant of A, and if b is a Boolean endomorphism of A such that $b\exists = \exists b \exists$ and such that either* (i) $ab = aba$ *or* (ii) $abp = 0$ *whenever* $ap = 0$, *then* $ab = ba$.

Proof. Under the assumptions of the lemma the conditions (i) and (ii) are equivalent to each other. Indeed, (i) obviously implies (ii). If, on the other hand, (ii) is true, then $ab(p \wedge ap') = 0$ for all p (since $a(p \wedge ap') = 0$), so that $abp \leqslant abap$. Applying this result to p' in place of p and combining the two inequalities so obtained, we conclude that (i) is true. The proof that (ii), together with the preceding assumptions of the lemma, implies the desired commutativity, is a simple calculation, as follows:

$$\begin{aligned} ab = aba = ab(\exists a) = a(b\exists)a = a(\exists b \exists)a = (a\exists)(b\exists a) \\ = \exists(b\exists a) = (\exists b \exists)a = (b\exists)a = ba. \end{aligned}$$

§ 14. Constants from endomorphisms

For the purpose of constructing constants an application of the definition itself is quite unwieldy. It is natural to guess, however, on the basis of intuitive considerations, that a constant c ought to be completely determined by the knowledge of $c(i)$ for any one i in I. The idea is that if we know how to replace one particular variable by a constant, then, using the transformation laws at our disposal, we can deduce the effect of replacing any other variable by that constant. Our next purpose is to make these indications precise by showing that a constant is indeed uniquely determined by one suitably selected endomorphism.

(14.1) THEOREM. *If A is a locally finite I-algebra, if f is a Boolean endomorphism of A, and if i is an element of I such that*

(i) $\exists(i)f = f,$

(ii) $f\exists(i) = \exists(i),$

(iii) $f\exists(j) = \exists(j)f$ *whenever* $j \neq i,$

then there exists a unique constant c of A such that $c(i) = f$.

Remark. Note that the sufficient conditions of this theorem are also obviously necessary, in this sense: if c is a constant of A and if $f = c(i)$, then f satisfies (i), (ii), and (iii).

Proof. Uniqueness is easy. Suppose, indeed, that c is a constant of A. It follows from (C_2) that

$$c(J)=\prod_{j\in J} c(j)$$

for every finite subset J of I. (Since, conventionally, the empty product of endomorphisms is the identity endomorphism, the situation is in harmony with (C_1).) It follows (cf. Lemma 12.1 and Theorem 12.3) that a constant c is uniquely determined by the endomorphisms $c(j)$, $j\in I$. Since $(i,j)^{-1}\{i\}=\{j\}$, it follows from (C_5) that

$$c(j)\,\boldsymbol{S}(i,j)=\boldsymbol{S}(i,j)c(i)$$

and hence that

$$c(j)=\boldsymbol{S}(i,j)c(i)\,\boldsymbol{S}(i,j)$$

for all j. This means that the endomorphisms $c(j)$, in turn, are uniquely determined by any one of them, and the uniqueness assertion is proved.

We shall need to make use of the fact that (i), (ii), and (iii) imply

$$\text{(iv)}\qquad f\boldsymbol{S}(j/i)f=f\boldsymbol{S}(j/i).$$

Since $f^2=f$ (cf. (1.9)), (iv) is obviously true when $j=i$. It follows that for every p in A and for every j in I the two Boolean endomorphisms in (iv) agree on fp, and that $f(p+fp)=0$. Since

$$\begin{aligned} f\boldsymbol{S}(j/i)(p+fp) &\leqslant f\boldsymbol{S}(j/i)\exists(j)(p+fp)\\ &=f\exists(j)(p+fp)=\exists(j)f(p+fp)=0, \end{aligned}$$

it follows also that the endomorphisms in (iv) agree on $p+fp$. Since $p=fp+(p+fp)$ for all p, the proof of (iv) is complete.

The proof of uniqueness suggests an approach to the proof of existence; given f, we must write

$$\text{(14.2)}\qquad c(J)=\prod_{j\in J}\boldsymbol{S}(i,j)f\boldsymbol{S}(i,j)$$

for every finite subset J of I. In order to be sure, however, that this product unambiguously defines something (independently of the order of the factors) we must prove first of all that its factors commute. To do this, we begin by writing

$$\text{(14.3)}\qquad a_j=\boldsymbol{S}(i,j)f\boldsymbol{S}(i,j)$$

whenever j is an element of I, and we go on to prove that a_j and a_k always commute. If $j=k$, then $a_j=a_k$ and the result is obvious. If $j\neq k$, we apply Lemma 13.4 with $A=A(j)$, $a=a_j$, and $b=a_k$. We proceed to verify the hypotheses of Lemma 13.4; in the course of the verification we shall make use, without any further reference, of the following special cases of (P_6):

$$S(i,j)\exists(i) = \exists(j)S(i,j),$$
$$S(i,j)\exists(j) = \exists(i)S(i,j),$$
$$S(i,j)\exists(k) = \exists(k)S(i,j),$$

the latter in case k is distinct from both i and j.

We have

$$(14.4)\quad \exists(j)a_j = \exists(j)S(i,j)fS(i,j) = S(i,j)\exists(i)fS(i,j) = S(i,j)fS(i,j) = a_j$$

and

$$(14.5)\quad a_j\exists(j) = S(i,j)fS(i,j)\exists(j) = S(i,j)f\exists(i)S(i,j)$$
$$= S(i,j)\exists(i)S(i,j) = \exists(j)S(i,j)S(i,j) = \exists(j);$$

this proves that a_j is a constant of $A(j)$.

We prove next that

$$(14.6)\quad a_k\exists(j) = \exists(j)a_k \quad \textit{whenever} \quad j \neq k.$$

The proof splits naturally into two parts. We have assumed that $j \neq k$; if also $j \neq i$, then

$$a_k\exists(j) = S(i,k)fS(i,k)\exists(j) = S(i,k)f\exists(j)S(i,k) = S(i,k)\exists(j)fS(i,k)$$
$$= \exists(j)S(i,k)fS(i,k) = \exists(j)a_k.$$

If, on the other hand, $j = i$, then we must have $i \neq k$ (since $j \neq k$), and therefore

$$a_k\exists(j) = a_k\exists(i) = S(i,k)fS(i,k)\exists(i) = S(i,k)f\exists(k)S(i,k)$$
$$= S(i,k)\exists(k)fS(i,k) = \exists(i)S(i,k)fS(i,k) = \exists(i)a_k = \exists(j)a_k.$$

This implies among other things that the hypothesis $b\exists = \exists b\exists$ holds in the present case.

If $a_jp = 0$, then $a_j\exists(k)p = \exists(k)a_jp = 0$ (by (14.6)). Since our assumptions are symmetric in j and k, we know from (14.4) and (14.5) that a_k is a constant of $A(k)$ and therefore that $a_kp \leqslant \exists(k)p$ for all p in A. Putting these two facts together, we conclude that if $a_jp = 0$, then $a_ja_kp \leqslant a_j\exists(k)p = 0$. This completes the verification of the hypotheses of Lemma 13.4; applying Lemma 13.4, we conclude that a_j and a_k commute. Since this holds for all j and k, we now know that all the factors of (14.2) commute and hence that the equation (14.2) unambigously defines a Boolean endomorphism $c(J)$ of A for every finite subset J of I. In view of Theorem 12.3, what remains to be done is to prove that the mapping c from finite subsets of I to Boolean endomorphisms of A satisfies (C_1)-(C_5)

whenever J and K are finite subsets of I and τ is a finite transformation on I.

The validity of (C_1) (*i. e.*, the fact that $c(\emptyset)$ is the identity mapping on $\boldsymbol{A}$) follows from the customary interpretation of the empty product. The validity of (C_2) (*i. e.*, $c(J\cup K)=c(J)c(K)$) follows from a straightforward application of the definition of c.

To prove (C_3) we recall that in view of (P_4)

$$\exists(K)=\prod_{k\in K}\exists(k)$$

for every finite subset K of I. It follows from repeated applications of (14.5) and (14.6) that

$$a_j\exists(K)=\exists(K) \quad \textit{whenever} \quad j\in K$$

and

$$a_j\exists(K)=\exists(K)a_j \quad \textit{whenever} \quad j\notin K.$$

Since (cf. (14.2) and (14.3))

$$c(J)=\prod_{j\in J}a_j,$$

we obtain

$$\begin{aligned} c(J)\exists(K)&=\prod_{j\in J-K}a_j\cdot\prod_{j\in J\cap K}a_j\cdot\exists(K)\\ &=\exists(K)\prod_{j\in J-K}a_j=\exists(K)c(J-K). \end{aligned}$$

The derivation of (C_4) proceeds similarly; indeed, since

$$\exists(j)c(K)=c(K) \quad \textit{whenever} \quad j\in K$$

and

$$\exists(j)c(K)=c(K)\exists(j) \quad \textit{whenever} \quad j\notin K,$$

it follows that

$$\begin{aligned} \exists(J)c(K)&=\prod_{j\in J-K}\exists(j)\cdot\prod_{j\in J\cap K}\exists(j)\cdot c(K)\\ &=c(K)\prod_{j\in J-K}\exists(j)=c(K)\exists(J-K). \end{aligned}$$

We begin the proof of (C_5) (*i. e.*, $c(J)\boldsymbol{S}(\tau)=\boldsymbol{S}(\tau)c(\tau^{-1}J)$) by assuming that J is a singleton, $J=\{j\}$, and that τ is a transposition or a replacement. If $\tau=(h,k)$ or if $\tau=(h/k)$, with $h\neq j$ and $k\neq j$, then $\tau^{-1}\{j\}=\{j\}$, and therefore the desired conclusion is that a_j and $\boldsymbol{S}(\tau)$ commute. The result can be derived from Lemma 13.4 with $\boldsymbol{A}=\boldsymbol{A}(j)$, $a=a_j$, and $b=\boldsymbol{S}(\tau)$. We already know that a_j is a constant of $\boldsymbol{A}(j)$ (cf. (14.4) and (14.5)). We know also that

$$a_j\exists(h)\exists(k)=\exists(h)\exists(k)a_j$$

(cf. (C_3)) and that

$$S(\tau)\exists(h)\exists(k)=\exists(h)\exists(k) \tag{14.7}$$

(cf. (P_5)). From (14.7) and the fact that $p\leqslant\exists(h)\exists(k)p$ for all p, it follows that $S(\tau)p\leqslant\exists(h)\exists(k)p$. If p is such that $a_jp=0$, then $\exists(h)\exists(k)a_jp=0$ and therefore

$$a_jS(\tau)p\leqslant a_j\exists(h)\exists(k)p=\exists(h)\exists(k)a_jp=0.$$

The fact that $S(\tau)\exists(j)=\exists(j)S(\tau)$ follows from (P_6). Consequently Lemma 13.4 is indeed applicable and we have proved (C_5) in this case.

In the next case $\tau=(j,k)$. Since $\tau^{-1}\{j\}=\{k\}$, the desired conclusion is

$$a_jS(j,k)=S(j,k)a_k. \tag{14.8}$$

If $j=k$, the conclusion is trivial; if j and k are distinct but one of them is equal to i, the conclusion is equivalent to the definition of the a's (cf. (14.3)). If both j and k are distinct from i, then

$$(i,j)(j,k)=(j,k)(i,k)$$

and therefore

$$a_jS(j,k)=S(i,j)fS(i,j)S(j,k)=S(i,j)fS(j,k)S(i,k)=S(i,j)S(j,k)fS(i,k)$$

by the preceding paragraph. It follows that

$$a_jS(j,k)=S(j,k)S(i,k)fS(i,k)=S(j,k)a_k.$$

If $\tau=(j/k)$ with $j\neq k$, then $\tau^{-1}\{j\}=\emptyset$ and the desired conclusion becomes

$$a_jS(j/k)=S(j/k). \tag{14.9}$$

The proof is simple:

$$\begin{aligned} a_jS(j/k)&=a_j\exists(j)S(j/k) && \text{(by } (P_6)) \\ &=\exists(j)S(j/k) && \text{(by (14.5))} \\ &=S(j/k). \end{aligned}$$

The only other replacement τ can be is (k/j) with $k\neq j$. Since $\tau^{-1}\{j\}=\{j,k\}$, the desired conclusion is

$$a_jS(k/j)=S(k/j)a_ja_k. \tag{14.10}$$

If $j=i$, then

$$\begin{aligned} S(k/j)a_ja_k&=S(k/i)fa_k=S(k/i)a_kf \\ &=S(k/i)S(i,k)fS(i,k)f=S(k/i)fS(k/i)f \\ &=S(k/i)\exists(k)fS(k/i)f=fS(k/i)f \\ &=fS(k/i)=a_jS(k/j); \end{aligned}$$

if $k=i$, then

$$(14.11)\qquad \begin{aligned} S(k/j)a_ja_k &= S(i/j)a_jf = S(i/j)S(i,j)fS(i,j)f \\ &= S(i/j)fS(j/i)f \\ &= S(i/j)fS(j/i) = S(i,j)fS(j/i) \\ &= S(i,j)fS(i,j)S(i/j) \\ &= a_jS(k/j). \end{aligned}$$

If both j and k are different from i (and $j \neq k$), then

$$(i,k)(i/j)(i,k) = (k/j)$$

and therefore

$$\begin{aligned} S(k/j)a_ja_k &= S(i,k)S(i/j)S(i,k)a_jS(i,k)fS(i,k) \\ &= S(i,k)S(i/j)a_jfS(i,k) \\ &= S(i,k)a_jS(i/j)S(i,k) \qquad \text{(by (14.11))} \\ &= S(i,k)a_jS(i,k)S(i,k)S(i/j)S(i,k) \\ &= a_jS(k/j). \end{aligned}$$

This completes the proof of (14.10) and hence of (C_5) in the special cases under consideration.

If J and K are finite subsets of I and τ is a finite transformation on I such that

$$c(J)S(\tau) = S(\tau)c(\tau^{-1}J)$$

and

$$c(K)S(\tau) = S(\tau)c(\tau^{-1}K),$$

then

$$\begin{aligned} c(J \cup K)S(\tau) &= c(J)c(K)S(\tau) = c(J)S(\tau)c(\tau^{-1}K) \\ &= S(\tau)c(\tau^{-1}J)c(\tau^{-1}K) = S(\tau)c\big(\tau^{-1}(J \cup K)\big). \end{aligned}$$

The fact that (C_5) holds whenever J is a singleton and τ is a transposition or a replacement, together with the result just obtained, implies that (C_5) holds whenever J is an arbitrary finite subset of I and τ is a transposition or a replacement.

If τ and σ are finite transformations on I such that

$$c(J)S(\tau) = S(\tau)c(\tau^{-1}J)$$

and

$$c(J)S(\sigma) = S(\sigma)c(\sigma^{-1}J)$$

for all finite subsets J of I, then

$$\begin{aligned} c(J)S(\sigma\tau) &= c(J)S(\sigma)S(\tau) = S(\sigma)c(\sigma^{-1}J)S(\tau) \\ &= S(\sigma)S(\tau)c\big(\tau^{-1}(\sigma^{-1}J)\big) = S(\sigma\tau)c\big((\sigma\tau)^{-1}J\big). \end{aligned}$$

The fact that (C_5) for finite sets holds for transpositions and replacements, together with the result just obtained, implies that (C_5) for finite sets holds for all finite products of transpositions and replacements. In view of Theorem 13.3 this means that (C_5) holds in all finite cases, and, therefore, completes the proof of Theorem 14.1.

§ 15. Examples of constants

It is instructive to see an example of a constant that is *prima facie* different from the ones that motivated the definition, *i. e.*, from those constants of a functional polyadic algebra that are defined in terms of the replacement of variables by some element of the domain. The idea behind the example is that if, given an arbitrary polyadic algebra, we hold a certain variable fixed, *i. e.*, if we steadfastly refuse to perform any transformations involving it, then that variable will act as a constant.

Suppose, to be precise, that I and I^+ are sets such that I is a proper subset of I^+, and suppose that $(\boldsymbol{A}^+, I^+, \boldsymbol{S}^+, \boldsymbol{\exists}^+)$ is a polyadic algebra. If we write $\boldsymbol{A}=\boldsymbol{A}^+$ and if we define $\boldsymbol{S}$ and $\boldsymbol{\exists}$ (for transformations on I and subsets of I) by (11.3) and (11.4), then it follows easily that $(\boldsymbol{A}, I, \boldsymbol{S}, \boldsymbol{\exists})$ is a polyadic algebra. The proof is, in fact, the same as the one given for compressions in § 11; for this reason it is all the more important to note (in order to avoid confusion) that the I-algebra $\boldsymbol{A}$ is not a compression of the I^+-algebra $\boldsymbol{A}^+$. (The part of the definition of compression that is not satisfied here is the requirement that $\boldsymbol{A}$ consist of all the elements of $\boldsymbol{A}^+$ independent of $I^+ - I$.) We shall say that the algebra $(\boldsymbol{A}, I, \boldsymbol{S}, \boldsymbol{\exists})$ is obtained from the algebra $(\boldsymbol{A}^+, I^+, \boldsymbol{S}^+, \boldsymbol{\exists}^+)$ by *fixing* the variables in $I^+ - I$. We note that if $(\boldsymbol{A}^+, I^+, \boldsymbol{S}^+, \boldsymbol{\exists}^+)$ is locally finite, then the same is true of $(\boldsymbol{A}, I, \boldsymbol{S}, \boldsymbol{\exists})$.

If, in the notation of the preceding paragraph, i is an element of $I^+ - I$, and if J is an arbitrary subset of I, we shall write (J/i) for the transformation τ such that $\tau j = i$ when $j \in J$ and $\tau j = j$ otherwise; we shall denote the corresponding endomorphism $S^+(\tau)$ by $\boldsymbol{S}^+(J/i)$. We assert now that if $c_i(J) = \boldsymbol{S}^+(J/i)$, then c_i is a constant of the I-algebra $\boldsymbol{A}$; it is in this sense that we may say that a fixed variable is (better: becomes or induces) a constant. The proof of our assertion is easy. Indeed (C_1) follows from (P_1) via the fact that $(\emptyset/i) = \delta$, and (C_2) follows from (P_2) via the fact that $(J/i)(K/i) = \big((J \cup K)/i\big)$. To prove (C_3) observe that $\big((J-K)/i\big)^{-1}K = K$; then apply (P_6) to get $\boldsymbol{S}^+\big((J-K)/i\big)\boldsymbol{\exists}^+(K) = \boldsymbol{\exists}^+(K)\,\boldsymbol{S}^+\big((J-K)/i\big)$ and apply (P_5) to get $\boldsymbol{S}^+(K/i)\boldsymbol{\exists}^+(K) = \boldsymbol{\exists}^+(K)$. (Recall that throughout this proof the only relevant sets are subsets of I and that $i \in I^+ - I$.) For (C_4), note that $(K/i)^{-1}J = J - K$ and apply (P_6).

The condition (C_5) follows immediately from (P_2) and the fact that $(J/i)\tau^+ = \tau^+(\tau^{-1}J/i)$ whenever τ is a transformation on I. The fact that each $c_i(J)$ is a Boolean endomorphism of $\boldsymbol{A}$ follows from the fact that $\boldsymbol{S}^+$ maps transformations to Boolean endomorphisms of $\boldsymbol{A}^+$.

Besides fixing variables, there are two other useful methods of constructing constants: lifting to dilations (Lemma 15.1) and transfer to quotients (Lemma 15.3).

(15.1) LEMMA. *If* $(\boldsymbol{A}, I, \boldsymbol{S}, \boldsymbol{\exists})$ *is a locally finite polyadic algebra of infinite degree, if* $(\boldsymbol{A}^+, I^+, \boldsymbol{S}^+, \boldsymbol{\exists}^+)$ *is a dilation of it, and if* c *is a constant of* $\boldsymbol{A}$, *then there exists a unique constant* c^+ *of* $\boldsymbol{A}^+$ *such that* $c^+(J)p = c(J)p$ *whenever* $J \subset I$ *and* $p \in \boldsymbol{A}$.

Proof. We shall prove the result only in the special case in which $I^+ = I \cup \{i^+\}$; the general case follows easily by transfinite induction. To construct c^+ we apply Theorem 14.1 with i^+ in the role of i. If $p^+ \in \boldsymbol{A}^+$, so that $p^+ = \boldsymbol{S}^+(i/i^+)p$ with $i \in I$ and $p \in \boldsymbol{A}$ (cf. Theorem 11.9), we write $fp^+ = c(i)p$. If

$$\boldsymbol{S}^+(i/i^+)p = \boldsymbol{S}^+(j/i^+)q, \tag{15.2}$$

we select an element k in I such that both p and q are independent of k and we apply to (15.2) first $\boldsymbol{S}^+(i^+/k)$, obtaining $\boldsymbol{S}(i/k)p = \boldsymbol{S}(j/k)q$, and then $c(k)$, obtaining (via (C_5))

$$c(i)p = \boldsymbol{S}(i/k)c(i)p = \boldsymbol{S}(j/k)c(j)q = c(j)q\,.$$

Thus fp^+ is unambiguously defined, independently of the representation of p^+. A routine verification shows that f is a Boolean endomorphism of $\boldsymbol{A}^+$; cf. especially the techniques used in the proof of (i) under Lemma 11.6.

We proceed now to verify the hypotheses of Theorem 14.1.

(i) If $p^+ = \boldsymbol{S}^+(i/i^+)p$, then $\boldsymbol{\exists}^+(i^+)fp^+ = \boldsymbol{\exists}^+(i^+)c(i)p = c(i)p$ (since $c(i)p \in \boldsymbol{A}) = fp^+$.

(ii) If $p^+ = \boldsymbol{S}^+(i/i^+)p$, then

$$\begin{aligned} f\boldsymbol{\exists}^+(i^+)p^+ &= f\boldsymbol{\exists}^+(i^+)\,\boldsymbol{S}^+(i,i^+)p = f\boldsymbol{S}^+(i,i^+)\boldsymbol{\exists}(i)p = f\boldsymbol{\exists}(i)p \\ &= \boldsymbol{\exists}(i)p = \boldsymbol{S}^+(i,i^+)\boldsymbol{\exists}(i)p = \boldsymbol{\exists}^+(i^+)\,\boldsymbol{S}^+(i,i^+)p = \boldsymbol{\exists}^+(i^+)p^+. \end{aligned}$$

(iii) If $j \in I$ and if $p^+ = \boldsymbol{S}^+(i/i^+)p$, we may assume that $i \neq j$ (cf. (11.7)). It follows that

$$\begin{aligned} f\boldsymbol{\exists}^+(j)p^+ &= f\boldsymbol{\exists}^+(j)\,\boldsymbol{S}^+(i,i^+)p = f\boldsymbol{S}^+(i,i^+)\boldsymbol{\exists}(j)p \\ &= c(i)\boldsymbol{\exists}(j)p = \boldsymbol{\exists}(j)c(i)p = \boldsymbol{\exists}^+(j)fp^+. \end{aligned}$$

From the preceding paragraph we may conclude (via Theorem 14.1) that there exists a (unique) constant c^+ of $\boldsymbol{A}^+$ such that $c^+(i^+)=f$. It follows that $c^+(i^+)p=p$ whenever $p \in \boldsymbol{A}$. If $i \in I$, then

$$\begin{aligned} c^+(i)p &= c^+(i)\boldsymbol{S}^+(i^+/i)p = \boldsymbol{S}^+(i^+/i)c^+(i)c^+(i^+)p \\ &= \boldsymbol{S}^+(i,i^+)c^+(i)c^+(i^+)p = \boldsymbol{S}^+(i,i^+)c^+(i)p \\ &= c^+(i^+)\boldsymbol{S}^+(i,i^+)p = f\boldsymbol{S}^+(i/i^+)p = c(i)p \end{aligned}$$

whenever $p \in \boldsymbol{A}$. This concludes the proof of the existence of c^+. If c^+ is a constant satisfying the conditions of the lemma, then

$$c^+(i^+)\boldsymbol{S}^+(i/i^+)p = \boldsymbol{S}^+(i,i^+)c^+(i)c^+(i^+)p = c^+(i)c^+(i^+)p = c^+(i)p = c(i)p.$$

This establishes uniqueness and completes the proof.

(15.3) LEMMA. *If f is a polyadic homomorphism from an I-algebra $\boldsymbol{A}$ onto an I-algebra $\boldsymbol{B}$, and if a is a constant of $\boldsymbol{A}$, then there exists a unique constant b of $\boldsymbol{B}$ such that $fa(J)=b(J)f$ for every subset J of I.*

Proof. If $p \in \boldsymbol{A}$, write $b(J)fp=fa(J)p$. If $fp=fq$, then $f(p+q)=0$ and therefore $\exists(J)f(p+q)=0$. It follows that $0=f\exists(J)(p+q) \geqslant fa(J)(p+q)$, and hence that $fa(J)p=fa(J)q$. This proves that $b(J)$ is unambiguously defined. Straighforward computations serve to prove that b is indeed a constant of $\boldsymbol{B}$. Uniqueness follows from the fact that f is onto.

The main difficulty with constants is that they do not always exist. To see this, we examine first of all a situation in intuitive logic. Suppose that X is a set with at least two elements; to be specific, write $X=\{0,1\}$. Let us consider a microcosmic logical system designed to make statements about the concept of equality in X and about nothing else. The system, in other words, contains two propositions (corresponding to truth and falsity), a dyadic predicate (corresponding to equality), and the negation p' of that predicate. If, in accordance with the intended interpretation, we write $x_0=x_1$ in place of $p(x_0,x_1)$ and $x_0 \neq x_1$ in place of $p'(x_0,x_1)$, the quantification and transformation rules of this system become obvious. Thus, for instance, the result of applying to p the existential quantifier corresponding to the variable in position 1 is the true proposition $(\exists x_1(x_0=x_1))$ and the result of replacing in p' the variable in position 1 by the variable in position 0 is the false proposition $(x_0 \neq x_0)$. The intuitively easily comprehended act of replacing the variable in position 1 by the "constant" 0 is not permissible; the system has no way of making a statement such as $x_0=0$.

In precise terms, the system discussed in the preceding paragraph is a functional dyadic algebra, *i. e.*, a functional polyadic algebra of degree 2. If, to be specific, we write $I=\{0,1\}$, then the pertinent algebra $\boldsymbol{A}$ can be described as an $\boldsymbol{O}$-valued I-algebra over X. (The triple role of

$\{0,1\}$ as $\boldsymbol{O}$, as X, and as I, is merely a notational simplification and is not likely to lead to confusion.) The algebra $\boldsymbol{A}$ consists therefore of functions from X^I into $\boldsymbol{O}$. It does not, however, consist of all sixteen such functions; its elements are the characteristic functions of the empty set, the full set X^I, the diagonal $\{(x_0,x_1)\colon x_0=x_1\}$, and the opposite diagonal $\{(x_0,x_1)\colon x_0\neq x_1\}$. The fact that the system is unable to say that $x_0=0$ is algebraically reflected by the fact that the algebra does not contain the characteristic function of the set $\{(x_0,x_1)\colon x_0=0\}$. (This example is essentially the same as the one described in the remark (d) following Theorem 10.1.)

It is now easy to prove that the algebra $\boldsymbol{A}$ does not have any constants. The reason is that the quantifiers $\exists(0)$ and $\exists(1)$ on $\boldsymbol{A}$ are equal to each other; they are both the simple quantifier on $\boldsymbol{A}$. If c were a constant of $\boldsymbol{A}$, then we should have $c(0)\exists(0)=\exists(0)$ and $c(0)\exists(1)=\exists(1)c(0)$, and therefore, since $\exists(0)=\exists(1)$, it would follow that $\exists(0)=\exists(0)c(0)=c(0)$. This is impossible because $c(0)$ must be a Boolean endomorphism of $\boldsymbol{A}$, whereas $\exists(0)$ is not.

The fact that the algebra in the above example is of degree 2 is not essential. Using the same basic idea (systems that can talk about equality only), we can construct constantless algebras of any degree whatever, excepting only the degree 1. For the corresponding facts about monadic algebras see [2].

If $\boldsymbol{A}$ had been the algebra of all functions from X^I into $\boldsymbol{O}$, it could not have been used to illustrate the same phenomenon. An examination of the motivating considerations that preceded our definition of constants shows that a "full" functional algebra always has many of them. Roughly speaking, we may say that the absence of constants for a $\boldsymbol{B}$-valued functional I-algebra over X indicates that the algebra is too small a subset of the set of all functions from X^I into $\boldsymbol{B}$. In the next section we shall partially justify this rough description by showing that all useful polyadic algebras can be embedded into algebras with many constants.

§ 16. Rich algebras

A finite set $\{c_1,\dots,c_n\}$ of constants of a locally finite I-algebra $\boldsymbol{A}$ will be called a *witness* to an element p of $\boldsymbol{A}$ if there exists a corresponding finite set $\{j_1,\dots,j_n\}$ of elements of I such that

(16.1) $$\exists(I)p=\big(c_1(j_1)\dots c_n(j_n)\big)p\,;$$

we shall also say that $\{c_1,\dots,c_n\}$ is a witness to p *with respect to* $\boldsymbol{A}$, or more simply, *in* $\boldsymbol{A}$. If $\boldsymbol{A}$ and $\boldsymbol{A}^*$ are I-algebras such that $\boldsymbol{A}$ is a polyadic subalgebra of A^* and such that every element of A has a witness in A^*, we shall say that A^* is a *rich extension* of A, or, more simply, that A^* is rich *for* A. A *rich algebra* is one that is rich for itself.

It is appropriate to remark at this point that the definition of rich algebras given above is not the same as the one that occurs in [1]. There it was required that for every p in A and for every j in I there exist a constant c of A such that $\exists(j)p = c(j)p$. This requirement is much too strong. Most polyadic algebras fail to satisfy it, and, in fact, most polyadic algebras cannot even be embedded into an algebra satisfying it. (Theorem 1 of [1] is essentially true nevertheless, for algebras of infinite degree, if "rich" is interpreted in the sense of the preceding paragraph; cf. Theorem 16.9 below.) To see what is wrong with the strong definition of a rich algebra, suppose that a polyadic algebra A contains an element p such that for some particular pair of variables, say j and k, it is true that $\exists(j)p = \exists(k)p' = 1$. (The constantless algebras given in § 15 satisfy these conditions.) Now we assert that there cannot be any constant c of A (or, for that matter, of any polyadic algebra including A as a polyadic subalgebra) such that $\exists(j)p = c(j)p$. Indeed, if such a constant did exist, then, since $c(j)p = 1$, it would follow that $c(j)p' = 0$. Hence we should have

$$0 = \exists(k)c(j)p' = c(j)\exists(k)p' = c(j)1 = 1,$$

and this is a contradiction.

There is a slight amount of unnaturality in connection with our present definition of a witness and the subsequent definitions of rich extensions and rich algebras, caused by the fact that the right term of (16.1) depends on the order in which the constants and the variables that enter into it are written down. Since, for several purposes, it is important to eliminate this unnaturality, we proceed to put on record a simple consequence of the commutativity lemma of § 13.

(16.2) LEMMA. *If a and b are (not necessarily distinct) constants of an I-algebra A and if J and K are disjoint subsets of I, then the Boolean endomorphisms $a(J)$ and $b(K)$ of A commute with each other.*

Proof. We apply Lemma 13.4 to the monadic algebra $A = A(J)$ with the quantifier $\exists(J)$ (see § 8); the roles of a and b are to be played by $a(J)$ and $b(K)$ respectively. Clearly $a(J)$ is a constant of $A(J)$ and $b(K)$ is a Boolean endomorphism of $A(J)$. If $a(J)p = 0$, then

$$a(J)b(K)p \leqslant a(J)\exists(K)p = \exists(K)a(J)p = 0,$$

so that $a(J)b(K)p = 0$. Since

$$b(K)\exists(J) = b(K)\exists(J)\exists(J) = \exists(J)b(K)\exists(J),$$

all the hypotheses of Lemma 13.4 are satisfied and the desired conclusion follows from the conclusion of that lemma.

(16.3) LEMMA. *If A is a locally finite I-algebra of infinite degree, then there exists a locally finite I-algebra A^* such that A^* is a rich extension of A and such that every constant of A can be extended to A^*.*

Remark. The statement about extending constants means that if c is a constant of A, then there exists a constant c^* of A^* such that $c^*(J)p = c(J)p$ whenever $J \subset I$ and $p \in A$.

For some purposes it is important to know the relation between the cardinal numbers of A and A^*. The proof will show that A^* can be constructed so that if A is finite, then A^* is countable, and if A is infinite, then the cardinal number of A^* is the same as that of A. We observe that it is perfectly feasible for a polyadic algebra of infinite degree to be finite. Thus, for example, if I is any set (finite or infinite), the Boolean algebra O can be converted into a polyadic I-algebra simply by setting both $S(\tau)$ and $\exists(J)$ equal to e, for all transformation τ on I and for all subsets J of I.

Proof. Let I^+ be a set including I such that if A is finite, then $I^+ - I$ is countably infinite, and if A is infinite, then the cardinal number of $I^+ - I$ is the same as that of A. By Theorem 11.9, there exists a locally finite polyadic algebra $(A^+, I^+, S^+, \exists^+)$ that is a dilation of the given algebra. We shall denote by $(A^+, I, S, \exists)$ the algebra obtained from $(A^+, I^+, S^+, \exists^+)$ by fixing the variables in $I^+ - I$.

It is easy to verify (cf. Lemma 11.6 and Theorem 11.9) that every element of A^+ has the form $S^+(\tau)p$, where $p \in A$ and τ is a transformation (on I^+) that sends some of the elements of a finite support of p (in I) into $I^+ - I$ and is equal to the identity otherwise. It follows that if A is finite, then A^+ is countable, and if A is infinite, then the cardinal number of A^+ is the same as that of A. Since the algebra $(A^*, I, S, \exists)$ that we shall construct will be defined as a quotient algebra of $(A^+, I, S, \exists)$, the remark about cardinal numbers is already established.

Since the algebra A is locally finite, it follows that to each element p of A there corresponds a finite subset J_p of I (a support of p) such that $\exists(I - J_p)p = p$. In order to avoid a needless unambiguity argument below, we shall assume that J_p is the minimal support of p, *i. e.*, that if $\exists(j)p = p$, then $j \notin J_p$. From what we know about the cardinality of $I^+ - I$ we can conclude that there exists a disjoint family $\{K_p\}$ of finite subsets of $I^+ - I$ such that K_p and J_p have the same number of elements for every p in A.

We know that every element of $I^+ - I$ induces in a natural way a constant of the algebra $(A^+, I, S, \exists)$. We shall find it convenient to denote an element of $I^+ - I$ and the constant it induces by the same symbol. Thus if $k \in I^+ - I$ and $J \subset I$, then $k(J)p = S^+(J/k)p$ for every element p of A^+.

Suppose now that p is any element of A: we shall make correspond to p a Boolean endomorphism f_p on A^+. If J_p is empty (*i. e.*, if p is closed), let f_p be e. If $J_p = \{j_1, \ldots, j_n\}$ and, correspondingly $K_p = \{k_1, \ldots, k_n\}$, write $f_p = k_1(j_1) \ldots k_n(j_n)$. (The indicated one-to-one correspondence between J_p and K_p is arbitrary, but fixed once and for all. The factors of f_p commute among each other by Lemma 16.2.) Roughly speaking, the endomorphism f_p has the effect of replacing by a constant each of the variables that p really depends on. The replacements are such that two distinct variables of the same p, and any two variables (distinct or not) of two distinct p's, are always replaced by distinct constants. The idea of the remainder of the proof is to force $f_p p$ to be equal to $\exists(J_p)p$, or, equivalently, to force $\exists(J_p)p - f_p p$ to be 0. Precisely speaking, we shall consider the polyadic ideal M, in the algebra $(A^+, I, S, \exists)$, generated by all elements that have the form $\exists(J_p)p - f_p p$ for some p in A, and we shall reduce A^+ modulo M. Essentially the only thing to worry about is the preservation of A in the passage from A^+ to A^+/M. Technically speaking, we shall have to show that the natural (projection) mapping from A^+ onto A^+/M is one-to-one on A, or, equivalently, that $A \cap M = \{0\}$. A remark is in order concerning our use of $\exists(J_p)p - f_p p$. Since

$$f_p p = k_1(j_1) \ldots k_n(j_n) p \leqslant \exists(j_1) \ldots \exists(j_n) p = \exists(J_p) p,$$

it follows that $\exists(J_p)p - f_p p$ is the same as the Boolean sum (or symmetric difference) $\exists(J_p)p + f_p p$. Since, furthermore, J_p supports p, we have $\exists(J_p)p = \exists(I)p$; it follows that identifying $\exists(J_p)p - f_p p$ with 0 has exactly the same effect as identifying $\exists(I)p$ with $f_p p$, and, in view of the definition of witnesses, the latter is what we must do.

Continuing with the notation of the preceding paragraph, we shall write $g_p = S^+(k_1/j_1) \ldots S^+(k_n/j_n)$. Clearly g_p is a Boolean endomorphism of A^+ such that $g_p f_p p = p$. For later purposes we observe that if $q \in A$, then $g_p q = q$; this follows from the fact that the elements of A are independent of $I^+ - I$.

An element p of A^+ belongs to the ideal M if and only if it is dominated by the supremum of a finite number of elements of the form $\exists(J_p)p - f_p p$ with $p \in A$. Since $\exists(J_p)p = \exists(I)p$, what we are trying to prove is that if q is an element of A such that

$$q \leqslant \big(\exists(I)p_1 - f_{p_1}p_1\big) \vee \ldots \vee \big(\exists(I)p_m - f_{p_m}p_m\big), \tag{16.4}$$

where m is a positive integer and $p_1, \ldots, p_m$ are distinct elements of A, then $q = 0$. If $m > 1$, we form the infimum of both sides of (16.4) with the complement of $(\exists(I)p_m - f_{p_m}p_m)$. The resulting inequality implies (after increasing its larger side by omitting from it the factor $(\exists(I)p_m - f_{p_m}p_m)'$) that

$$(16.5)\quad q \wedge (\exists(I)p_m - f_{p_m}p_m)' \leqslant (\exists(I)p_1 - f_{p_1}p_1) \vee \ldots \vee (\exists(I)p_{m-1} - f_{p_{m-1}}p_{m-1}).$$

The smaller side of (16.5) belongs to the algebra obtained from $(A^+, I^+, S^+, \exists^+)$ by fixing the variables in $I^+ - (I \cup K_{p_m})$, whereas the larger side of (16.5) is independent of $I^+ - (I \cup K_{p_m}.)$ (Here is where the disjointness of the sets K_p comes in.) Thus it is only a matter of change of notation (replace I by $I \cup K_{p_m}$) to reduce the parameter m in the statement to be proved (*i. e.*, the statement that if $q \in A$ and (16.4) holds, then $q=0$) to $m-1$. It is therefore sufficient to prove that statement for $m=1$, *i. e.*, to prove that if p and q are in A and if

$$(16.6)\qquad q \leqslant \exists(I)p - f_p p,$$

then $q=0$. Since $\exists(I)f_p p = f_p p$, the right term of (16.6) is invariant under $\exists(I)$; consequently we may and do assume that q is closed. Since $g_p f_p p = p$, an application of g_p to (16.6) yields

$$(16.7)\qquad q \leqslant \exists(I)p - p.$$

Forming the infimum of both sides of (16.7) with p we obtain

$$(16.8)\qquad q \wedge p = 0.$$

Applying $\exists(I)$ and using the fact that q is closed, we conclude that $q \wedge \exists(I)p = 0$. Since, however, $q \leqslant \exists(I)p$ (by (16.7)), we can go on to conclude that $q=0$, as desired.

From what we have proved it follows that the I-algebra $A^* = A^+/M$ includes A as a polyadic subalgebra. Since, by Lemma 15.3, the constants k in $I^+ - I$ induce constants on A^*, the fact that A^* is a rich extension of A follows from the fact that if $p \in A$, then the elements $\exists(I)p$ and $f_p p$ of A^+ are congruent modulo M. The assertion about the extendability of constants from A to A^* follows in two steps. By Lemma 15.1 every constant of A can be lifted to a constant of A^+, and, by Lemma 15.3, every constant of A^+ can be transferred to A^*. The proof of Lemma 16.3 is complete.

(16.9) THEOREM. *Every locally finite polyadic algebra of infinite degree is a polyadic subalgebra of a locally finite rich algebra.*

Remark. As in the case of Lemma 16.3, the proof will show that the extension algebra A^* of the given algebra A can be constructed so that if A if finite, then A^* is countable, and if A is infinite, then the cardinal number of A^* is the same as that of A.

Proof. Given an algebra A, we write $A_0 = A$, and, given A_n, we let A_{n+1} be a rich extension of A_n ($n = 0, 1, 2, \ldots$) obtained by an application of Lemma 16.3. If A^* is the set-theoretic union of the increasing sequence of polyadic algebras so obtained, then A^* is in an obvious way a polyadic algebra that includes A as a polyadic subalgebra. Since an inductive application of Lemma 16.3 implies that every constant of each A_n can be extended to a constant of A^*, it follows that A^* has enough constants to bear witness to every element of A^*, *i. e.*, that A^* is rich.

§ 17. Representation

We have now reached the culmination of our work. In this section we shall prove the principal representation theorem of the theory of polyadic algebras and from it we shall deduce the representation theorem for simple algebras needed to guarantee that Theorem 9.3 is indeed the algebraic version of Gödel's completeness theorem.

We shall say that a locally finite functional I-algebra A, with domain X, say, is *functionally rich* if the elements of X constitute a sufficient supply of constants for A. In order to make this definition precise, we recall that to each finite subset J of I and to each element c of X there corresponds in a natural way a mapping, to be denoted by $c(J)$, that sends every function from X^I into B onto another such function. If p is a function from X^I into B and if $x \in X^I$, the value of $c(J)p$ at x is the value of p at the point that x becomes when x_i is replaced by c for each i in J (cf. § 12). To say that A is functionally rich means that A is closed under each of the operations $c(J)$ so obtained (so that c is, from this point of view, a constant of A), and that each element of A has a witness $\{c_1, \ldots, c_n\}$ such that the constants $c_1, \ldots, c_n$ belong to X in the sense just described. (It is clear that a functionally rich algebra is rich in the sense of § 16.) It is a consequence of this definition that if A is functionally rich, then the suprema defining $\exists(I)$ are always attained, or, more explicitly, that if $p \in A$, then there exists a point y in X^I such that $\exists(I)p(x) = p(y)$ for all x in X^I.

(17.1) THEOREM. *Every rich, locally finite polyadic I-algebra A of infinite degree is isomorphic to a functionally rich functional algebra. The domain X of the functional algebra can be chosen so that if A is finite, then X is finite and if A is infinite, then the cardinal number of X is less than or equal to that of A.*

Proof. The value-algebra B of the functional algebra that we shall construct is defined as the Boolean algebra of all closed elements in A. For the domain X we select a sufficiently large set of constants of A;

this means simply that each element of A is to have a witness $\{c_1,\dots,c_n\}$ such that $c_1,\dots,c_n$ belong to X. The assertion concerning the cardinal number of X follows immediately from this description of how X is to be chosen.

We are now ready to define a mapping f that associates with each element p of A a function $\tilde{p}$ from X^I into B; the mapping f will turn out to be one of the isomorphisms whose existence the theorem asserts. If $x \in X^I$ and if J is a finite subset of I, we shall write $x(J)=\prod_{j\in J} x_j(j)$; clearly $x(J)$ is a Boolean endomorphism of A. Let J be a finite support of the given element p; the value of $\tilde{p}$ at x is, by definition, $x(J)p$. Since J supports p, it follows that $x(J)p$ is closed, so that $\tilde{p}(x) \in B$; we must still prove, however, that $\tilde{p}(x)$ is unambiguously defined, independently of the choice of the support J. To prove this, suppose that K is another finite support of p. We observe that $x(J)=x(J\cap K)x(J-K)$. Since K supports p, it follows that p is independent of $J-K$, and hence that $x(J-K)p=p$. We conclude that $x(J)p=x(J\cap K)p$; since, similarly, $x(K)p=x(J\cap K)p$, the proof of unambiguity is complete.

To prove that f is a Boolean isomorphism, observe first that

$$(fp')(x)=x(J)p'=\big(x(J)p\big)'=\big(fp(x)\big)';$$

this proves that f preserves complementation. To prove that f preserves suprema, given p and q in A, let J be a finite set that supports them both; then

$$f(p\vee q)(x)=x(J)(p\vee q)=x(J)p\vee x(J)q=fp(x)\vee fq(x).$$

Suppose finally that $fp=0$. Let $\{c_1,\dots,c_n\}$ be a witness to p, so that $\exists(I)p=c_1(j_1)\dots c_n(j_n)p$ for a suitable finite subset $J=\{j_1,\dots,j_n\}$ of I. If x is a point of X^I such that $x_{j_1}=c_1,\dots,x_{j_n}=c_n$, then

$$0=fp(x)=x(J)p=c_1(j_1)\dots c_n(j_n)p=\exists(I)p$$

and therefore $p=0$. This proves that f is indeed a Boolean isomorphism.

We show next that f commutes with $S(\tau)$. By § 8, it is sufficient to treat the case in which τ is a finite transformation on I. Given an element p in A, let $J=\{j_1,\dots,j_n\}$ be a finite subset of I such that J supports both p and τ and such that $\tau^{-1}J\subset J$. Since, by Lemma 6.14, τJ supports $S(\tau)p$, it follows that

$$fS(\tau)p(x)=x(\tau J)S(\tau)p.$$

If $k\in\tau J$, then $\tau^{-1}\{k\}$ is a certain subset of J; say, for simplicity, $\tau^{-1}\{k\}=\{j_1,\dots,j_m\}$. Then

$$\begin{aligned}x_k(k)\,S(\tau)&=S(\tau)x_k(\tau^{-1}\{k\})=S(\tau)x_k(\{j_1,\dots,j_m\})\\&=S(\tau)x_{\tau j_1}(j_1)\dots x_{\tau j_m}(j_m).\end{aligned}$$

Since, similarly, $S(\tau)$ can be pulled past each $x_k(k)$ with $k \in \tau J$, it follows that

$$\begin{aligned} fS(\tau)p(x) &= S(\tau)x_{\tau j_1}(j_1)\dots x_{\tau j_n}(j_n)p \\ &= S(\tau)(\tau_* x)_{j_1}(j_1)\dots(\tau_* x)_{j_n}(j_n)p \\ &= S(\tau)(\tau_* x)(J)p = S(\tau)\big(fp(\tau_* x)\big). \end{aligned}$$

Since $fp(\tau_* x) \in \mathbf{B}$, it is unaffected by the application of $S(\tau)$, so that

$$fS(\tau)p(x) = fp(\tau_* x) = S(\tau)fp(x).$$

In order to conclude that f is a polyadic isomorphism we must still show that f commutes with $\exists(K)$; by § 8, it is sufficient to treat the case in which $K = \{k_1, \dots, k_m\}$ is a finite subset of I. We are to prove, in other words, that if $p \in A$ and $x \in X^I$, then the set $\{fp(y): xK_* y\}$ has a supremum and that supremum is equal to $f\exists(K)p(x)$. Let J be a finite support of p such that $K \subset J$ and write $q = x(J-K)p$. Since K supports q (cf. Lemma 6.10), it follows from the definition of a rich algebra that there exist constants $c_1, \dots, c_m$ in X such that

$$\exists(I)q = \exists(K)q = c_1(k_1)\dots c_m(k_m)q\,.$$

Let $\bar{y}$ be the element of X^I for which $\bar{y}_{k_1} = c_1, \dots, \bar{y}_{k_m} = c_m$, and $\bar{y}_i = x_i$ when $i \notin K$. It follows that $xK_*\bar{y}$ and that

$$fp(\bar{y}) = \bar{y}(J)p = \bar{y}(K)x(J-K)p = \bar{y}(K)q = \exists(K)q\,.$$

If $y \in X^I$ and $xK_* y$, then, similarly,

$$fp(y) = y(J)p = y(K)x(J-K)p = y(K)q \leqslant \exists(K)q = fp(\bar{y})\,.$$

Thus the set $\{fp(y): xK_* y\}$ has a largest element, namely $fp(\bar{y})$; it follows that the supremum of that set certainly exists, and, of course, is equal to $fp(\bar{y})$. What we must show therefore is that $fp(\bar{y}) = f\exists(K)p(x)$, or, since $J-K$ supports $\exists(K)p$, that $\bar{y}(J)p = x(J-K)\exists(K)p$. We already know that $\bar{y}(J)p = \exists(K)q$; the fact that also $x(J-K)\exists(K)p = \exists(K)q$ follows from the fact that a multiple application of (C_3) allows us to pull $\exists(K)$ past $x(J-K)$.

To prove that the image of A under f is functionally rich it suffices to observe that f carries an adequate supply of constants of A into (a necessarily adequate supply of) "functional" constants of the image algebra. The proof of the principal representation theorem for polyadic algebras is complete.

(17.2) Lemma. *If A is a functionally rich, locally finite, $\mathbf{B}$-valued functional I-algebra over X, if f_0 is a Boolean homomorphism from $\mathbf{B}$ into a Boolean algebra $\mathbf{B}_0$, and if $(fp)(x) = f_0(p(x))$ whenever $p \in A$ and $x \in X^I$, then f is a polyadic homomorphism from A onto a B_0-valued functional I-algebra over X.*

Proof. The only non-trivial part of this assertion is that f commutes with $\exists(K)$ for every finite subset K of I. The proof is similar in almost all details to the proof of the corresponding part of Theorem 17.1. (It is in fact possible, but not worth while, to formulate a highly technical lemma that could be cited in both places.) If $x \in X^I$, $J \subset I$, and $p \in A$, we denote by $x(J)p$ the function (in A) whose value at y is the value of p at the point that y becomes when y_j is replaced by x for each j in J. Suppose that $p \in A$ and that K is a finite subset of I; we are to prove that, for each fixed x in X^I, the set $\{fp(y): xK_*y\}$ has a supremum and that supremum is equal to $f\exists(K)p(x)$. Let J be a finite support of p such that $K \subset J$ and write $q = x(J-K)p$. It follows, as in the proof of Theorem 17.1, that there exists a point $\bar{y}$ in X^I such that $xK_*\bar{y}$ and such that $\bar{y}(K)q = \exists(K)q$. If $y \in X^I$ and xK_*y, then

$$fp(y) = f_0(p(y)) = f_0(y(K)p(x)) = f_0(y(K)x(J-K)p(x)) = f_0(y(K)q(x)) .$$

It follows that $fp(\bar{y}) = f_0(\bar{y}(K)q(x)) = f_0(\exists(K)q(x))$ and hence that $fp(y) \leq fp(\bar{y})$. Thus the set $\{fp(y): xK_*y\}$ has $fp(\bar{y})$ as its largest element. Since

$$fp(\bar{y}) = f_0(\exists(K)q(x)) = f_0(\exists(K)x(J-K)p(x)) = f_0(\exists(K)p(x)) = f\exists(K)p((x) ,$$

the proof of the lemma is complete.

(17.3) THEOREM. *Every simple, locally finite polyadic I-algebra A of infinite degree is (isomorphic to) a model, i. e., an O-valued functional polyadic algebra. The domain X of the model can be chosen so that if A is finite, then X is countable, and if A is infinite, then the cardinal number of X is less than or equal to that of A.*

Proof. By Theorem 16.9, A is a polyadic subalgebra of a locally finite, rich I-algebra A^*, such that if A is finite, then A^* is countable and such that otherwise A and A^* have the same cardinal number. By Theorem 17.1, we may assume that A^* is a functionally rich functional algebra, with value-algebra B, say, and domain X; the information that Theorem 17.1 gives about the cardinality of X implies the second assertion of the present theorem. Let f_0 be an arbitrary Boolean homomorphism from B into (and therefore onto) O, and write $(fp)(x) = f_0(p(x))$ whenever $p \in A^*$ and $x \in X^I$. It follows from Lemma 17.2 (with A^* and O playing the roles of A and B_0, respectively) that f is a polyadic homomorphism from A^* onto a model with domain X. If g is the restriction of f to A, then g is a polyadic homomorphism from A onto such a model (in general a subalgebra of the model onto which f maps A^*). Since the kernel of a polyadic homomorphism is a proper ideal, and since the only proper ideal of a simple algebra is trivial, it follows that q is an isomorphism, and the proof of the theorem is complete.

The cardinality assertions of Theorem 17.3 constitute the algebraic version of the general Skolem-Löwenheim theorem.

Appendix

The purpose of this appendix is to put on record some minor additions to the first paper in this series (*i. e.*, [2]).

The additions are as follows:

(a) In order to apply Lemma 9 in the proof of Theorem 9, it is necessary to prove first that the product of two Boolean relations is again a Boolean relation. This result, in turn, follows from the fact that the direct image of a closed set under a Boolean relation is again closed. For the proof, suppose that φ is a Boolean relation from Y to X, and let ξ denote the projection from $Y \times X$ onto X. If F is a subset of Y, then $\varphi F = \xi\big((F \times X) \cap \varphi\big)$; it is, consequently, sufficient to prove that *every Boolean relation is closed*, *i. e.*, that φ is a closed subset of $Y \times X$. If $(y_0, x_0) \notin \varphi$, then it is false that $p(x_0) \leq fp(y_0)$ for all p in the dual algebra A of X; let p_0 be an element of A such that $p_0(x_0) = 1$ and $fp_0(y_0) = 0$. The set

$$\{(y, x)\colon\ p_0(x) = 1,\ fp_0(y) = 0\}$$

is a neighborhood of (y_0, x_0) disjoint from φ; this proves that the complement of φ is open. From this result it follows also that the inverse image of a closed set under a Boolean relation is again closed; all that is needed is to observe that if η is the projection from $Y \times X$ onto Y, and if E is a subset of X, then $\varphi^{-1}E = \eta\big((Y \times E) \cap \varphi\big)$. All the results in this paragraph are due to R. J. Blattner.

(b) The first part of the proof of Lemma 18 must be completed by establishing that the mapping σ there constructed is indeed a cross section of π, *i. e.*, that $\pi\sigma y = y$ for all y in Y. The argument runs as follows. Since $c\exists p(x) = \exists p(x)$, it follows that $\exists p(\gamma x) = \exists p(x)$ for all p and all x, and hence that $\pi\gamma x = \pi x$ for all x. If $y \in Y$, then $y = \pi x$ for some x in X, and $\sigma y = \gamma x$; the desideratum follows from the result of the preceding sentence by making the obvious substitutions.

(c) In the last paragraph of the proof of Lemma 19 it must be proved that if $q \in A$ and if

$$q^+ \leq (\exists^+ p_0^+ - c_0^+ p_0^+) + r\,,$$

where $r \in I^+$, then $q^+ \in I^+$. There is no loss of generality in assuming that $p_0 \neq 0$; let x_0 be a point of X such that $p_0(x_0) = 1$. Forming the infimum of both sides of the assumed inequality with $c_0^+ p_0^+$, we see that it is sufficient to prove that if $q \in \boldsymbol{A}$ and if $q^+ \wedge c_0^+ p_0^+ \in \boldsymbol{I}^+$, then $q \in \boldsymbol{I}^+$. To say that $q^+ \wedge c_0^+ p_0^+ \in \boldsymbol{I}^+$ means that

$$q^+ \wedge c_0^+ p_0^+ \leqslant \bigvee_{i=1}^{n} (\exists^+ p_i + c_0^+ \exists^+ p_i)$$

with $p_i \in \boldsymbol{A}^+$, $i=1,\ldots,n$. If we evaluate both terms of this inequality at (x_0, y), and if we recall that $q^+(x,y) = q^+(x_0,y) = q(y)$ for all x and y, we obtain

$$q^+(x,y) \leqslant \bigvee_{i=1}^{n} \left(\exists^+ p_i(x_0,y) + c_0^+ \exists^+ p_i(x_0,y)\right).$$

If we write $r_i(y) = p_i(x_0,y)$, then $r_i \in \boldsymbol{A}$, and

$$\begin{aligned} \exists^+ p_i(x_0,y) &= \bigvee \{p_i(x,z)\colon\ x = x_0,\ y \exists^* z\} \\ &= \bigvee \{p_i(x_0,z)\colon\ y \exists^* z\} \\ &= \bigvee \{r_i(z)\colon\ y \exists^* z\} \\ &= \exists r_i(y) = \exists^+ r_i^+(x,y). \end{aligned}$$

It follows that

$$q^+ \leqslant \bigvee_{i=1}^{n} (\exists^+ r_i^+ + c_0^+ \exists^+ r_i^+),$$

and hence that $q^+ \in \boldsymbol{I}^+$, as asserted.

References

[1] P. R. Halmos, *Polyadic Boolean algebras*, Proc. Nat. Acad. Sci. U. S. A. 40 (1954), p. 296-301.

[2] — *Algebraic logic, I. Monadic Boolean algebras*, Compositio Math., 12 (1955), p. 217-249.

[3] L. Henkin, *The completeness of the first-order functional calculus*, J. Symbolic Logic 14 (1949), p. 159-166.

[4] B. Jónsson and A. Tarski, *Boolean algebras with operators*, Amer. J. Math., 73 (1951), p. 891-939.

[5] H. Rasiowa and R. Sikorski, *A proof of the completeness theorem of Gödel*, Fund. Math. 37 (1950), p. 193-200.

[6] A. Tarski and F. B. Thompson, *Some general properties of cylindric algebras*, Bull. Amer. Math. Soc. 58 (1952), p. 65.

[7] A. Tarski, *A representation theorem for cylindric algebras*, Bull. Amer. Math. Soc. 58 (1952), p. 65-66.

[8] H. Wang, *Logic of many-sorted theories*, J. Symbolic Logic 17 (1952), p. 105-116.

UNIVERSITY OF CHICAGO

Reçu par la Rédaction le 5.4.1955

VI

TERMS

ALGEBRAIC LOGIC, III. PREDICATES, TERMS, AND OPERATIONS IN POLYADIC ALGEBRAS

Introduction. The theory of polyadic algebras is an algebraic photograph of the logical theory of first-order functional calculi. It is known, for instance, that the Gödel completeness theorem can be formulated in algebraic language as a representation theorem for a large class of simple polyadic algebras, together with the statement that every polyadic algebra is semisimple. The next desideratum is an algebraic study of the celebrated Gödel incompleteness theorem. Before that can be achieved, it is necessary to investigate the algebraic counterparts of some fundamental logical concepts (such as the ones mentioned in the title above). The purpose of this paper is to report the results of such an investigation(1).

Although this is the third of a sequence of papers on algebraic logic, its development does not lean very heavily on the first two papers. The main purpose of the first paper was to study the topological properties of quantification, via the Stone duality theory for Boolean algebras [*Algebraic logic* I, Compositio Math. vol. 12 (1955) pp. 217–249]. The purpose of the second paper was to study the algebraic properties of polyadic algebras, with main emphasis on their representation theory [*Algebraic logic* II, to appear in Fund. Math.](2). Since the methods and results of this paper might be described as combinatorial, neither the duality theory nor the representation theory plays any role; a sympathetic understanding of the basic definitions and of their elementary consequences is sufficient for present purposes. For the convenience of the reader, the basic definitions and theorems are summarized in §1 below.

The most difficult concept introduced in this paper is the concept of a term (§4), and the most difficult theorems are the ones that describe properties of terms (§§5–8) and the ones that give methods of constructing them (§§9, 11, and 13). The climax is reached in §9; the existence theorem of that section (despite its somewhat complicated statement) turns out to be a most efficient tool for constructing terms satisfying various conditions. The remaining sections are devoted to auxiliary matters and to the more easily

Received by the editors January 5, 1956.

(1) The work on this paper was sponsored in part by the National Science Foundation, NSF grant 2266. Most of these results were announced in a note in the Proceedings of the National Academy of Sciences; cf. vol. 42 (1956) pp. 130–136.

(2) These papers will be referred to by the corresponding Roman numerals.

treated algebraic theory of predicates and operations. The applications (e.g., to the theory of equality in polyadic algebras and, eventually, to the algebraic version of the incompleteness theorem) will be published in subsequent papers of the sequence.

1. **Polyadic algebras.** The set-theoretic notation to be used (e.g., $\cup$, $\cap$, $\subset$, and ϵ) is the standard one. The empty set is denoted by $\varnothing$, and nonbelonging by ϵ', so that, for instance, $x \epsilon' \varnothing$ for all x. A singleton (i.e., a set consisting of a single element, say x) is denoted by $\{x\}$; more generally, braces are employed, as always, to indicate the set of elements described within them. Sometimes, for the sake of notational simplicity, the braces are omitted from a symbol such as $\{x\}$. This is done mostly on the occasions when $\{x\}$ appears as an argument of a function; it is only rarely that the practice of writing $f(x)$ for $f(\{x\})$ can lead to confusion.

We shall be working with a fixed set I, and, unless there is explicit warning to the contrary, we shall assume that *set* means a subset of I and *transformation* means a transformation (not necessarily one-to-one and not necessarily onto) from I into itself. If a transformation does map I onto itself in a one-to-one manner, we shall call it a *permutation*. Certain special transformations recur sufficiently often to deserve special symbols. Thus the identity transformation will be denoted by δ, so that $\delta i = i$ for all i in I. If i and j are in I, the *transposition* that interchanges them will be denoted by (i, j), so that (i, j) is the transformation such that $(i, j)i = j$, $(i, j)j = i$, and $(i, j)k = k$ whenever $k \epsilon' \{i, j\}$. Clearly δ and (i, j) are permutations.

If $i_1, \cdots, i_n$ are distinct elements of I, and if $j_1, \cdots, j_n$ are arbitrary (not necessarily distinct) elements of I, the symbol $(i_1, \cdots, i_n/j_1, \cdots, j_n)$ denotes the transformation that maps each i onto the corresponding j and everything else onto itself. Thus, in particular, if i and j are in I, then (i/j) is the transformation such that $(i/j)i = j$ and $(i/j)k = k$ whenever $k \neq i$. This special case can be generalized in a slightly different direction: if $K \subset I$ and $j \epsilon I$, we shall write (K/j) for the transformation that maps every element of K onto j and every element of $I - K$ onto itself.

We shall say that a transformation τ *lives on* a set J if $\tau J \subset J$ and if $\tau = \delta$ outside J. A transformation is *finite* if it lives on some finite set. Sometimes we shall consider a transformation, say τ^-, on a subset I^- of I (i.e., a transformation of I^- into itself); we shall then say that the *canonical extension* of τ^- is the transformation τ on I such that $\tau = \tau^-$ on I^- and $\tau = \delta$ on $I - I^-$.

The following notation will be used for every Boolean algebra $\boldsymbol{A}$: the supremum of two elements p and q of $\boldsymbol{A}$ is $p \vee q$, the infimum of p and q is $p \wedge q$, the complement of p is p', the zero element of $\boldsymbol{A}$ is 0, and the unit element of $\boldsymbol{A}$ is 1. The natural order relation is denoted by $\leqq$, so that $p \leqq q$ means that $p \vee q = q$ (or, equivalently, that $p \wedge q = p$).

A *quantifier* (more precisely, an *existential quantifier*) on a Boolean algebra $\boldsymbol{A}$ is a mapping $\exists$ of $\boldsymbol{A}$ into itself such that

$$\exists 0 = 0, \tag{Q1}$$

$$p \leqq \exists p, \tag{Q2}$$

$$\exists(p \wedge \exists q) = \exists p \wedge \exists q, \tag{Q3}$$

for all p and q in A. Suppose that A is a Boolean algebra, I is a set, $\mathbf{S}$ is a mapping that associates a Boolean endomorphism $\mathbf{S}(\tau)$ of A with every transformation τ on I, and $\exists$ is a mapping that associates a quantifier $\exists(J)$ on A with every subset J of I. The quadruple $(A, I, \mathbf{S}, \exists)$ is a *polyadic algebra* if

(P1) $\exists(\varnothing)$ *is the identity mapping on* A,

(P2) $\exists(J \cup K) = \exists(J)\exists(K)$ *for all* J *and* K,

(P3) $\mathbf{S}(\delta)$ *is the identity mapping on* A,

(P4) $\mathbf{S}(\sigma\tau) = \mathbf{S}(\sigma)\mathbf{S}(\tau)$ *for all* σ *and* τ,

(P5) $\mathbf{S}(\sigma)\exists(J) = \mathbf{S}(\tau)\exists(J)$ *whenever* $\sigma = \tau$ *outside* J,

(P6) $\exists(J)\mathbf{S}(\tau) = \mathbf{S}(\tau)\exists(\tau^{-1}J)$ *whenever* τ *is such that it never maps two distinct elements of* I *onto the same element of* J.

It is often convenient to be slightly elliptical, and, instead of saying that $(A, I, \mathbf{S}, \exists)$ is a polyadic algebra, to say that A is a polyadic algebra, or, alternatively, to say that A is an I-algebra. An element of I is called a *variable* of the algebra A. The *degree* of A is the cardinal number of the set of its variables.

An element p of an I-algebra A is *independent* of a subset J of I if $\exists(J)p = p$; the set J is a *support* of p (or J *supports* p) if p is independent of $I - J$. The following facts about supports are easy to prove; we shall often use them without any explicit reference. (i) If J supports p and if $J \subset K$, then K supports p. (ii) If J supports p, then J supports p'; if J supports p and q, then J supports $p \vee q$. (iii) If J supports p, then $\exists(K)p = \exists(K \cap J)p$ for all K. (iv) If J supports p, then τJ supports $\mathbf{S}(\tau)p$ for all τ.

An I-algebra A is *locally finite* if each of its elements has a finite support. It is a fortunate circumstance that the logically most important polyadic algebras are the ones most amenable to algebraic treatment. These algebras are characterized by two conditions that pull against each other in a certain sense. The first condition is that their degree be infinite (i.e., that there be infinitely many variables); the second condition is that they be locally finite (i.e., that no single element depend on more than finitely many variables). We hereby establish a notation and an assumption that will be used throughout: *we shall always assume that* $(A, I, \mathbf{S}, \exists)$ *is a locally finite polyadic algebra of infinite degree.*

Suppose that c is a mapping that associates a Boolean endomorphism $c(J)$ of A with every subset J of I. The mapping c is a *constant* of the algebra A if

$$\text{(C1)} \qquad c(\varnothing) \textit{ is the identity mapping on } \mathbf{A},$$

$$\text{(C2)} \qquad c(J \cup K) = c(J)c(K),$$

$$\text{(C3)} \qquad c(J)\exists(K) = \exists(K)c(J - K),$$

$$\text{(C4)} \qquad \exists(J)c(K) = c(K)\exists(J - K),$$

$$\text{(C5)} \qquad c(J)\mathbf{S}(\tau) = \mathbf{S}(\tau)c(\tau^{-1}J),$$

whenever J and K are subsets of I and τ is a transformation on I. We shall need to make use of the following three facts about constants.

(1.1) THEOREM. [II, (12.3)]. *If c is a mapping from all finite subsets of I into Boolean endomorphisms of $\mathbf{A}$ satisfying* (C1)–(C5) *whenever J and K are finite subsets of I and τ is a finite transformation on I, then there exists a unique constant $\bar{c}$ of $\mathbf{A}$ such that $\bar{c}(J) = c(J)$ for all finite subsets J of I.*

(1.2) THEOREM. [II, (14.1)]. *If f is a Boolean endomorphism of $\mathbf{A}$ and if i is an element of I such that*

$$\text{(1)} \qquad \exists(i)f = f,$$

$$\text{(2)} \qquad f\exists(i) = \exists(i),$$

$$\text{(3)} \qquad f\exists(j) = \exists(j)f \qquad \textit{whenever } j \neq i,$$

then there exists a unique constant c of $\mathbf{A}$ such that $c(i) = f$.

(1.3) THEOREM. [II, (16.2)]. *If b and c are (not necessarily distinct) constants and if J and K are disjoint subsets of I, then the Boolean endomorphisms $b(J)$ and $c(K)$ commute with each other.*

Associated with every subset I^- of I there is a polyadic algebra $(\mathbf{A}^-, I^-, \mathbf{S}^-, \exists^-)$ that is said to be obtained from $\mathbf{A}$ by *fixing* the variables of $I - I^-$. The Boolean algebra $\mathbf{A}^-$ is (by definition) the same as the Boolean algebra $\mathbf{A}$. To define $\mathbf{S}^-$, suppose that τ^- is a transformation on I^-, let τ be its canonical extension to I, and write $\mathbf{S}^-(\tau^-) = \mathbf{S}(\tau)$; to define $\exists^-$, suppose that J is a subset of I^-, and write $\exists^-(J) = \exists(J)$. The following result establishes a connection between this concept and the preceding one; it says, in effect, that a fixed variable is a constant.

(1.4) THEOREM. [II, §15]. *Suppose that $I^- \subset I$ and that $(\mathbf{A}^-, I^-, \mathbf{S}^-, \exists^-)$ is the algebra obtained from $\mathbf{A}$ by fixing the variables of $I - I^-$. If $j \in I - I^-$ and if $c^-(J) = \mathbf{S}(J/j)$ for every subset J of I^-, then c^- is a constant of $\mathbf{A}^-$.*

2. **Predicates.** The elements of a polyadic algebra are the algebraic counterparts of propositional functions. In a suggestive and reasonably flexible terminology, an element with support J will be called a *J-proposition.* If J is empty, a J-proposition (i.e., a $\varnothing$-proposition) will be called a *closed* proposition, or, simply a *proposition.*

There is a distinction, in ordinary logical parlance, between a "propositional function" and a "predicate"; that distinction can be profitably imitated in the theory of polyadic algebras. In the usual treatment of first-order functional calculi a predicate symbol all by itself is not a well-formed formula; it becomes one, however, when it is followed by the appropriate number of individual variables. This suggests that in the algebraic theory a predicate should be a function whose arguments are variables and whose values are elements of a polyadic algebra. Not every such function deserves to be called a predicate; the general definition must somehow take account of special facts such as that if in $x=y$ the variables x and y are replaced by u and v respectively, then the result is $u=v$. These considerations motivate the following definition.

An n-place (or n-ary) *predicate* ($n=1, 2, 3, \cdots$) of an I-algebra $\boldsymbol{A}$ is a function P from I^n into $\boldsymbol{A}$ such that if $(i_1, \cdots, i_n) \in I^n$ and if τ is a transformation on I, then

$$\text{(2.1)} \qquad \mathbf{S}(\tau)P(i_1, \cdots, i_n) = P(\tau i_1, \cdots, \tau i_n).$$

We shall sometimes say that an n-place predicate is a predicate of *degree n*. The basic theory of predicates is quite easy; there are only two main facts that it is useful to know. Theorem (2.2) asserts that the variables that a particular value of a predicate depends on are apparent at a glance. An n-place predicate and an element of I^n together define an element of $\boldsymbol{A}$; Lemma (2.3) gives a sort of converse for this assertion.

(2.2) THEOREM. *If P is an n-place predicate and if $(i_1, \cdots, i_n) \in I^n$, then $\{i_1, \cdots, i_n\}$ supports $P(i_1, \cdots, i_n)$.*

Proof. In view of the local finiteness of $\boldsymbol{A}$, it is sufficient to show that $\exists(j)P(i_1, \cdots, i_n) = P(i_1, \cdots, i_n)$ whenever $j \in' \{i_1, \cdots, i_n\}$. If k is any variable distinct from j, then

$$\begin{aligned}\exists(j)P(i_1, \cdots, i_n) &= \exists(j)\mathbf{S}(j/k)P(i_1, \cdots, i_n) \ [\text{by (2.1)}] \\ &= \mathbf{S}(j/k)P(i_1, \cdots, i_n) \ [\text{since } j \neq k] = P(i_1, \cdots, i_n) \ [\text{by (2.1)}].\end{aligned}$$

(2.3) LEMMA. *If $p \in \boldsymbol{A}$ and if $j_1, \cdots, j_n$ are distinct elements of I such that $\{j_1, \cdots, j_n\}$ supports p, then there exists a unique n-place predicate P of $\boldsymbol{A}$ such that $P(j_1, \cdots, j_n) = p$.*

Proof. The predicate P is defined by

$$P(i_1, \cdots, i_n) = \mathbf{S}(j_1, \cdots, j_n/i_1, \cdots, i_n)p.$$

To prove that the function P is indeed a predicate, let τ be an arbitrary transformation on I and note that if $j \in \{j_1, \cdots, j_n\}$, then

$$\tau \cdot (j_1, \cdots, j_n/i_1, \cdots, i_n)j = (j_1, \cdots, j_n/\tau i_1, \cdots, \tau i_n)j.$$

Since, by assumption, $\{j_1, \cdots, j_n\}$ supports p, it follows that

$$\mathsf{S}(\tau)P(i_1, \cdots, i_n) = \mathsf{S}(\tau)\mathsf{S}(j_1, \cdots, j_n/i_1, \cdots, i_n)p$$
$$= \mathsf{S}(j_1, \cdots, j_n/\tau i_1, \cdots, \tau i_n)p = P(\tau i_1, \cdots, \tau i_n).$$

The uniqueness of P is an immediate consequence of (2.1), with

$$(j_1, \cdots, j_n/i_1, \cdots, i_n)$$

in the role of τ.

(2.4) COROLLARY. *Every element of* $\boldsymbol{A}$ *is in the range of some predicate of* $\boldsymbol{A}$.

Whereas (2.2) guarantees that a value of a predicate depends on no more variables than the visible ones, it may well happen that it depends on fewer. Suppose, for example, that Q is a 1-place (*unary*) predicate and write

$$P(i, j) = Q(i)$$

whenever i and j are variables. If τ is an arbitrary transformation on I, then

$$\mathsf{S}(\tau)P(i, j) = \mathsf{S}(\tau)Q(i) = Q(\tau i) = P(\tau i, \tau j),$$

so that P is a 2-place (*binary*) predicate. By (2.2), $\{i, j\}$ supports $P(i, j)$ for each i and j; since, however, $P(i, j) = Q(i)$, an application to (2.2) to Q shows that in fact $\{i\}$ also supports $P(i, j)$.

The preceding paragraph suggests a possible generalization of the concept of predicate: we may ask what happens if the roles of P and Q are interchanged. If, in other words, Q is a binary predicate, and if j is some particular variable, we may write

$$P(i) = Q(i, j)$$

for every variable i. If τ is a transformation on I such that $\tau j = j$, then

$$\mathsf{S}(\tau)P(i) = \mathsf{S}(\tau)Q(i, j) = Q(\tau i, j) = P(\tau i).$$

If, however, $\tau j \neq j$, then $\mathsf{S}(\tau)P(i)$ is not necessarily the same as $P(\tau i)$. If, in particular, $\tau = (i, j)$, then $\mathsf{S}(\tau)P(i) = Q(j, i)$, whereas $P(\tau i) = Q(j, j)$, and, in general, $Q(j, i) \neq Q(j, j)$. If, similarly, $\tau = (j/i)$, then $\mathsf{S}(\tau)P(j) = Q(i, i)$, whereas $P(\tau j) = Q(i, j)$, and, in general, $Q(i, i) \neq Q(i, j)$.

A study of this phenomenon shows that the appropriate general concept is defined as follows. Let J be an arbitrary finite subset of I, write $I^- = I - J$, and let $(\boldsymbol{A}^-, I^-, \mathsf{S}^-, \exists^-)$ be the polyadic algebra obtained from $\boldsymbol{A}$ by fixing the variables of J; define an n-place J-*predicate* of $\boldsymbol{A}$ as an n-place predicate of $\boldsymbol{A}^-$. Equivalently, an n-place J-predicate of $\boldsymbol{A}$ is a function P from $(I-J)^n$ into $\boldsymbol{A}$ with the following property: if τ^- is a transformation on $I-J$ and if τ is its canonical extension to I, then

$$\mathsf{S}(\tau)P(i_1, \cdots, i_n) = P(\tau i_1, \cdots, \tau i_n)$$

for every element $(i_1, \cdots, i_n)$ of $(I-J)^n$. The extent to which J-predicates generalize predicates is clear: the special concept is obtained from the general one by setting J equal to $\varnothing$. In other words, what was called a predicate before, is a $\varnothing$-predicate in the present terminology. Note also that the relation of J-predicates to predicates is the same as the relation of J-propositions to propositions.

The following result has the effect of reducing the theory of J-predicates to the theory of ordinary predicates.

(2.5) THEOREM. *Suppose that J is a finite subset of I; let $j_1, \cdots, j_m$ be the distinct elements of J. If Q is an $(n+m)$-place predicate, and if*

$$P(i_1, \cdots, i_n) = Q(i_1, \cdots, i_n, j_1, \cdots, j_m) \tag{2.6}$$

whenever $(i_1, \cdots, i_n) \in (I-j)^n$, then P is an n-place J-predicate. If, conversely, P is an n-place J-predicate, then there exists a unique $(n+m)$-place predicate Q such that (2.6) *holds for all $(i_1, \cdots, i_n)$ in $(I-J)^n$.*

Proof. The first conclusion is an immediate consequence of the definition of a J-predicate; it is necessary to prove the converse only. For this purpose, write $I^- = I-J$ and let $(A^-, I^-, \mathbf{S}^-, \exists^-)$ be the algebra obtained from $\boldsymbol{A}$ by fixing the variables of J. If $(i_1, \cdots, i_n) \in (I-J)^n$, then it follows from (2.2) that $\{i_1, \cdots, i_n\}$ supports $P(i_1, \cdots, i_n)$ in $\boldsymbol{A}^-$. This means that if $i \in I-J$ and $i \in' \{i_1, \cdots, i_n\}$, then $\exists(i)P(i_1, \cdots, i_n) = P(i_1, \cdots, i_n)$, and this, in turn, implies that $\{i_1, \cdots, i_n, j_1, \cdots, j_m\}$ supports $P(i_1, \cdots, i_n)$ in $\boldsymbol{A}$. If, in particular, $k_1, \cdots, k_n$ are distinct elements of I-J, then it follows from (2.3) that there exists a unique $(n+m)$-place predicate Q such that

$$P(k_1, \cdots, k_n) = Q(k_1, \cdots, k_n, j_1, \cdots, j_m).$$

Suppose now that $(i_1, \cdots, i_n) \in (I-J)^n$ and let τ^- be the transformation $(k_1, \cdots, k_n/i_1, \cdots, i_n)$ on $I-J$. If τ is the canonical extension of τ^- to I, then

$$\begin{aligned} Q(i_1, \cdots, i_n, j_1, \cdots, j_m) &= Q(\tau k_1, \cdots, \tau k_n, \tau j_1, \cdots, \tau j_m) \\ &= \mathbf{S}(\tau)Q(k_1, \cdots, k_n, j_1, \cdots, j_m) = \mathbf{S}(\tau)P(k_1, \cdots, k_n) \\ &= \mathbf{S}^-(\tau^-)P(k_1, \cdots, k_n) = P(\tau^- k_1, \cdots, \tau^- k_n) = P(i_1, \cdots i_n). \end{aligned}$$

This completes the proof of the theorem.

3. **Transformations of type** (M, N). The reason for restricting the discussion of polyadic algebras to algebras of infinite degree is that the availability of a large supply of variables often makes it possible to avoid unpleasant notational collisions. In this section we introduce a special class of transformations designed to exploit the availability of infinitely many variables; using these transformations we shall be able to avoid most collisions in an almost automatic manner.

Suppose that M and N are finite subsets of I. A transformation σ on I will be said to be of *type* (M, N) if

$$(3.1)\qquad \begin{cases} \sigma \text{ is one-to-one on } M, \\ \sigma M \subset I - (M \cup N), \\ \sigma = \delta \text{ outside } M. \end{cases}$$

If σ is of type (M, N), we shall denote by $\tilde{\sigma}$ the transformation that acts as the inverse of σ in σM and is equal to δ outside σM; in other words,

$$(3.2)\qquad \begin{cases} \tilde{\sigma}\sigma i = i & \text{when} \quad i \in M, \\ \tilde{\sigma} i = i & \text{when} \quad i \in' \sigma M. \end{cases}$$

A straightforward application of the definitions (3.1) and (3.2) shows that

$$(3.3)\qquad \tilde{\sigma}\sigma = \tilde{\sigma}.$$

When a transformation is used to avoid variable collisions, it is often necessary to prove that the result is independent of the particular choice of that transformation. The following lemma is the most useful tool in such unambiguity proofs.

(3.4) Lemma. *If both σ and τ are transformations of type (M, N), then there exists a permutation π on I such that*

$$(3.5)\qquad \pi\tau = \sigma\pi,$$

$$(3.6)\qquad \pi = \delta \quad \text{outside} \quad \tau M \cup \sigma M.$$

If π is a permutation satisfying (3.5) *and* (3.6), *then*

$$(3.7)\qquad \pi\tilde{\tau} = \tilde{\sigma}\pi.$$

Proof. Since $\sigma M - \tau M$ has the same (finite) number of elements as $\tau M - \sigma M$, there exists a one-to-one mapping π from $\sigma M - \tau M$ onto $\tau M - \sigma M$; extend π to I by writing

$$\pi = \sigma\tilde{\tau} \text{ on } \tau M \text{ and } \pi = \delta \text{ outside } \sigma M \cup \tau M.$$

(i) If $j \in M$, then $\pi\tau j = \sigma\tilde{\tau}\tau j = \sigma j$ and $\sigma\pi j = \sigma j$. (ii) If $j \in \tau M$, then $\pi\tau j = \pi j = \sigma\tilde{\tau} j$ and $\sigma\pi j = \sigma\sigma\tilde{\tau} j = \sigma\tilde{\tau} j$. (iii) If $j \in \sigma M - \tau M$, then $\pi\tau j = \pi j$ and $\sigma\pi j = \pi j$. (iv) If, finally, $j \in' M \cup \sigma M \cup \tau M$, then $\pi\tau j = \pi j = j$ and $\sigma\pi j = \sigma j = j$. This proves the first assertion of the lemma.

To prove the second assertion, suppose first that $j \in \tau M$, so that $j = \tau i$ with i in M. It follows that $\pi\tilde{\tau} j = \pi\tilde{\tau}\tau i = \pi i = i$ and $\tilde{\sigma}\pi j = \tilde{\sigma}\pi\tau i = \tilde{\sigma}\sigma\pi i = \tilde{\sigma}\sigma i = i$. If, on the other hand, $j \in' \tau M$, then $\pi j \in' \pi\tau M = \sigma\pi M = \sigma M$. Since $j \in' \tau M$, it follows that $\pi\tilde{\tau} j = \pi j$; since $\pi j \in' \sigma M$, it follows that $\tilde{\sigma}\pi j = \pi j$. This completes the proof of the lemma.

4. **Terms.** Propositions, J-propositions, predicates, and J-predicates all have to do with the elements of a polyadic algebra. In many treatments they are even denoted by similar symbols, namely by functional symbols that must be followed by a certain number of arguments before they take on a specific

meaning. They are thus, roughly speaking, the concepts whose symbols stand outside the parentheses in the functional notation. What are the things that stand inside those parentheses? By what, in other words, may we replace the variables that a J-proposition or a predicate depends on? One answer is: by constants; another answer is: by variables. There is still another answer. There is a concept that includes both constants and variables, a general concept that stands in the same relation to constants as J-propositions stand to propositions. The new concept is that of a term, or, more explicitly, a J-term.

To understand the definition of a term, it is advisable first to introduce the auxiliary concept of a J-constant, or, roughly speaking, a "variable constant." Suppose that J is a finite subset of I. If $I^- = I - J$, and if $\boldsymbol{A}^-$ is the I^--algebra obtained from $\boldsymbol{A}$ by fixing the variables of J, then a *J-constant* of $\boldsymbol{A}$ is defined to be a constant of the algebra $\boldsymbol{A}^-$. Observe that a $\varnothing$-constant is simply a constant.

If c is a J-constant, we shall denote the value of c at a subset K of $I - J$ by $\mathsf{S}(K/c)$. (This notation is different from the one used before for constants; the innovation has a very beneficial unifying and simplifying effect.) The symbol $\mathsf{S}(K/c)$ denotes thus a Boolean endomorphism of $\boldsymbol{A}$. Using this notation, and recalling the definitions of a constant and of the process of fixing variables, we may express the definition of a J-constant as follows. A J-constant is a mapping c that associates with every subset K of $I - J$ a Boolean endomorphism of $\boldsymbol{A}$, to be denoted by $\mathsf{S}(K/c)$, so that

(T1) $\quad \mathsf{S}(\varnothing/c)$ *is the identity mapping on* $\boldsymbol{A}$,

(T2) $$\mathsf{S}(H \cup K/c) = \mathsf{S}(H/c)\mathsf{S}(K/c),$$

(T3) $$\mathsf{S}(H/c)\exists(K) = \exists(K)\mathsf{S}(H - K/c),$$

(T4) $$\exists(H)\mathsf{S}(K/c) = \mathsf{S}(K/c)\exists(H - K),$$

(T5) $$\mathsf{S}(K/c)\mathsf{S}(\tau) = \mathsf{S}(\tau)\mathsf{S}(\tau^{-1}K/c),$$

whenever H and K are subsets of $I - J$ and τ is a transformation that lives on $I - J$.

We have already observed that every constant is a $\varnothing$-constant. Constants can also be used to obtain examples of J-constants; in fact it might be said that a constant is automatically a J-constant (for every finite set J). This is not quite accurate; what should be said is that the restriction of a constant to the subsets of $I - J$ is a J-constant. These examples are rather trivial. A nontrivial example is obtained from variables. If j is an element of I and if c_j denotes the mapping (from subsets of $I - \{j\}$ to Boolean endomorphisms of $\boldsymbol{A}$) that assigns to a subset K of $I - \{j\}$ the value $\mathsf{S}(K/j)$, then c_j is a j-constant (i.e., a $\{j\}$-constant); cf. (1.4)). The distinction between j and c_j is not always easy to maintain; we shall sometimes speak as if j itself were a j-constant.

If c is a J-constant, then $\mathbf{S}(K/c)$ is not defined for every subset K of I; this is the main point in which the concept of a J-term differs from that of a J-constant. A *J-term* (where J is a finite subset of I) is a mapping t that associates with every subset K of I a Boolean endomorphism $\mathbf{S}(K/t)$ of $\boldsymbol{A}$, so that the restriction of t to the subsets of $I-J$ is a J-constant of $\boldsymbol{A}$, and so that

$$\mathbf{S}(K/t)p = \mathbf{S}(\sigma K/t)\mathbf{S}(\sigma)p \tag{4.1}$$

whenever p is an element of $\boldsymbol{A}$ with finite support L, K is a finite subset of I, and σ is a transformation of type $(K, L \cup J)$. The reason for the restriction (4.1) can be illustrated by borrowing the notation from the subsequent theory of "terms in predicates." If P is a binary predicate and if $L = \{i, j\}$, then L supports $p = P(i, j)$. If $K = \{j, k\}$ (with $i \notin K$), and if σ is a transformation of type $(K, L \cup J)$, then $\sigma i = i$. Since both $\mathbf{S}(K/t)P(i,j)$ and $\mathbf{S}(\sigma K/t)\mathbf{S}(\sigma)P(i,j)$ are equal to $P(i, t)$, the assertion of equation (4.1) agrees with our intuitive expectations. The idea is simply this: to replace the variables of K by t in p, first transform them into some neutral variables far from the scene of action (and, in particular, not in J), and then replace those new variables by the J-constant that t reduces to out there.

It is sometimes useful to know that if J_1 and J_2 are finite subsets of I such that $J_1 \subset J_2$, then every J_1-term is a J_2-term. Suppose, indeed, that t is a J_1-term. Since $I - J_2 \subset I - J_1$, it follows that the restriction of t to the subsets of $I - J_2$ is a J_2-constant. If, moreover, K and L are finite subsets of I, then every transformation of type $(K, L \cup J_2)$ is automatically a transformation of type $(K, L \cup J_1)$; the assertion that t is a J_2-term now follows immediately from the definition.

The condition (4.1) in the definition of a J-term is very strong; *ab initio* it is not even clear that any nontrivial J-terms exist. Trivial examples can be obtained from constants. A constant c is a J-term (for every finite set J), but even this takes some proving. The proof of (4.1) runs as follows. If p has finite support L, if K is a finite subset of I, and if σ is a transformation of type $(K, L \cup J)$, then

$$\begin{aligned}
\mathbf{S}(\sigma K/c)\mathbf{S}(\sigma)p &= \mathbf{S}(\sigma)\mathbf{S}(\sigma^{-1}\sigma K/c)p \text{ [by (C5)]} \\
&= \mathbf{S}(\sigma)\mathbf{S}(K \cup \sigma K/c)p \text{ [by (3.1)]} = \mathbf{S}(\sigma)\mathbf{S}(K/c)\mathbf{S}(\sigma K/c)p \text{ [by (C2)]} \\
&= \mathbf{S}(\sigma)\mathbf{S}(K/c)p \text{ [by (3.1), (C3), and (C1)]} \\
&= \mathbf{S}(K/c)p \text{ [by (3.1) and (C4)]}.
\end{aligned}$$

The strength of the definition will be useful to us; once we know that something is a J-term, we shall be able to make powerful use of (4.1). We shall presently prove the existence of many J-terms; we shall show, in fact, that every J-constant has a unique extension to a J-term. For this purpose we turn now to a systematic study of some of the elementary properties of J-constants.

5. *J*-**constants.**

(5.1) LEMMA. *If c is a J-constant, if L supports p, and if $K \subset I-J$, then $J \cup (L-K)$ supports $\mathbf{S}(K/c)p$.*

Proof. If $i \epsilon' J \cup (L-K)$, then either $i \epsilon' J \cup K \cup L$ or $i \epsilon K$. If $i \epsilon' J \cup K \cup L$, then

$$\exists(i)\mathbf{S}(K/c)p = \mathbf{S}(K/c)\exists(i)p \text{ [by (T4)]} = \mathbf{S}(K/c)p;$$

if $i \epsilon K$, then

$$\exists(i)\mathbf{S}(K/c)p = \mathbf{S}(K/c)p \text{ [by (T4)]}.$$

(5.2) LEMMA. *If c is a J-constant, if L supports p, and if $K \subset I-J$, then*

$$\mathbf{S}(K/c)p = \mathbf{S}(K \cap L/c)p.$$

Proof. Since p is independent of $K-L$, it follows that

$$\mathbf{S}(K/c)p = \mathbf{S}(K/c)\exists(K - L)p,$$

and hence, by (T3), that

$$\mathbf{S}(K/c)p = \exists(K - L)\mathbf{S}(K \cap L/c)p.$$

By (5.1), $J \cup (L-K)$ supports $\mathbf{S}(K \cap L/c)p$; since this set is disjoint from $K-L$, it follows that $\mathbf{S}(K \cap L/c)p$ is independent of $K-L$, and hence that

$$\mathbf{S}(K/c)p = \mathbf{S}(K \cap L/c)p.$$

(5.3) LEMMA. *If c is a J-constant, if L is a finite support of p, if K is a finite subset of I, and if σ is a transformation of type $(K, L \cup J)$, then $J \cup (L-K)$ supports $\mathbf{S}(\sigma K/c)\mathbf{S}(\sigma)p$.*

Proof. Since L supports p, it follows that σL supports $\mathbf{S}(\sigma)p$, and therefore, by (5.1), that $J \cup (\sigma L-\sigma K)$ supports $\mathbf{S}(\sigma K/c)\mathbf{S}(\sigma)p$. Since

$$\sigma L = \sigma(L \cap K) \cup \sigma(L - K) \subset \sigma K \cup (L - K),$$

it follows that

$$\sigma L - \sigma K \subset L - K.$$

(5.4) LEMMA. *If c is a J-constant, if L is a finite support of p, if K is a finite subset of I, and if both σ and τ are transformations of type $(K, L \cup J)$, then*

$$\mathbf{S}(\sigma K/c)\mathbf{S}(\sigma)p = \mathbf{S}(\tau K/c)\mathbf{S}(\tau)p.$$

Proof. Apply (3.4) to find a permutation π with the properties there described, and replace $\mathbf{S}(\tau)$ in $\mathbf{S}(\tau K/c)\mathbf{S}(\tau)p$ by $\mathbf{S}(\pi^{-1})\mathbf{S}(\sigma)\mathbf{S}(\pi)$. Since $\pi=\delta$ in L and since L supports p, the factor $\mathbf{S}(\pi)$ may be erased from the result. Since π^{-1} lives on $I-J$ (in fact on $I-(L \cup J)$), (T5) is applicable; it yields

$$\mathbf{S}(\tau K/c)\mathbf{S}(\pi^{-1}) = \mathbf{S}(\pi^{-1})\mathbf{S}(\pi\tau K/c).$$

Since $\pi\tau K = \sigma\pi K = \sigma K$, it follows that

$$\mathbf{S}(\tau K/c)\mathbf{S}(\tau)p = \mathbf{S}(\pi^{-1})\mathbf{S}(\sigma K/c)\mathbf{S}(\sigma)p.$$

Since, by (5.3), $J \cup L$ supports $\mathbf{S}(\sigma K/c)\mathbf{S}(\sigma)p$, and since $\pi^{-1} = \delta$ on $J \cup L$, the proof is complete.

(5.5) Lemma. *If c is a J-constant, if L_1 and L_2 are finite supports of p, if K is a finite subset of I, and if σ_1 and σ_2 are transformations of types $(K, L_1 \cup J)$ and $(K, L_2 \cup J)$ respectively, then*

$$\mathbf{S}(\sigma_1 K/c)\mathbf{S}(\sigma_1)p = \mathbf{S}(\sigma_2 K/c)\mathbf{S}(\sigma_2)p.$$

Proof. If $L = L_1 \cup L_2$, then L is a finite support of p; let τ be an arbitrary transformation of type $(K, L \cup J)$. (The proof that such a τ exists uses the assumption that I is infinite.) It follows that τ is a transformation of type $(K, L_1 \cup J)$ and also a transformation of type $(K, L_2 \cup J)$; the desired result is now an immediate consequence of (5.4).

6. The extension of J-constants. Since a constant is determined by its values on finite sets, the same is true of a J-constant, and, therefore, of a J-term. For convenience of reference, we restate this informal comment as a lemma; the proof of the lemma is an easy application of the corresponding result (1.1) for constants.

(6.1) Lemma. *If J is a finite subset of I and if t is a mapping that associates with every finite subset K of I a Boolean endomorphism $\mathbf{S}(K/t)$ of $\mathbf{A}$ so that* (T1)–(T5) *are satisfied for finite subsets H and K of $I - J$ and for finite transformations τ that live on $I - J$ and so that $\mathbf{S}(K/t) = \mathbf{S}(\sigma K/t)\mathbf{S}(\sigma)p$ whenever p is an element of $\mathbf{A}$ with finite support L, K is a finite subset of I, and σ is a transformation of type $(K, L \cup J)$, then there exists a unique J-term $\bar{t}$ such that $\mathbf{S}(K/\bar{t}) = \mathbf{S}(K/t)$ for all finite subsets K of I.*

Once this result is available, there is not much point in maintaining the distinction between t and $\bar{t}$; in the sequel we shall use the same symbol (namely t) for both, and, correspondingly, we shall say that t *is* a J-term (instead of saying that t can be extended to infinite sets so as to become one).

Suppose now that c is a J-constant; it is natural to try to extend c to a J-term as follows. Given any element p of A and any finite subset K of I, let L be a finite support of p, let σ be a transformation of type $(K, L \cup J)$, and write

(6.2) $$\mathbf{S}(K/t)p = \mathbf{S}(\sigma K/c)\mathbf{S}(\sigma)p.$$

It follows from (5.5) that this definition is unambiguous, i.e., that it does not depend on the particular choice of L and of σ. We indicate the dependence of t on c by writing

(6.3) $$t = \bar{c};$$

we proceed to prove that t has the properties it should have.

(6.4) LEMMA. *If c is a J-constant, if $t = \bar{c}$, and if K is a finite subset of I, then $\mathbf{S}(K/t)$ is a Boolean endomorphism of A.*

Proof. If $p \in A$, and if L is a finite support of p, then L supports p' also; it follows that if σ is a transformation of type $(K, L \cup J)$, then

$$\mathbf{S}(K/t)p' = \mathbf{S}(\sigma K/c)\mathbf{S}(\sigma)p' = (\mathbf{S}(\sigma K/c)\mathbf{S}(\sigma)p)' = (\mathbf{S}(K/t)p)'.$$

If p and q are in A, let L be a finite set that supports both of them, and let σ be a transformation of type $(K, L \cup J)$. Since L supports $p \vee q$, it follows that

$$\begin{aligned}\mathbf{S}(K/t)(p \vee q) &= \mathbf{S}(\sigma K/c)\mathbf{S}(\sigma)(p \vee q)\\ &= \mathbf{S}(\sigma K/c)\mathbf{S}(\sigma)p \vee \mathbf{S}(\sigma K/c)\mathbf{S}(\sigma)q = \mathbf{S}(K/t)p \vee \mathbf{S}(K/t)q.\end{aligned}$$

(6.5) LEMMA. *If c is a J-constant, if $t = \bar{c}$, and if K is a finite subset of $I - J$, then $\mathbf{S}(K/t) = \mathbf{S}(K/c)$.*

Proof. This is similar to the proof that a constant is a J-term. If $p \in A$, if L is a finite support of p, and if σ is a transformation of type $(K, L \cup J)$, then

$$\begin{aligned}\mathbf{S}(K/t)p &= \mathbf{S}(\sigma K/c)\mathbf{S}(\sigma)p \text{ [by (6.2)]} = \mathbf{S}(\sigma)\mathbf{S}(\sigma^{-1}\sigma K/c)p \text{ [by (T5)]}\\ &= \mathbf{S}(\sigma)\mathbf{S}(K \cup \sigma K/c)p \text{ [by (3.1)]} = \mathbf{S}(\sigma)\mathbf{S}(K/c)\mathbf{S}(\sigma K/c)p \text{ [by (T2)]}\\ &= \mathbf{S}(\sigma)\mathbf{S}(K/c)p \text{ [by (3.1), (T3), and (T1)]} = \mathbf{S}(K/c)p\\ &\qquad \text{[by (3.1) and (T4)]}.\end{aligned}$$

(6.6) LEMMA. *If c is a J-constant, if $t = \bar{c}$, if p is an element of A with finite support L, if K is a finite subset of I, and if σ is a transformation of type $(K, L \cup J)$, then*

$$\mathbf{S}(K/t)p = \mathbf{S}(\sigma K/t)\mathbf{S}(\sigma)p.$$

Proof. By definition, $\mathbf{S}(K/t)p = \mathbf{S}(\sigma K/c)\mathbf{S}(\sigma)p$; the desired result follows from (6.5) (with σK in the role of K).

(6.7) THEOREM. *If c is a J-constant, then there exists a unique J-term t (namely, $t = \bar{c}$) such that*

$$\mathbf{S}(K/t) = \mathbf{S}(K/c)$$

whenever $K \subset I - J$.

Proof. The existence of t is contained in the preceding four lemmas. Uniqueness follows from the fact that a term is uniquely determined by its values at finite sets, together with the fact that the value of a J-term at each finite set can be calculated, via (4.1), from its values on finite subsets of $I-J$.

7. **Properties of J-terms.** If t is a J-term, let c be the J-constant obtained by restricting t to the subsets of $I-J$; we shall indicate the dependence of c on t by writing

$$c = t_J. \tag{7.1}$$

The mapping that sends t onto t_J is a one-to-one mapping from all J-terms onto all J-constants; its inverse is the mapping that sends c onto $\bar{c}$ (cf. (6.3)). In other words, if t is a J-term, and if $c=t_J$, then $t=\bar{c}$; similarly, if c is a J-constant, and if $t=\bar{c}$, then $c=t_J$. All the properties of J-terms can be studied by means of this correspondence between J-terms and J-constants.

(7.2) LEMMA. *If t is a J-term, then $\mathbf{S}(\varnothing/t)$ is the identity mapping on $\boldsymbol{A}$.*

Proof. By (T1), $\mathbf{S}(\varnothing/t_J)p=p$ for all p; the assertion follows from the fact that if $K\subset I-J$ (and, in particular, if $K=\varnothing$), then $\mathbf{S}(K/t)=\mathbf{S}(K/t_J)$.

(7.3) LEMMA. *If t is a J-term and if $K\subset I$, then*

$$\mathbf{S}(K/t) = \mathbf{S}(K-J/t)\mathbf{S}(K\cap J/t).$$

Proof. Suppose first that K is finite. Let p be an element of $\boldsymbol{A}$ with finite support L and let τ be a transformation of type $(K, L\cup J)$. If $\rho=\tau$ on $K-J$ and $\rho=\delta$ outside $K-J$, and if $\sigma=\tau$ on $K\cap J$ and $\sigma=\delta$ outside $K\cap J$, then the transformations ρ and σ are of types $(K-J, L\cup J)$ and $(K\cap J, L\cup J)$ respectively, and $\rho\sigma=\tau$. By (5.3), $J\cup L$ supports $q=\mathbf{S}(\sigma(K\cap J)/t)\mathbf{S}(\sigma)p$; it follows that

$$\begin{aligned}
\mathbf{S}(K-J/t)\mathbf{S}(K\cap J/t)p &= \mathbf{S}(K-J/t)q\\
&= \mathbf{S}(\rho(K-J)/t)\mathbf{S}(\rho)\mathbf{S}(\sigma(K\cap J)/t)\mathbf{S}(\sigma)p\\
&= \mathbf{S}(\rho(K-J)/t)\mathbf{S}(\sigma(K\cap J)/t)\mathbf{S}(\rho)\mathbf{S}(\sigma)p \text{ [by (T5)]}\\
&= \mathbf{S}(\tau(K-J)/t)\mathbf{S}(\tau(K\cap J)/t)\mathbf{S}(\tau)p\\
&= \mathbf{S}(\tau K/t)\mathbf{S}(\tau)p \text{ [by (T2)]} = \mathbf{S}(K/t)p.
\end{aligned}$$

In the general case, let u be the mapping from subsets K of I into operators $\mathbf{S}(K/u)$ on $\boldsymbol{A}$ defined by

$$\mathbf{S}(K/u) = \mathbf{S}(K-J/t)\mathbf{S}(K\cap J/t).$$

Clearly (i) $\mathbf{S}(K/u)$ is a Boolean endomorphism of $\boldsymbol{A}$ for each subset K of I. If $K\subset I-J$, then

$$\mathbf{S}(K/u) = \mathbf{S}(K/t)\mathbf{S}(\varnothing/t) = \mathbf{S}(K/t) \text{ [by (7.2)]};$$

this implies that (ii) the restriction of u to the subsets of $I-J$ is a J-constant (namely t_J). Finally (iii) if p is an element of $\boldsymbol{A}$ with finite support L, if K is a finite subset of I, and if σ is a transformation of type $(K, L\cup J)$, then

$$\begin{aligned}\mathbf{S}(K/u)p &= \mathbf{S}(K - J/t)\mathbf{S}(K \cap J/t) = \mathbf{S}(K/t) \text{ [by the preceding paragraph]}\\ &= \mathbf{S}(\sigma K/t)\mathbf{S}(\sigma)p = \mathbf{S}(\sigma K - J/t)\mathbf{S}(\sigma K \cap J/t)\mathbf{S}(\sigma)p\\ &= \mathbf{S}(\sigma K/u)\mathbf{S}(\sigma)p.\end{aligned}$$

The assertions (i), (ii), and (iii) add up to the fact that u is a J-term; since $u_J = t_J$, it follows that $u = t$.

(7.4) Lemma. *If t is a J-term and if H and K are subsets of I, then*

$$\mathbf{S}(H \cup K/t) = \mathbf{S}(H - J/t)\mathbf{S}(K - J/t)\mathbf{S}((H \cup K) \cap J/t).$$

Proof. By (7.3)

$$\mathbf{S}(H \cup K/t) = \mathbf{S}((H \cup K) - J/t)\mathbf{S}((H \cup K) \cap J/t);$$

the conclusion follows from (T2).

(7.5) Lemma. *If t is a J-term, if L supports p, and if $K \subset I$, then $J \cup (L - K)$ supports $\mathbf{S}(K/t)p$.*

Proof. Let L_0 be a finite support of p such that $L_0 \subset L$, and let σ be a transformation of type $(K \cap J,\ L \cup J)$. Since, by (7.3),

$$\mathbf{S}(K/t)p = \mathbf{S}(K - J/t)\mathbf{S}(K \cap J/t)p = \mathbf{S}(K - J/t)\mathbf{S}(\sigma(K \cap J)/t)\mathbf{S}(\sigma)p,$$

and since, by (5.3), the set

$$J \cup (L_0 - (K \cap J)) \quad (= J \cup (L_0 - K) \cup (L_0 - J))$$

supports $\mathbf{S}(\sigma(K \cap J)/t)\mathbf{S}(\sigma)p$, it follows, again by (5.3), that the set

$$J \cup [((L_0 - K) \cup (L_0 - J)) - (K - J)]$$

supports $\mathbf{S}(K/t)p$. The conclusion follows from the fact that

$$((L_0 - K) \cup (L_0 - J)) - (K - J) \subset L_0 - K \subset L - K.$$

(7.6) Lemma. *If t is a J-term, if L supports p, and if $K \subset I$, then*

$$\mathbf{S}(K/t)p = \mathbf{S}(K \cap L/t)p.$$

Proof. Suppose first that $K \subset J$. Let L_0 be a finite support of p such that $L_0 \subset L$ and write $L_1 = L_0 \cup (K \cap L)$. It follows that L_1 is a finite support of p and that $K \cap L_1 = K \cap L$. Consequently it is sufficient, in this case, to prove the conclusion under the added assumption that L is finite. We make this assumption. Let σ be a transformation of type $(K,\ L \cup J)$; write $\tau = \sigma$ on $K \cap L$ and $\tau = \delta$ outside $K \cap L$. It follows that τ is a transformation of type $(K \cap L,\ L \cup J)$, and hence that

$$\mathbf{S}(K \cap L/t)p = \mathbf{S}(\tau(K \cap L)/t)\mathbf{S}(\tau)p.$$

We have, by (T2),

$$\mathsf{S}(K/t)p = \mathsf{S}(\sigma K/t)\mathsf{S}(\sigma)p = \mathsf{S}(\sigma(K \cap L)/t)\mathsf{S}(\sigma(K - L)/t)\mathsf{S}(\sigma)p.$$

Since σL supports $\mathsf{S}(\sigma)p$ and is disjoint from $\sigma(K-L)$, we have also

$$\begin{aligned}\mathsf{S}(\sigma(K - L)/t)\mathsf{S}(\sigma)p &= \mathsf{S}(\sigma(K - L)/t)\exists(I - \sigma L)\mathsf{S}(\sigma)p \\ &= \mathsf{S}(\sigma(K - L)/t)\exists((I - J) - \sigma L)\exists(J \cap \sigma L)\mathsf{S}(\sigma)p \\ &= \exists((I - J) - \sigma L)\exists(J \cap \sigma L)\mathsf{S}(\sigma)p \text{ [by (T3) and (T1)]} \\ &= \exists(I - \sigma L)\mathsf{S}(\sigma)p = \mathsf{S}(\sigma)p.\end{aligned}$$

The preceding three equations together imply that

$$\begin{aligned}\mathsf{S}(K/t)p &= \mathsf{S}(\sigma(K \cap L)/t)\mathsf{S}(\sigma)p \\ &= \mathsf{S}(\tau(K \cap L)/t)\mathsf{S}(\tau)p \text{ [since } \tau = \sigma \text{ on } L] \\ &= \mathsf{S}(K \cap L/t)p.\end{aligned}$$

In the general case

$$\begin{aligned}\mathsf{S}(K/t)p &= \mathsf{S}(K - J/t)\mathsf{S}(K \cap J/t)p \text{ [by (7.3)]} \\ &= \mathsf{S}(K - J/t)\mathsf{S}(K \cap J \cap L/t)p \text{ [by the preceding paragraph]} \\ &= \mathsf{S}((K - J) \cap L/t)\mathsf{S}(K \cap J \cap L/t)p \text{ [by (5.1)]} \\ &= \mathsf{S}(K \cap L/t)p \text{ [by (7.3)]};\end{aligned}$$

this completes the proof of Lemma (7.6).

(7.7) LEMMA. *If t is a J-term and if H and K are subsets of I, then*

$$\mathsf{S}(H/t)\exists(K) = \exists(K - J)\mathsf{S}(H - K/t)\exists(K \cap J).$$

Proof. We shall first prove the special case of the lemma in which $H \subset J$ and K is a finite subset of $I-J$; in that case the desired conclusion takes the form

$$\mathsf{S}(H/t)\exists(K) = \exists(K)\mathsf{S}(H/t). \tag{7.8}$$

Let p be an element of A with finite support L, and let σ be a transformation of type $(H, L \cup J \cup K)$. It follows that σ is of type $(H, L \cup J)$, and therefore

$$\mathsf{S}(H/t)p = \mathsf{S}(\sigma H/t)\mathsf{S}(\sigma)p$$

and

$$\mathsf{S}(H/t)\exists(K)p = \mathsf{S}(\sigma H/t)\mathsf{S}(\sigma)\exists(K)p.$$

Since $K \cap L \subset K \subset I-J$, and since σH is disjoint from K, it follows that σ lives outside $K \cap L$, and hence that

$$\mathsf{S}(\sigma)\exists(K \cap L) = \exists(K \cap L)\mathsf{S}(\sigma).$$

The same reasons imply, via (T3), that

$$\mathbf{S}(\sigma H/t)\exists(K \cap L) = \exists(K \cap L)\mathbf{S}(\sigma H/t).$$

It follows, since $\exists(K)p = \exists(K \cap L)p$, that

$$\mathbf{S}(H/t)\exists(K)p = \exists(K \cap L)\mathbf{S}(\sigma H/t)\mathbf{S}(\sigma)p.$$

Since $J \cup L$ supports $\mathbf{S}(H/t)p$ and since $K-L$ is disjoint from $J \cup L$, it follows that $\exists(K-L)$ leaves $\mathbf{S}(H/t)p$ invariant, and hence that

$$\mathbf{S}(H/t)\exists(K)p = \exists(K)\mathbf{S}(\sigma H/t)\mathbf{S}(\sigma)p = \exists(K)\mathbf{S}(H/t)p,$$

as desired.

Suppose next that K is an arbitrary (not necessarily finite) subset of $I-J$. If p is an element of $\boldsymbol{A}$ with finite support L, and if $K_0 = K \cap L$, then

$$\begin{aligned}\mathbf{S}(H/t)\exists(K)p &= \mathbf{S}(H/t)\exists(K \cap L)p = \mathbf{S}(H/t)\exists(K_0)p \\ &= \exists(K_0)\mathbf{S}(H/t)p \ [\text{by } (7.8)] = \exists(K \cap L)\mathbf{S}(H/t)p.\end{aligned}$$

(We are still assuming that $H \subset J$.) Since $J \cup L$ supports $\mathbf{S}(H/t)p$ and is disjoint from $K-L$, it follows that $\exists(K-L)$ leaves $\mathbf{S}(H/t)p$ invariant; this, together with the preceding equation, implies that

$$\mathbf{S}(H/t)\exists(K) = \exists(K)\mathbf{S}(H/t) \tag{7.9}$$

whenever $H \subset J$ and $K \subset I-J$.

We assert next that if H and K are quite arbitrary subsets of I, then

$$\mathbf{S}(H/t)\exists(K) = \exists(K-J)\mathbf{S}(H/t)\exists(K \cap J). \tag{7.10}$$

Indeed:

$$\begin{aligned}\mathbf{S}(H/t)\exists(K) &= \mathbf{S}(H-J/t)\mathbf{S}(H \cap J/t)\exists(K-J)\exists(K \cap J) \ [\text{by } (7.3)] \\ &= \mathbf{S}(H-J/t)\exists(K-J)\mathbf{S}(H \cap J/t)\exists(K \cap J) \ [\text{by } (7.9)] \\ &= \exists(K-J)\mathbf{S}(H-J/t)\mathbf{S}(H \cap J/t)\exists(K \cap J) \ [\text{by } (\text{T}3)] \\ &= \exists(K-J)\mathbf{S}(H/t)\exists(K \cap J) \ [\text{by } (7.3)].\end{aligned}$$

Replacing H by $H-K$ in (7.10) we obtain

$$\mathbf{S}(H-K/t)\exists(K) = \exists(K-J)\mathbf{S}(H-K/t)\exists(K \cap J). \tag{7.11}$$

Since $I-K$ supports $\exists(K)p$ for every p in $\boldsymbol{A}$, it follows that

$$\mathbf{S}(H/t)\exists(K)p = \mathbf{S}(H-K/t)\exists(K)p \ [\text{by } (7.6)]; \tag{7.12}$$

(7.10), (7.12), and (7.11) imply the desired conclusion.

(7.13) LEMMA. *If t is a J-term and if H and K are subsets of I, then*

$$\exists(H)\mathbf{S}(K/t) = \exists(H \cap J)\mathbf{S}(K/t)\exists(H-(J \cup K)).$$

Proof. Compute, as follows:

$$\begin{aligned}\exists(H)\mathbf{S}(K/t) &= \exists(H \cap J)\exists(H - J)\mathbf{S}(K - J/t)\mathbf{S}(K \cap J/t) \ [\text{by } (7.3)] \\ &= \exists(H \cap J)\mathbf{S}(K - J/t)\exists((H - J) - (K - J))\mathbf{S}(K \cap J/t) \ [\text{by (T4)}] \\ &= \exists(H \cap J)\mathbf{S}(K - J/t)\exists(H - (J \cup K))\mathbf{S}(K \cap J/t) \\ &= \exists(H \cap J)\mathbf{S}(K - J/t)\mathbf{S}(K \cap J/t)\exists(H - (J \cup K)) \ [\text{by } (7.8)] \\ &= \exists(H \cap J)\mathbf{S}(K/t)\exists(H - (J \cup K)) \ [\text{by } (7.3)].\end{aligned}$$

8. **Terms and transformations.** In the preceding section we discussed what might be called the commutation relations involving terms and the quantifier structure of A; in this section we do the same thing for the transformation structure of A.

It is convenient to begin by introducing a new item of notation and making one or two elementary remarks on it. If τ is a transformation on I and if J is a subset of I, we shall denote by τ_J the transformation such that

$$\tau_J = \tau \text{ on } \tau^{-1}J \quad \text{and} \quad \tau_J = \delta \text{ outside } \tau^{-1}J.$$

Observe that if τ happens to live on J, then $\tau^{-1}J = J$, and therefore $\tau_J = \tau$ and $\tau_{I-J} = \delta$; if, on the other hand, τ lives on $I - J$, then $\tau_J = \delta$ and $\tau_{I-J} = \tau$.

(8.1) LEMMA. *If $\tau = \delta$ on J, then τ_{I-J} lives on $I - J$ and $\tau_{I-J}\tau_J = \tau$.*

Proof. If $i \in J$, then $\tau i = i \in J$, and therefore $\tau_{I-J}i = i$. If $i \in' J$, then either $\tau i \in J$, in which case $\tau_{I-J}i = i \in' J$, or $\tau i \in' J$, in which case $\tau_{I-J}i = \tau i \in' J$. This proves that τ_{I-J} lives on $I - J$. To prove the asserted equation, observe that if $i \in \tau^{-1}J$, then (since $\tau = \delta$ on J) $\tau\tau i = \tau i \in J$ and therefore $\tau_{I-J}\tau_J i = \tau_{I-J}\tau i = \tau i$, while if $i \in' \tau^{-1}J$, then $\tau_{I-J}\tau_J i = \tau_{I-J} i = \tau i$.

(8.2) LEMMA. *If t is a J-term, if τ is a transformation such that $\tau = \delta$ on J, and if $K \subset I$, then*

$$\mathbf{S}(K/t)\mathbf{S}(\tau) = \mathbf{S}(\tau_{I-J})\mathbf{S}((\tau_{I-I})^{-1}K/t)\mathbf{S}(\tau_J).$$

Proof. Suppose first that the transformation τ is finite, and let H be a finite set such that τ lives on H. Given an element p of A, let L be a finite support of p. If σ is a transformation of type $(K \cap J, L \cup J \cup H)$, then, since τL supports $\mathbf{S}(\tau)p$ and since $\tau L \subset L \cup J \cup H$, it follows (via (7.3)) that

$$\mathbf{S}(K/t)\mathbf{S}(\tau)p = \mathbf{S}(K - J/t)\mathbf{S}(\sigma(K \cap J)/t)\mathbf{S}(\sigma)\mathbf{S}(\tau)p.$$

Since $\tau_{I-J}i$ is always equal either to τi or to i, it follows that τ_{I-J} lives on H (along with τ), and hence (by (8.1)) that it lives on $H \cap (I - J) = H - J$. Since σ lives on $(K \cap J) \cup \sigma(K \cap J)$, and since that set is disjoint from $H - J$, it follows that σ commutes with τ_{I-J}. Consequently (via another use of (8.1))

$$\begin{aligned}\mathbf{S}(K/t)\mathbf{S}(\tau)p &= \mathbf{S}(K - J/t)\mathbf{S}(\sigma(K \cap J)/t)\mathbf{S}(\tau_{I-J})\mathbf{S}(\sigma)\mathbf{S}(\tau_J)p \\ &= \mathbf{S}(K - J/t)\mathbf{S}(\tau_{I-J})\mathbf{S}((\tau_{I-J})^{-1}\sigma(K \cap J)/t)\mathbf{S}(\sigma)\mathbf{S}(\tau_J)p.\end{aligned} \tag{8.3}$$

Since τ_{I-J} lives on $I - J$ and since $K - J \subset I - J$, it follows that $(\tau_{I-J})^{-1}(K - J)$

$\subset I-J$. Since τ_{I-J} lives on H and since $\sigma(K \cap J) \subset I-H$, it follows that $(\tau_{I-J})^{-1}\sigma(K \cap J) = \sigma(K \cap J)$. These two remarks and (8.3) imply that

$$(8.4)\quad \mathbf{S}(K/t)\mathbf{S}(\tau)p = \mathbf{S}(\tau_{I-J})\mathbf{S}((\tau_{I-J})^{-1}(K-J)/t)\mathbf{S}(\sigma(K \cap J)/t)\mathbf{S}(\sigma)\mathbf{S}(\tau_J)p.$$

Observe next that (since $\tau = \delta$ on J)

$$(\tau_{I-J})^{-1}K - J = (\tau_{I-J})^{-1}(K-J) \quad \text{and} \quad (\tau_{I-J})^{-1}K \cap J = K \cap J.$$

Since $\tau_J L$ supports $\mathbf{S}(\tau_J)p$ and $\tau_J L \subset L \cup J \cup H$, it follows from (8.4) that

$$\mathbf{S}(K/t)\mathbf{S}(\tau)p = \mathbf{S}(\tau_{I-J})\mathbf{S}((\tau_{I-J})^{-1}K - J/t)\mathbf{S}((\tau_{I-J})^{-1}K \cap J/t)\mathbf{S}(\tau_J)p;$$

the proof (in the case that τ is finite) is complete.

Suppose next that τ lives on $I-J$ and that $K \subset J$; in this case the desideratum takes the form

$$(8.5)\qquad \mathbf{S}(K/t)\mathbf{S}(\tau) = \mathbf{S}(\tau)\mathbf{S}(\tau^{-1}K/t).$$

Given an element p of A, let L be a finite support of p; write $\rho = \tau$ on L and $\rho = \delta$ outside L. The transformation ρ is finite and $\mathbf{S}(\tau)p = \mathbf{S}(\rho)p$. Since ρ lives on $I-J$ (along with τ), it follows that $\rho_{I-J} = \rho$ and $\rho_J = \delta$; the preceding paragraph implies therefore that

$$\mathbf{S}(K/t)\mathbf{S}(\tau)p = \mathbf{S}(K/t)\mathbf{S}(\rho)p = \mathbf{S}(\rho)\mathbf{S}(\rho^{-1}K/t)p.$$

Since $L \cup J$ supports $\mathbf{S}(\rho^{-1}K/t)p$ and since $\rho = \tau$ on $L \cup J$, it follows that

$$\mathbf{S}(K/t)\mathbf{S}(\tau)p = \mathbf{S}(\tau)\mathbf{S}(\rho^{-1}K/t)p.$$

By (7.6),

$$\mathbf{S}(\tau^{-1}K/t)p = \mathbf{S}(\tau^{-1}K \cap L/t)p \quad \text{and} \quad \mathbf{S}(\rho^{-1}K/t)p = \mathbf{S}(\rho^{-1}K \cap L/t)p.$$

Now $\tau^{-1}K \cap L = \rho^{-1}K \cap L$ (by the definition of ρ), and the proof of (8.5) is complete.

The general case is deduced from the special case just treated, as follows:

$$\begin{aligned}
\mathbf{S}(K/t)\mathbf{S}(\tau) &= \mathbf{S}(K-J/t)\mathbf{S}(K \cap J/t)\mathbf{S}(\tau_{I-J})\mathbf{S}(\tau_J) \ \text{[by (7.3) and (8.1)]}\\
&= \mathbf{S}(K-J/t)\mathbf{S}(\tau_{I-J})\mathbf{S}((\tau_{I-J})^{-1}(K \cap J)/t)\mathbf{S}(\tau_J) \ \text{[by (8.5)]}\\
&= \mathbf{S}(\tau_{I-J})\mathbf{S}((\tau_{I-J})^{-1}(K-J)/t)\mathbf{S}((\tau_{I-J})^{-1}(K \cap J)/t)\mathbf{S}(\tau_J) \ \text{[by (T5)]}\\
&= \mathbf{S}(\tau_{I-J})\mathbf{S}((\tau_{I-J})^{-1}K - J/t)\mathbf{S}((\tau_{I-J})^{-1}K \cap J/t)\mathbf{S}(\tau_J) \ \text{[by (8.1)]}\\
&= \mathbf{S}(\tau_{I-J})\mathbf{S}((\tau_{I-J})^{-1}K/t)\mathbf{S}(\tau_J) \ \text{[by (7.3)]}.
\end{aligned}$$

This concludes the proof of (8.2).

(8.6) Lemma. *If t is a J-term, if τ is a transformation such that $\tau = \delta$ on J and $\tau = \delta$ outside $\tau^{-1}J$, and if $H \subset I - \tau^{-1}J$, then $\mathbf{S}(H/t)$ and $\mathbf{S}(\tau)$ commute.*

Proof. Assume first that the transformation τ is finite; it follows that the set $\tau^{-1}J - J$ is finite. If that set is empty, then $\tau = \delta$ and the conclusion is trivial. In the remaining case, let $j_1, \cdots, j_n$ be the distinct elements of $\tau^{-1}J-J$; it is easy to verify that

$$\tau = (j_1/\tau j_1) \cdots (j_n/\tau j_n),$$

and that the factors of this product commute among themselves. Since the sets $\{j_1\}, \cdots, \{j_n\}$, and H are pairwise disjoint, it follows from the commutativity theorem (1.3) for constants (applied to the J-constants induced by the variables $\tau j_1, \cdots, \tau j_n$ and by the J-term t) that the Boolean endomorphisms $\mathbf{S}(j_1/\tau j_1), \cdots, \mathbf{S}(j_n/\tau j_n)$, and $\mathbf{S}(H/t)$ commute among themselves. Since

$$\mathbf{S}(\tau) = \mathbf{S}(j_1/\tau j_1) \cdots \mathbf{S}(j_n/\tau j_n),$$

the desideratum is proved in this case.

In the general case we argue separately for each element of A. Given p in A, let L be a finite support of p; write $\tau_0=\tau$ on L and $\tau_0=\delta$ outside L. Since $\tau_0=\delta$ whenever $\tau=\delta$, it follows that $\tau_0=\tau$ on $J\cup L$, and hence that $\tau_0=\tau$ on a support of $\mathbf{S}(H/t)p$. The preceding paragraph implies that

$$\mathbf{S}(H/t)\mathbf{S}(\tau)p = \mathbf{S}(H/t)\mathbf{S}(\tau_0)p = \mathbf{S}(\tau_0)\mathbf{S}(H/t)p = \mathbf{S}(\tau)\mathbf{S}(H/t)p;$$

this completes the proof of the lemma.

(8.7) LEMMA. *If t is a J-term, if τ is a transformation such that $\tau=\delta$ on J, and if $K\subset I-\tau^{-1}J$, then*

$$\mathbf{S}(K/t)\mathbf{S}(\tau) = \mathbf{S}(\tau)\mathbf{S}(\tau^{-1}K/t).$$

Proof. The assumptions on τ and K imply that

$$(\tau_{I-J})^{-1}K = \tau^{-1}K. \tag{8.8}$$

Indeed, if $\tau i\in K$ (i.e., $i\in\tau^{-1}K$), then $i\in\tau^{-1}(I-J)$, so that $\tau_{I-J}i=\tau i$, and therefore $\tau_{I-J}i\in K$. If, on the other hand, $\tau_{I-J}i\in K$, then $\tau i\in I-J$ (i.e., $i\in\tau^{-1}(I-J)$). The reason is that if $\tau i\notin I-J$, then $\tau_{I-J}i=i$, so that $i\in K$ and $\tau i\in J$; this contradicts the fact that $K\cap\tau^{-1}J=\varnothing$. It follows that $\tau_{I-J}i=\tau i$, and, therefore, that $\tau i\in K$.

By (8.2) and (8.8),

$$\mathbf{S}(K/t)\mathbf{S}(\tau) = \mathbf{S}(\tau_{I-J})\mathbf{S}(\tau^{-1}K/t)\mathbf{S}(\tau_J). \tag{8.9}$$

Since $K\subset I-\tau^{-1}J\subset I-J$, it follows that $\tau^{-1}K\subset I-\tau^{-1}J$, so that (8.6) is applicable to τ_J and $\tau^{-1}K$ in place of τ and H, respectively. The desired conclusion follows from (8.9) and (8.1).

For convenience of reference we proceed to summarize the most important ones among the results about J-terms obtained above.

(8.10) THEOREM. *If t is a J-term, if H and K are subsets of I, and if τ is a transformation such that $\tau=\delta$ on J, then*

$(\overline{\mathrm{T}}1)$ $\quad \mathbf{S}(\varnothing/t)$ *is the identity mapping on* A,

$(\overline{\mathrm{T}}2)$ $\quad \mathbf{S}(H \cup K/t) = \mathbf{S}(H - J/t)\mathbf{S}(K - J/t)\mathbf{S}((H \cup K) \cap J/t)$,

$(\overline{\mathrm{T}}3)$ $\quad \mathbf{S}(H/t)\exists(K) = \exists(K - J)\mathbf{S}(H - K/t)\exists(K \cap J)$,

$(\overline{\mathrm{T}}4)$ $\quad \exists(H)\mathbf{S}(K/t) = \exists(H \cap J)\mathbf{S}(K/t)\exists(H - (J \cap K))$,

$(\overline{\mathrm{T}}5)$ $\quad \mathbf{S}(K/t)\mathbf{S}(\tau) = \mathbf{S}(\tau_{I-J})\mathbf{S}((\tau_{I-J})^{-1}K/t)\mathbf{S}(\tau_J)$.

We conclude this section by a brief description of how its results are connected with the corresponding results for J-constants. In view of the extension theorem (6.7), $(\overline{\mathrm{T}}1)$ says the same thing as (T1). (The extension theorem is frequently used, above and below, without explicit reference.) If H and K are subsets of $I-J$, $(\overline{\mathrm{T}}2)$ reduces to (T2). If $K \subset I-J$, then $(\overline{\mathrm{T}}3)$ reduces to a generalization of (T3); the assumption $H \subset I-J$ is not necessary. If $H \subset I-J$, then $(\overline{\mathrm{T}}4)$ reduces to a generalization of (T4); the assumption $K \subset I-J$ is not necessary. If τ lives on $I-J$, then $(\overline{\mathrm{T}}5)$ reduces to a generalization of (T5); the assumption $K \subset I-J$ is not necessary. The assertion (8.7) is also a generalization of (T5), in a different direction. It asserts that if K is suitably specialized, then it is not necessary to assume that τ lives on $I-J$; it is sufficient to assume merely that $\tau = \delta$ on J.

9. **The construction of terms.** To construct terms, it is useful to know that a term, like a constant, is uniquely determined by a single Boolean endomorphism.

(9.1) LEMMA. *If J is a finite subset of I, if f is a Boolean endomorphism of A, and if i is an element of $I-J$ such that*

(1) $\quad \exists(i)f = f$,

(2) $\quad f\exists(i) = \exists(i)$,

(3) $\quad f\exists(j) = \exists(j)f$ *whenever* $j \neq i$ *and* $j \in I - J$,

then there exists a unique J-term t of A such that $\mathbf{S}(i/t) = f$.

This lemma is an immediate consequence of the corresponding result (1.2) for constants (applied to the algebra obtained from A by fixing the variables of J) together with the theorem (6.7) on the extension of J-constants to J-terms.

In the applications it often happens that it is easy to define a Boolean homomorphism g from a part of A into A so that g satisfies the conditions of (9.1) as far as the limitations of its domain allow. The domains that arise are "finite compressions" of A, i.e., they are obtained by taking a finite subset K of I and considering only those elements of A that are independent of K. In other words, such a domain consists of all those elements p of A for which $\exists(K)p = p$, or, equivalently, it is the range of the quantifier $\exists(K)$. Our next result says that under a relatively mild additional restriction (involving the transformation structure of A) even such a partial endomorphism yields a term. The result is a generalization of (9.1); the proof is based on an application of that lemma. To prepare the ground we apply a slight refinement of the techniques that were used to extend J-constants to J-terms. The idea of the proof is to extend the given partial endomorphism g to an honest endomorphism f to which (9.1) applies. To do this, we first transform the variables of each element of A so as to obtain an element in the domain of g, then we apply g, and then, finally, we undo the effect of the first transformation. All this is, of course, a vague qualitative description; we proceed now to the precise details.

(9.2) THEOREM. *If J and K are finite subsets of I, if g is a Boolean homomorphism from the range of $\exists(K)$ into A, and if i is an element of $I-(J\cup K)$ such that*

(1) $\exists(i)g\exists(K) = g\exists(K)$,

(2) $g\exists(K)\exists(i) = \exists(K)\exists(i)$,

(3) $g\exists(K)\exists(j) = \exists(j)g\exists(K)$ *whenever* $j \neq i$ *and* $j \in I - J$,

(4) $g\exists(K)\mathbf{S}(\tau) = \mathbf{S}(\tau)g\exists(K)$ *whenever τ is a finite transformation that lives on* $I - (\{i\} \cup J \cup K)$,

then there exists a J-term t of A such that $\mathbf{S}(i/t)\exists(K) = g\exists(K)$.

REMARK. If $K=\varnothing$, then (1), (2), and (3) coincide with the corresponding conditions of (9.1), and therefore, in that case, (4) is superfluous. In the general case it is an immediate consequence of (T5) that the condition (4) is necessary for the existence of t.

Proof. Given an element p of A with finite support L, let σ be a transformation of type $(K, \{i\} \cup L \cup J)$, and define an operator f on A by writing

$$fp = \mathbf{S}(\bar{\sigma})g\mathbf{S}(\sigma)p. \tag{9.3}$$

(For the definition of $\bar{\sigma}$, see (3.2).) We observe that since σL supports $\mathbf{S}(\sigma)p$ and is disjoint from K, the element $\mathbf{S}(\sigma)p$ is independent of K and, consequently, belongs to the domain of g. Thus the definition (9.3) makes sense, but it must, of course, be supported by an unambiguity proof.

If both σ and τ are transformations of type $(K, \{i\} \cup L \cup J)$, apply (3.4) to find a permutation π such that $\pi\tau=\sigma\pi$, $\pi\bar{\tau}=\bar{\sigma}\pi$, and $\pi=\delta$ in $\{i\} \cup J \cup K \cup L$. Form $\mathbf{S}(\bar{\tau})g\mathbf{S}(\tau)p$, and then replace $\mathbf{S}(\tau)$ by $\mathbf{S}(\pi^{-1})\mathbf{S}(\sigma)\mathbf{S}(\pi)$ and $\mathbf{S}(\bar{\tau})$ by $\mathbf{S}(\pi^{-1})\mathbf{S}(\bar{\sigma})\mathbf{S}(\pi)$. Since $\pi=\delta$ in L and since L supports p, the rightmost $\mathbf{S}(\pi)$ may be erased from the resulting equation. Since π lies on $I-(\{i\} \cup J \cup K)$, it follows from (9.2)(4) that the rightmost $\mathbf{S}(\pi^{-1})$ can be pulled to the left of g. As a consequence of these manipulations we obtain

$$\mathbf{S}(\bar{\tau})g\mathbf{S}(\tau)p = \mathbf{S}(\pi^{-1})\mathbf{S}(\bar{\sigma})g\mathbf{S}(\sigma)p. \tag{9.4}$$

In order to prove now that the factor $\mathbf{S}(\pi^{-1})$ can be erased from (9.4), we investigate the supports of some of the constituents of that equation. We know that σL supports $\mathbf{S}(\sigma)p$; we assert that

(9.5) $\sigma L \cup J$ *supports* $g\mathbf{S}(\sigma)p$.

If $j \epsilon' \sigma L \cup J$, then either $j \neq i$, in which case

$$\exists(j)g\mathbf{S}(\sigma)p = g\exists(j)\mathbf{S}(\sigma)p \text{ [by (9.2)(3)] } = g\mathbf{S}(\sigma)p \text{ [since } j \epsilon' \sigma L],$$

or $j = i$, in which case

$$\exists(j)g\mathbf{S}(\sigma)p = \exists(i)g\mathbf{S}(\sigma)p = g\mathbf{S}(\sigma)p \text{ [by (9.2)(1)]}.$$

In any case, therefore, if $j \epsilon' \sigma L \cup J$, then $g\mathbf{S}(\sigma)p$ is independent of j; this proves (9.5). Since

$$\bar{\sigma}(\sigma L \cup J) = \bar{\sigma}L \cup \bar{\sigma}J = L \cup J,$$

it follows that

(9.6) $L \cup J$ *supports* $\mathbf{S}(\bar{\sigma})g\mathbf{S}(\sigma)p$.

Since $\pi^{-1} = \delta$ on $L \cup J$, it follows from (9.4) and (9.6) that

(9.7) $\mathbf{S}(\bar{\tau})g\mathbf{S}(\tau)p = \mathbf{S}(\bar{\sigma})g\mathbf{S}(\sigma)p$.

This does not quite finish the unambiguity proof; we must still show that the choice of L is immaterial. For this purpose, suppose that both L_1 and L_2 are finite supports of p and that σ_1 and σ_2 are transformations of types $(K, \{i\} \cup L_1 \cup J)$ and $(K, \{i\} \cup L_2 \cup J)$ respectively. If $L = L_1 \cup L_2$, then L is a finite support of p. Let τ be an arbitrary transformation of type $(K, \{i\} \cup L \cup J)$. Two applications of (9.7), first to σ_1 and τ and then to σ_2 and τ, imply that

$$\mathbf{S}(\bar{\sigma}_1)g\mathbf{S}(\sigma_1)p = \mathbf{S}(\bar{\sigma}_2)g\mathbf{S}(\sigma_2)p,$$

as desired.

A straightforward argument (cf. (6.4)) shows that the mapping f defined by (9.3) is a Boolean endomorphism of $\boldsymbol{A}$. The next thing to do is to verify that f satisfies the conditions (1), (2) and (3) of (9.1).

With p, L, and σ as before, we have

$$\begin{aligned}\exists(i)fp &= \exists(i)\mathbf{S}(\bar{\sigma})g\mathbf{S}(\sigma)p = \mathbf{S}(\bar{\sigma})\exists(i)g\mathbf{S}(\sigma)p \text{ [since } \bar{\sigma}^{-1}\{i\} = \{i\}] \\ &= \mathbf{S}(\bar{\sigma})g\mathbf{S}(\sigma)p \text{ [by (9.2)(1)] } = fp,\end{aligned}$$

and

$$\begin{aligned}f\exists(i)p &= \mathbf{S}(\bar{\sigma})g\mathbf{S}(\sigma)\exists(i)p = \mathbf{S}(\bar{\sigma})g\exists(i)\mathbf{S}(\sigma)p \text{ [since } \sigma^{-1}\{i\} = \{i\}] \\ &= \mathbf{S}(\bar{\sigma})\exists(i)\mathbf{S}(\sigma)p \text{ [by (9.2)(2)] } = \mathbf{S}(\bar{\sigma})\mathbf{S}(\sigma)\exists(i)p \\ &= \mathbf{S}(\bar{\sigma}\sigma)\exists(i)p = \mathbf{S}(\bar{\sigma})\exists(i)p \\ &= \exists(i)p \text{ [since } \bar{\sigma} = \delta \text{ outside } \sigma K \text{ and } p \text{ is independent of } \sigma K];\end{aligned}$$

this settles (9.1)(1) and (9.1)(2).

The proof of (9.1)(3) is harder. We continue to use p, L, and σ as before, and we consider an element j of $I-J$ such that $j \neq i$. If $j \epsilon' L$, then $j \epsilon' L \cup J$ and therefore, since (by (9.6)) $L \cup J$ supports fp,

$$\exists(j)fp = fp \quad \text{and} \quad f\exists(j)p = fp.$$

If $j \epsilon L-K$, then $\sigma^{-1}\{j\} = \tilde{\sigma}^{-1}\{j\} = \{j\}$ (since $j \epsilon' \sigma K$), and therefore

$$\begin{aligned} \exists(j)fp &= \exists(j)\mathbf{S}(\tilde{\sigma})g\mathbf{S}(\sigma)p = \mathbf{S}(\tilde{\sigma})\exists(j)g\mathbf{S}(\sigma)p \\ &= \mathbf{S}(\tilde{\sigma})g\exists(j)\mathbf{S}(\sigma)p \ [\text{by } (9.2)(3)] = \mathbf{S}(\tilde{\sigma})g\mathbf{S}(\sigma)\exists(j)p = f\exists(j)p. \end{aligned}$$

The only remaining case is the one in which $j \epsilon L \cap K$.

Let σ_- be the transformation that agrees with σ except at σj (where $\sigma = \delta$) and that sends σj onto j. Since both p and $\exists(j)p$ are independent of σj, it follows that

$$\mathbf{S}(\sigma)p = \mathbf{S}(\sigma_-)p \quad \text{and} \quad \mathbf{S}(\sigma)\exists(j)p = \mathbf{S}(\sigma_-)\exists(j)p.$$

Note, for later use, that $\sigma_-^{-1}\{\sigma j\} = \{j\}$. Let σ_+ be the transformation that agrees with $\tilde{\sigma}$ except at j (where $\tilde{\sigma} = \delta$) and that sends j onto σj. Since both $g\mathbf{S}(\sigma)p$ and $g\mathbf{S}(\sigma)\exists(j)p$ are independent of j (by (9.5)), it follows that

$$\mathbf{S}(\tilde{\sigma})g\mathbf{S}(\sigma)p = \mathbf{S}(\sigma_+)g\mathbf{S}(\sigma)p$$

and

$$\mathbf{S}(\tilde{\sigma})g\mathbf{S}(\sigma)\exists(j)p = \mathbf{S}(\sigma_+)g\mathbf{S}(\sigma)\exists(j)p.$$

The proof of the remaining case of (9.1)(3) is now a matter of computation, as follows:

$$\begin{aligned} f\exists(j)p &= \mathbf{S}(\tilde{\sigma})g\mathbf{S}(\sigma)\exists(j)p = \mathbf{S}(\sigma_+)g\mathbf{S}(\sigma_-)\exists(j)p \\ &= \mathbf{S}(\sigma_+)g\exists(\sigma j)\mathbf{S}(\sigma_-)p = \mathbf{S}(\sigma_+)\exists(\sigma j)g\mathbf{S}(\sigma_-)p \\ &= \exists(j)\mathbf{S}(\sigma_+)g\mathbf{S}(\sigma_-)p = \exists(j)\mathbf{S}(\tilde{\sigma})g\mathbf{S}(\sigma)p = \exists(j)fp. \end{aligned}$$

An application of (9.1) yields the existence of a J-term t such that $\mathbf{S}(i/t) = f$. Suppose now that an element p of A is independent of K. Let L be a finite support of p such that L is disjoint from K, and let σ be a transformation of type $(K, \{i\} \cup L \cup J)$. Since $\sigma = \delta$ in L, we have $\sigma L = L$ and $\mathbf{S}(\sigma)p = p$. By (9.5), $L \cup J$ supports gp; since $\tilde{\sigma} = \delta$ in $L \cup J$, it follows that

$$\mathbf{S}(i/t)p = fp = \mathbf{S}(\tilde{\sigma})g\mathbf{S}(\sigma)p = \mathbf{S}(\tilde{\sigma})gp = gp.$$

The proof of Theorem (9.2) is complete.

We prove next that the term t of (9.2) is unique.

(9.8) Theorem. *If J and K are finite subsets of I, if s and t are J-terms, and if i is an element of $I-(J \cup K)$ such that*

(9.9) $$\mathbf{S}(i/s)\exists(K) = \mathbf{S}(i/t)\exists(K),$$

then $s=t$.

Proof. We shall show that $\mathbf{S}(i/s)=\mathbf{S}(i/t)$; the desired result will then follow from the uniqueness assertion of (9.1). If p is an element of A with finite support L and if σ is a transformation of type $(K, \{i\} \cup L \cup J)$, then $\mathbf{S}(\sigma)p$ is independent of K; it follows from (9.9) that

$$\mathbf{S}(i/s)\mathbf{S}(\sigma)p = \mathbf{S}(i/t)\mathbf{S}(\sigma)p$$

and hence that

$$\mathbf{S}(\bar{\sigma})\mathbf{S}(i/s)\mathbf{S}(\sigma)p = \mathbf{S}(\bar{\sigma})\mathbf{S}(i/t)\mathbf{S}(\sigma)p.$$

By (8.7), first with s, $\bar{\sigma}$, and $\{i\}$ and then with t, $\bar{\sigma}$, and $\{i\}$ in the roles of t, τ, and K, respectively, $\mathbf{S}(\bar{\sigma})$ commutes with both $\mathbf{S}(i/s)$ and $\mathbf{S}(i/t)$; since $\bar{\sigma}\sigma=\bar{\sigma}$, it follows that

$$\mathbf{S}(i/s)\mathbf{S}(\bar{\sigma})p = \mathbf{S}(i/t)\mathbf{S}(\bar{\sigma})p.$$

Since $\sigma=\delta$ outside σK and since p is independent of σK, it follows that $\mathbf{S}(i/s)p=\mathbf{S}(i/t)p$, and the proof of the uniqueness theorem is complete.

10. **Terms in predicates.** Suppose that P is an n-place predicate and that $t_1, \cdots, t_n$ are terms; the purpose of this section is to show how it is possible to assign a sensible meaning to the expression $P(t_1, \cdots, t_n)$.

There is an obvious way to proceed: simply select n variables $i_1, \cdots, i_n$ and write

$$P(t_1, \cdots, t_n) = \mathbf{S}(i_1/t_1) \cdots \mathbf{S}(i_n/t_n)P(i_1, \cdots, i_n).$$

This procedure is open to several objections. The main difficulty is that the choice of the variables $i_1, \cdots, i_n$ must not be completely arbitrary. If, for instance, they are not all distinct, then the result is not the expected one, not even in the simplest case in which the roles of the terms are played by other variables. (Compare $\mathbf{S}(i/j)\mathbf{S}(i/k)P(i, i)$ with $P(j, k)$.) If the auxiliary variables are distinct, there could still be trouble. Suppose, for instance, that u is a J-term, v is a K-term, and P is a binary predicate. If we tried to form $P(u, v)$ by first applying $\mathbf{S}(k/v)$ to $P(j, k)$ and then applying $\mathbf{S}(j/u)$ to the result, we could encounter an unexpected collision among the variables at the end of the first step. The point is that the term v could depend on the auxiliary variable j (i.e., j could belong to K), and consequently $P(j, v)$ (i.e., $\mathbf{S}(k/v)P(j, k)$) could involve j in a way in which $P(j, k)$ did not involve it. Here, as above, the difficulty may already arise in the case in which the roles of the terms are played by variables. (Compare $\mathbf{S}(j/i)\mathbf{S}(k/j)P(j, k)$ with $P(i, j)$.)

The difficulties of the preceding paragraph can be avoided by a judicious choice of the auxiliary variables, but there is still another objection to the proposed definition of $P(t_1, \cdots, t_n)$. The objection is that even if the auxiliary variables are chosen carefully, their choice might still influence the result; the proposed definition is not obviously unambiguous. We proceed to show how all these obstacles can be overcome.

(10.1) LEMMA. *Suppose that $J_1, \cdots, J_n$ are finite subsets of I, and that t_1 is a J_1-term, $\cdots$, t_n is a J_n-term; write $J = J_1 \cup \cdots \cup J_n$. If $i_1, \cdots, i_n$, $k_1, \cdots, k_n$ are $2n$ distinct variables in $I - J$, and if p is an element of A independent of $\{i_1, \cdots, i_n\}$, then*

$$(10.2) \quad \mathbf{S}(i_1/t_1) \cdots \mathbf{S}(i_n/t_n)\mathbf{S}(k_1/i_1) \cdots \mathbf{S}(k_n/i_n)p = \mathbf{S}(k_1/t_1) \cdots \mathbf{S}(k_n/t_n)p.$$

Proof. If J_0 is a finite subset of I, if t_0 is a J_0-term, and if i, j, and k are distinct elements of $I - J_0$, then the transformation (k/j) lives on $I - J_0$, and therefore, by (T5), the endomorphisms $\mathbf{S}(i/t_0)$ and $\mathbf{S}(k/j)$ commute with each other. By an inductive application of this comment to the left side of (10.2), the desideratum reduces to

$$(10.3) \quad \mathbf{S}(i_1/t_1)\mathbf{S}(k_1/i_1) \cdots \mathbf{S}(i_n/t_n)\mathbf{S}(k_n/i_n)p = \mathbf{S}(k_1/t_1) \cdots \mathbf{S}(k_n/t_n)p.$$

Let L be a finite support of p disjoint from $\{i_1, \cdots, i_n\}$. Since (k_n/i_n) is a transformation of type $(\{k_n\}, L \cup J_n)$, and since $(k_n/i_n)k_n = i_n$, it follows from (4.1) that

$$\mathbf{S}(i_n/t_n)\mathbf{S}(k_n/i_n)p = \mathbf{S}(k_n/t_n)p.$$

By (7.5), $J_n \cup L$ supports $\mathbf{S}(k_n/t_n)p$, and, consequently, $\mathbf{S}(k_n/t_n)p$ is independent of $\{i_1, \cdots, i_n\}$. Since this comment prepares the ground for the induction step, (10.3) follows from an n-fold repetition of the argument just given.

(10.4) LEMMA. *Suppose that $J_1, \cdots, J_n$ are finite subsets of I, that t_1 is a J_1-term, $\cdots$, t_n is a J_n-term, and that P is an n-place predicate; write $J = J_1 \cup \cdots \cup J_n$. If $i_1, \cdots, i_n$ are distinct variables in $I - J$, and $j_1, \cdots, j_n$ are distinct variables in $I - J$, then*

$$(10.5) \quad \mathbf{S}(i_1/t_1) \cdots \mathbf{S}(i_n/t_n)P(i_1, \cdots, i_n) = \mathbf{S}(j_1/t_1) \cdots \mathbf{S}(j_n/t_n)P(j_1, \cdots, j_n).$$

Proof. Let $k_1, \cdots, k_n$ be distinct variables in $I - (J \cup \{i_1, \cdots, i_n\} \cup \{j_1, \cdots, j_n\})$, and write $P(k_1, \cdots, k_n) = p$. Since $(k_1/i_1) \cdots (k_n/i_n) = (k_1, \cdots, k_n/i_1, \cdots, i_n)$, it follows from (2.1) that

$$(10.6) \qquad P(i_1, \cdots, i_n) = \mathbf{S}(k_1/i_1) \cdots \mathbf{S}(k_n/i_n)p.$$

By (10.6) and (10.1), the left side of (10.5) is equal to $\mathbf{S}(k_1/t_1) \cdots \mathbf{S}(k_n/t_n)p$; the same argument with $j_1, \cdots, j_n$ in place $i_1, \cdots, i_n$ shows that the right side of (10.5) is equal to the same thing.

We are now in a position to make the promised definition. If P is an n-place predicate, and if t_1 is a J_1-term, $\cdots$, t_n is a J_n-term, we write

$$(10.7) \qquad P(t_1, \cdots, t_n) = \mathbf{S}(i_1/t_1) \cdots \mathbf{S}(i_n/t_n) P(i_1, \cdots, i_n)$$

whenever $i_1, \cdots, i_n$ are distinct variables in $I-(J_1 \cup \cdots \cup J_n)$; Lemma (10.4) guarantees that this is an unambiguous definition. We shall obtain some of the properties of $P(t_1, \cdots, t_n)$ later; for the time being we conclude the discussion by calling attention to two useful special cases.

Since every constant is a term (a $\varnothing$-term, in fact), (10.7) assigns a meaning to $P(c_1, \cdots, c_n)$ whenever P is an n-place predicate and $c_1, \cdots, c_n$ are constants. The proviso accompanying the definition is, of course, simpler than in the general case; we have

$$(10.8) \qquad P(c_1, \cdots, c_n) = \mathbf{S}(i_1/c_1) \cdots \mathbf{S}(i_n/c_n) P(i_1, \cdots, i_n)$$

whenever $i_1, \cdots, i_n$ are distinct variables. The general definition allows us to consider mixed cases also; if, for instance, P is a 4-place predicate, then $P(a, b, s, t)$ makes sense whenever a and b are constants and s and t are terms.

The second useful special case is the one in which some of the terms are variables. We recall (see §4) that to every variable j there corresponds a j-constant, denoted by c_j, such that

$$\mathbf{S}(K/c_j) = \mathbf{S}(K/j)$$

for every subset K of $I-\{j\}$. It follows (cf. (6.3)) that to every variable j there corresponds a j-term (i.e., a $\{j\}$-term), which we shall denote by t_j, such that

$$(10.9) \qquad \mathbf{S}(K/t_j) = \mathbf{S}(K/j)$$

for every subset K of $I-\{j\}$. (Here is a situation in which it is important not to identify a singleton with its single element; the term t_j is not the same as the j-constant $t_{\{j\}}$ defined by (7.1). Note that, in accordance with the definition (6.3), $t_j = \bar{c}_j$). It is worth observing that (10.9) holds not only for subsets of $I-\{j\}$, but in fact for all finite sets (and hence for all subsets of I). Suppose, indeed, that p is an element of A with finite support L, K is a finite subset of I, and σ is a transformation of type $(K, L \cup \{j\})$. By (4.1)

$$\mathbf{S}(K/t_j)p = \mathbf{S}(\sigma K/t_j)\mathbf{S}(\sigma)p = \mathbf{S}(\sigma K/j)\mathbf{S}(\sigma)p.$$

If $i \in L \cap K$, then $(\sigma K/j)\sigma i = j$ and $(K/j)i = j$; if $i \in L-K$, then $(\sigma K/j)\sigma i = i$ and $(K/j)i = i$. This proves that the transformations $(\sigma K/j)\sigma$ and (K/j) agree on L (in fact they agree outside σK), and this, in turn, implies that (10.9) is true for K.

If the transformation law (2.1) for predicates, the definition (10.7), and the equation (10.9) are combined, the result is that $P(t_1, \cdots, t_n)$ means what it should mean in case some of the t's are variables. If, in particular, $j_1, \cdots, j_n$ are variables, then

$$P(t_{j_1}, \cdots, t_{j_n}) = P(j_1, \cdots, j_n).$$

The general definition allows us to consider mixed cases also; if, for instance, P is a 4-place predicate, then $P(t_i, t_j, s, t)$ makes sense whenever i and j are variables and s and t are terms. There is no loss of rigor in simplifying the notation; we shall generally write $P(i, j, s, t)$ instead of $P(t_i, t_j, s, t)$. It is easy to verify that if s is a J-term and t is a K-term, then

$$P(i, j, s, t) = \mathbf{S}(h/s)\mathbf{S}(k/t)P(i, j, h, k)$$

whenever h and k are distinct variables in $I-(J \cup K \cup \{i, j\})$; we shall make use of such facts without any explicit reference.

11. **Transforms of terms.** Given a term t and a transformation τ, we shall define a term, to be denoted by τt, that may be thought of as the result of transforming the variables of t by τ. It is to be expected, of course, that τt will depend on τ and t in just about the same way as τj depends on τ and j when j is a variable. We shall see, in particular, that if $t = t_j$ for some variable j (cf. (10.9)) and if $\tau j = k$, then $\tau t = t_k$; in other words, $\tau t_j = t_{\tau j}$. We shall be able to prove also that if P is an n-place predicate and if $t_1, \cdots, t_n$ are terms, then

$$\mathbf{S}(\tau)P(t_1, \cdots, t_n) = P(\tau t_1, \cdots, \tau t_n).$$

These surely are the minimal decency conditions that should be satisfied by any reasonable definition of τt.

The clue to the desired definition is in the equation

$$\mathbf{S}(K/\tau j)\mathbf{S}(\tau) = \mathbf{S}(\tau)\mathbf{S}(\tau^{-1}K/j), \tag{11.1}$$

which holds whenever $j \in I$, $K \subset I$, and τ is a transformation on I. (In fact $(K/\tau j) = \tau(\tau^{-1}K/j)$. Proof: $(K/\tau j)\tau$ sends i onto τj or τi according as $i \in \tau^{-1}K$ or $i \in' \tau^{-1}K$, and the same is true of $\tau(\tau^{-1}K/j)$.) If i is an element of I such that $\tau^{-1}\{i\} = \{i\}$, and if we replace K by $\{i\}$ in (11.1), we obtain

$$\mathbf{S}(i/\tau j)\mathbf{S}(\tau) = \mathbf{S}(\tau)\mathbf{S}(i/j). \tag{11.2}$$

If p is an element of A such that $\tau = \delta$ on some support of p, then $\mathbf{S}(\tau)p = p$, and therefore (11.2) implies that

$$\mathbf{S}(i/\tau j)p = \mathbf{S}(\tau)\mathbf{S}(i/j)p. \tag{11.3}$$

Verbally (11.3) says (provided τ, i, and p satisfy the stated conditions) that if we first replace i by j in p and then transform the variables by τ, the result is the same as if we had simply replaced i by τj. On intuitive grounds this seems just as reasonable if instead of j we speak of an arbitrary term t; we may conjecture, in other words, that if τt were properly defined, we should have

$$\mathbf{S}(i/\tau t)p = \mathbf{S}(\tau)\mathbf{S}(i/t)p \tag{11.4}$$

whenever $\tau^{-1}\{i\} = \{i\}$ and $\tau = \delta$ on some support of p. Our definition of τt will be guided by (11.4).

One more preliminary comment is needed. If t is a J-term, what kind of a term should τt be? The answer is clear: if t depends on the variables in J, and on no others, and if those variables are transformed by τ, the result should depend on the variables in τJ, and on no others. If t is a J-term, it is to be expected that τt will be a τJ-term.

(11.5) Lemma. *If t is a J-term, if τ is a transformation that lives on a finite subset K of I, and if i is an element of $I-(J \cup K)$, then there exists a unique τJ-term s such that*

$$\mathbf{S}(i/s)p = \mathbf{S}(\tau)\mathbf{S}(i/t)p \tag{11.6}$$

whenever p is independent of K.

Proof. The construction of s is based on (9.2); the set τJ will play the role of what was there denoted by J. (Observe that since $i \in I-K$, it follows that $\tau^{-1}\{i\} = \{i\}$; since also $i \in I-J$, it follows that $i \in I-\tau J$.) If p is independent of K, we write

$$gp = \mathbf{S}(\tau)\mathbf{S}(i/t)p.$$

It is clear that g is a Boolean homomorphism from the range of $\exists(K)$ into $\boldsymbol{A}$; we proceed to verify that g satisfies the conditions (1)–(4) of (9.2) (with τJ in place of J).

We have

$$\begin{aligned}\exists(i)g\exists(K) &= \exists(i)\mathbf{S}(\tau)\mathbf{S}(i/t)\exists(K) = \mathbf{S}(\tau)\exists(i)\mathbf{S}(i/t)\exists(K)\\ &= \mathbf{S}(\tau)\mathbf{S}(i/t)\exists(K)\ [\text{by (T4)}] = g\exists(K),\end{aligned}$$

and

$$\begin{aligned}g\exists(K)\exists(i) &= \mathbf{S}(\tau)\mathbf{S}(i/t)\exists(K)\exists(i) = \mathbf{S}(\tau)\mathbf{S}(i/t)\exists(i)\exists(K)\\ &= \mathbf{S}(\tau)\exists(i)\exists(K)\ [\text{by (T3) and (T1)}] = \exists(i)\mathbf{S}(\tau)\exists(K)\\ &= \exists(i)\exists(K)\ [\text{since } \tau \text{ lives on } K] = \exists(K)\exists(i);\end{aligned}$$

this settles (9.2)(1) and (9.2)(2).

Suppose next that j is an element of $I-\tau J$ such that $j \neq i$. We split the discussion into two cases, according as $j \in K$ or $j \notin K$. If $j \in K$, and if p is an arbitrary element of $\boldsymbol{A}$, then $I-K$ supports $\exists(K)p$ and therefore $J \cup (I-K)$ supports $\mathbf{S}(i/t)\exists(K)p$ (7.5). It follows that $\tau(J \cup (I-K))$ supports $g\exists(K)p$. Since $\tau(J \cup (I-K)) = \tau J \cup (I-K)$, and since j does not belong to this set, $g\exists(K)p$ is independent of j; consequently

$$g\exists(K)\exists(j)p = g\exists(K)p\ [\text{since } j \in K] = \exists(j)g\exists(K)p.$$

If $j \notin K$, then $\tau j = j$ and therefore (since $j \notin \tau J$) $j \notin J$. It follows that

$$g\exists(K)\exists(j) = \mathbf{S}(\tau)\mathbf{S}(i/t)\exists(K)\exists(j) = \mathbf{S}(\tau)\mathbf{S}(i/t)\exists(j)\exists(K)$$
$$= \mathbf{S}(\tau)\exists(j)\mathbf{S}(i/t)\exists(K) \text{ [by (T3)]}$$
$$= \exists(j)\mathbf{S}(\tau)\mathbf{S}(i/t)\exists(K) \text{ [since } j \epsilon' K] = \exists(j)g\exists(K);$$

this concludes the proof of (9.2)(3).

Suppose finally that σ is a finite transformation that lives on $I-(\{i\} \cup \tau J \cup K)$. It follows from (8.7) (with σ and $\{i\}$ in the roles of what are there denoted by τ and K) that $\mathbf{S}(i/t)$ commutes with $\mathbf{S}(\sigma)$. Since τ lives on K, and since this implies that τ commutes with σ, it follows that

$$g\exists(K)\mathbf{S}(\sigma) = \mathbf{S}(\tau)\mathbf{S}(i/t)\exists(K)\mathbf{S}(\sigma)$$
$$= \mathbf{S}(\tau)S(i/t)\mathbf{S}(\sigma)\exists(K) \text{ [since } \sigma \text{ lives on } I-K]$$
$$= \mathbf{S}(\sigma)\mathbf{S}(\tau)\mathbf{S}(i/t)\exists(K) = \mathbf{S}(\sigma)g\exists(K).$$

The existence of the τJ-term s satisfying (11.6) is now an immediate consequence of (9.2); uniqueness is guaranteed by (9.8).

We prove next that the term s of (11.5) does not depend on the choice of i and K.

(11.7) LEMMA. *Suppose that t is a J-term, that τ is a transformation that lives on both H and K (where H and K are finite subsets of I), and that i and j are elements of $I-(J \cup H)$ and $I-(J \cup K)$ respectively. If u and v are τJ-terms such that*

$$\mathbf{S}(i/u)p = \mathbf{S}(\tau)\mathbf{S}(i/t)p$$

whenever p is independent of H and

$$\mathbf{S}(j/v)p = \mathbf{S}(\tau)\mathbf{S}(j/t)p$$

whenever p is independent of K, then $u=v$.

Proof. Since τ lives on $H \cap K$ and since i and j are in $I-(H \cap K)$, it follows that τ commutes with the transformation (i, j). We observe that (i, j) lives on both $I-J$ and $I-\tau J$, and that if p is independent of $H \cup K$ (so that, in particular, p is independent of both H and K), then $\mathbf{S}(i, j)p = p$.[1] It follows that

$$\mathbf{S}(i/u)p = \mathbf{S}(\tau)\mathbf{S}(i/t)p = \mathbf{S}(\tau)\mathbf{S}(i/t)\mathbf{S}(i, j)p$$
$$= \mathbf{S}(\tau)\mathbf{S}(i, j)\mathbf{S}(j/t)p \text{ [by (T5)]} = \mathbf{S}(i, j)\mathbf{S}(\tau)\mathbf{S}(j/t)p$$
$$= \mathbf{S}(i, j)\mathbf{S}(j/v)p = \mathbf{S}(i/v)\mathbf{S}(i, j)p \text{ [by (T5)]} = \mathbf{S}(i/v)p;$$

the desired equality is now a consequence of (9.8).

We are now prepared to define τt whenever t is a J-term and τ is a finite transformation. Let K be a finite set such that τ lives on K and let i be an element of $I-(J \cup K)$. We define τt to be the unique τJ-term obtained by applying (11.5) to this situation, so that

[1] See p. 239.

(11.8) $$\mathbf{S}(i/\tau t)p = \mathbf{S}(\tau)\mathbf{S}(i/t)p$$

whenever p is independent of K. Lemma (11.7) guarantees that the definition is unambiguous (i.e., that as long as K and i satisfy the stated conditions, it does not matter how they are chosen). We proceed to prove that this definition has all the reasonable properties we promised for it.

(11.9) LEMMA. *If t is a J-term and if σ and τ are finite transformations such that $\sigma = \tau$ on J, then $\sigma t = \tau t$.*

Proof. Select a finite set K such that both σ and τ live on K and let i be an element of $I-(J \cup K)$. If p is independent of K, then

$$\mathbf{S}(i/\sigma t)p = \mathbf{S}(\sigma)\mathbf{S}(i/t)p \quad \text{and} \quad \mathbf{S}(i/\tau t)p = \mathbf{S}(\tau)\mathbf{S}(i/t)p.$$

Since $J \cup (I-K)$ supports $\mathbf{S}(i/t)p$ and since $\sigma = \tau$ on $J \cup (I-K)$, it follows that $\mathbf{S}(i/\sigma t)p = \mathbf{S}(i/\tau t)p$; the desired equality is now a consequence of (9.8).

Using (11.9), we can extend the definition of τt to the case of a not necessarily finite τ, as follows. If t is a J-term and if τ is a transformation, let σ be an arbitrary finite transformation such that $\tau = \sigma$ on J, and write $\tau t = \sigma t$. If σ_1 and σ_2 are finite transformations such that $\tau = \sigma_1$ in J and $\tau = \sigma_2$ in J, then $\sigma_1 t = \sigma_2 t$ (by (11.9)) and therefore the definition of τt is unambiguous. If, in particular, τ happens to be finite, the new definition of τt is compatible with the old one. Observe that since $\sigma J = \tau J$, the term τt is always a τJ-term.

(11.10) THEOREM. *If t is a term, then*

(11.11) $$\delta t = t;$$

if σ and τ are transformations, then

(11.12) $$(\sigma\tau)t = \sigma(\tau t).$$

If t is a J-term, then

(11.13) $$\sigma t = \tau t \text{ whenever } \sigma = \tau \text{ on } J.$$

Proof. Suppose that t is a J-term. Let i be an element of $I-J$. Since δ lives on $\varnothing$, it follows that

$$\mathbf{S}(i/\delta t)p = \mathbf{S}(\delta)\mathbf{S}(i/t)p = \mathbf{S}(i/t)p$$

for all p, and hence (9.8) that $\delta t = t$.

To prove (11.12), write $\rho = \sigma\tau$ and define the auxiliary transformations ρ_0, σ_0, and τ_0, as follows:

$$\rho_0 = \rho \text{ on } J, \qquad \rho_0 = \delta \text{ on } I - J,$$

$$\sigma_0 = \sigma \text{ on } \tau J, \qquad \sigma_0 = \delta \text{ on } I - \tau J,$$

$$\tau_0 = \tau \text{ on } J, \qquad \tau_0 = \delta \text{ on } I - J.$$

The transformations ρ_0, σ_0, and τ_0 are finite; it follows from the definition of the action of a not necessarily finite transformation that $\rho t=\rho_0 t$, $\tau t=\tau_0 t$, and $\sigma(\tau t)=\sigma_0(\tau t)$.

Let K be a finite set such that ρ_0, σ_0, and τ_0 live on K, and let i be an element of $I-(J \cup K)$. Since $\sigma_0\tau_0$ lives on K, it follows that if p is independent of K, then

$$\mathbf{S}(i/(\sigma_0\tau_0)t)p = \mathbf{S}(\sigma_0\tau_0)\mathbf{S}(i/t)p = \mathbf{S}(\sigma_0)\mathbf{S}(\tau_0)\mathbf{S}(i/t)p$$
$$= \mathbf{S}(\sigma_0)\mathbf{S}(i/\tau_0 t)p = \mathbf{S}(i/\sigma_0(\tau_0 t))p.$$

The uniqueness theorem (9.8) implies now that $(\sigma_0\tau_0)t=\sigma_0(\tau_0 t)$. The transformations ρ_0 and $\sigma_0\tau_0$ are not necessarily equal, but it is easy to verify that they agree on J; it follows that

$$(\sigma\tau)t = \rho t = \rho_0 t = (\sigma_0\tau_0)t = \sigma_0(\tau_0 t) = \sigma_0(\tau t) = \sigma(\tau t).$$

The last assertion of the theorem (i.e., (11.13)) is immediate from the definitions.

(11.14) THEOREM. *If τ is a transformation and if j is a variable, then* $\tau t_j=t_{\tau j}$.

Proof. It is sufficient to consider finite transformations only. Let K be a finite set such that τ lives on K. If $i \in I-(J \cup K)$ and if $k \in I-K$, then either $k=i$, in which case

$$(i/\tau j)k = (i/\tau j)i = \tau j = \tau(i/j)i = \tau(i/j)k,$$

or else $k\neq i$, in which case

$$(i/\tau j)k = k = \tau k = \tau(i/j)k.$$

Hence, in either case, $(i/\tau j)=\tau(i/j)$ outside K; it follows that if p is independent of K, then $\mathbf{S}(i/\tau j)p=\mathbf{S}(\tau)\mathbf{S}(i/j)p$.

(11.15) THEOREM. *If t is a term, if τ is a transformation, and if i is an element of I such that $\tau^{-1}\{i\}=\{i\}$, then*

$$\mathbf{S}(i/\tau t)\mathrm{S}(\tau) = \mathbf{S}(\tau)\mathbf{S}(i/t).$$

Proof. Suppose that t is a J-term. Let p be an element of $\boldsymbol{A}$ with finite support L; observe that $L \cup J$ supports $\mathbf{S}(i/t)p$. If $\tau_0=\tau$ on $L \cup J$ and $\tau_0=\delta$ outside $L \cup J$, then τ_0 is a finite transformation such that $\tau_0^{-1}\{i\}=\{i\}$, $\tau_0 t=\tau t$, $\mathbf{S}(\tau_0)p=\mathbf{S}(\tau)p$, and $\mathbf{S}(\tau_0)\mathbf{S}(i/t)p=\mathbf{S}(\tau)\mathbf{S}(i/t)p$. It follows that there is no loss of generality in assuming (as we do from now on) that the given transformation τ is finite. Let K be a finite set not containing i such that τ lives on K.

Assume now that $i \in I-J$. Let σ be a transformation of type $(K, \{i\} \cup J \cup L)$. If ρ is the transformation such that $\rho=\delta$ outside σK and $\rho k=\tau\bar{\sigma}k$ when $k \in \sigma K$, then $\tau\bar{\sigma}=\rho\tau$. By (8.7) (first with τt, τJ, ρ, and $\{i\}$, and then with t, J, $\bar{\sigma}$, and $\{i\}$ in the roles of t, J, τ, and K, respectively) it follows that $\mathbf{S}(\rho)$

commutes with $\mathbf{S}(i/\tau t)$ and $\mathbf{S}(\bar{\sigma})$ commutes with $\mathbf{S}(i/t)$. Consequently

$$
\begin{aligned}
\mathbf{S}(i/\tau t)\mathbf{S}(\tau)p &= \mathbf{S}(i/\tau t)\mathbf{S}(\tau)\mathbf{S}(\bar{\sigma})p \\
&= \mathbf{S}(i/\tau t)\mathbf{S}(\tau)\mathbf{S}(\bar{\sigma})\mathbf{S}(\sigma)p \quad [\text{by } (3.3)] \\
&= \mathbf{S}(i/\tau t)\mathbf{S}(\rho)\mathbf{S}(\tau)\mathbf{S}(\sigma)p = \mathbf{S}(\rho)\mathbf{S}(i/\tau t)\mathbf{S}(\tau)\mathbf{S}(\sigma)p \\
&= \mathbf{S}(\rho)\mathbf{S}(\tau)\mathbf{S}(i/t)\mathbf{S}(\sigma)p \quad [\text{since } \mathbf{S}(\sigma)p \text{ is independent of } K] \\
&= \mathbf{S}(\tau)\mathbf{S}(\bar{\sigma})\mathbf{S}(i/t)\mathbf{S}(\sigma)p = \mathbf{S}(\tau)\mathbf{S}(i/t)\mathbf{S}(\bar{\sigma})\mathbf{S}(\sigma)p \\
&= \mathbf{S}(\tau)\mathbf{S}(i/t)p;
\end{aligned}
$$

this completes the proof in case $i \in I-J$.

If $i \in J$, let j be an element of $I-(L \cup J \cup K)$. The transformation (i/j) is of type $(\{i\}, L \cup J)$ and also of type $(\{i\}, \tau L \cup \tau J)$; it follows from (4.1) that

$$\mathbf{S}(i/t)p = \mathbf{S}(j/t)\mathbf{S}(i/j)p \tag{11.16}$$

and

$$
\begin{aligned}
\mathbf{S}(i/\tau t)\mathbf{S}(\tau)p &= \mathbf{S}(i/\tau t)\mathbf{S}(i/j)\mathbf{S}(\tau)p \\
&= \mathbf{S}(j/\tau t)\mathbf{S}(\tau)\mathbf{S}(i/j)p \quad [\text{since } \{i, j\} \subset I - K] \\
&= \mathbf{S}(\tau)\mathbf{S}(j/t)\mathbf{S}(i/j)p \quad [\text{since } j \in I - J] \\
&= \mathbf{S}(\tau)\mathbf{S}(i/t)p \quad [\text{by } (11.16)].
\end{aligned}
$$

This completes the proof of Theorem (11.15).

We show finally that the process of replacing the arguments of a predicate by terms is decently related to the action of transformations on terms.

(11.17) THEOREM. *If P is an n-place predicate, if $t_1, \cdots, t_n$ are terms, and if τ is a transformation, then*

$$\mathbf{S}(\tau)P(t_1, \cdots, t_n) = P(\tau t_1, \cdots, \tau t_n).$$

Proof. Suppose that t_1 is a J_1-term, $\cdots$, t_n is a J_n-term, and write $J = J_1 \cup \cdots \cup J_n$. Let τ_0 be the transformation that agrees with τ on J and that is equal to δ outside J. Since the desired conclusion is the same for τ as for τ_0, we may henceforth assume that the transformation τ is finite to begin with.

If $i_1, \cdots, i_n$ are distinct elements of $I-(J \cup \tau J)$, such that $\tau^{-1}\{i_1\} = \{i_1\}, \cdots, \tau^{-1}\{i_n\} = \{i_n\}$, then

$$
\begin{aligned}
\mathbf{S}(\tau)P(t_1, \cdots, t_n) &= \mathbf{S}(\tau)\mathbf{S}(i_1/t_1) \cdots \mathbf{S}(i_n/t_n)P(i_1, \cdots, i_n) \quad [\text{by } (10.7)] \\
&= \mathbf{S}(i_1/\tau t_1) \cdots \mathbf{S}(i_n/\tau t_n)\mathbf{S}(\tau)P(i_1, \cdots, i_n) \quad [\text{by } (11.15)] \\
&= \mathbf{S}(i_1/\tau t_1) \cdots \mathbf{S}(i_n/\tau t_n)P(i_1, \cdots, i_n) \quad [\text{by } (2.1)] \\
&= P(\tau t_1, \cdots, \tau t_n) \quad [\text{by } (10.7)].
\end{aligned}
$$

This concludes our discussion of the transforms of terms.

12. **Operations.** An operation (as we shall define it) is something that converts variables into J-terms, in about the same way as a predicate converts variables into J-propositions. Equality, for instance, is a binary predicate; suitably combined with the variables x and y it becomes the $\{x, y\}$-proposition $x=y$. Similarly, addition is a binary operation; suitably combined with the variables x and y it becomes the $\{x, y\}$-term $x+y$. Using the facts about terms and their transforms derived in the preceding sections, we can obtain the basic theory of operations quite easily; the techniques are minor modifications of the ones used at the corresponding parts of the theory of predicates.

We begin with an auxiliary result (a corollary of (9.2)) that enables us to tell when a term is a J-term.

(12.1) LEMMA. *If J and K are finite subsets of I, if t is a $(J \cup K)$-term, and if i is an element of $I-(J \cup K)$ such that*

$$(12.2)\quad \mathbf{S}(i/t)\exists(K)\exists(j) = \exists(j)\mathbf{S}(i/t)\exists(K) \text{ whenever } j \neq i \text{ and } j \in I-J,$$

then t is a J-term.

Proof. Write $gp=\mathbf{S}(i/t)p$ whenever p is independent of K. Clearly g is a Boolean homomorphism from the range of $\exists(K)$ into $\boldsymbol{A}$; we shall prove that g satisfies the conditions (1)–(4) of (9.2). We have

$$\exists(i)g\exists(K) = \exists(i)\mathbf{S}(i/t)\exists(K) = \mathbf{S}(i/t)\exists(K) \text{ [by (T4)]} = g\exists(K),$$

and

$$\begin{aligned} g\exists(K)\exists(i) &= \mathbf{S}(i/t)\exists(K)\exists(i) = \mathbf{S}(i/t)\exists(i)\exists(K) \\ &= \exists(i)(K) \text{ [by (T3) and (T1)]} = \exists(K)\exists(i); \end{aligned}$$

this, together with (12.2), settles (9.2)(1), (9.2)(2), and (9.2)(3). If τ is a finite transformation that lives on $I-(\{i\} \cup J \cup K)$, then

$$\begin{aligned} g\exists(K)\mathbf{S}(\tau) &= \mathbf{S}(i/t)\exists(K)\mathbf{S}(\tau) = \mathbf{S}(i/t)\mathbf{S}(\tau)\exists(K) \text{ [since } \tau^{-1}K = K] \\ &= \mathbf{S}(\tau)\mathbf{S}(i/t)\exists(K) \text{ [by (T5), since } \tau^{-1}\{i\} = \{i\}] = \mathbf{S}(\tau)g\exists(K); \end{aligned}$$

this settles (9.2)(4). It follows from (9.2) that there exists a J-term s such that $\mathbf{S}(i/s)p=\mathbf{S}(i/t)p$ whenever p is independent of K. Since $s=t$ (by (9.8)), it follows that t is a J-term, as asserted.

An n-place (or n-ary) *operation* ($n=1, 2, 3, \cdots$) of an I-algebra $\boldsymbol{A}$ is a function T on I^n whose values are terms of $\boldsymbol{A}$, such that if $(i_1, \cdots, i_n) \in I^n$ and if τ is a transformation on I, then

$$(12.3)\qquad \tau T(i_1, \cdots, i_n) = T(\tau i_1, \cdots, \tau i_n).$$

We shall sometimes say that an n-place operation is an operation of *degree n.*

We prove now that the analogues of (2.2), (2.3), and (2.4) are valid for operations.

(12.4) THEOREM. *If T is an n-place operation and if $(i_1, \cdots, i_n) \epsilon I^n$, then $T(i_1, \cdots, i_n)$ is an $\{i_1, \cdots, i_n\}$-term.*

Proof. Write $t = T(i_1, \cdots, i_n)$. We know that t is a term; let us say that t is a $\bar{J}$-term. If $J = \{i_1, \cdots, i_n\}$ and $K = \bar{J} - J$, then $\bar{J} \subset J \cup K$ and therefore t is a $(J \cup K)$-term. We shall prove that t is a J-term by applying (12.1). Suppose, accordingly, that $i \epsilon I - (J \cup K)$ and that j is an element of $I - J$ distinct from i; it is sufficient to prove (cf. (12.2)) that

$$\mathbf{S}(i/t)\exists(j) = \exists(j)\mathbf{S}(i/t).$$

For this purpose, let k be an element of $I - (J \cup K)$ distinct from both i and j. Since $(j, k)^{-1}\{i\} = \{i\}$, and since (by (12.3)) $(j, k)t = t$, it follows from (11.15) that

$$\mathbf{S}(i/t)\mathbf{S}(j, k) = \mathbf{S}(j, k)\mathbf{S}(i/t). \tag{12.5}$$

Consequently

$$\begin{aligned}\mathbf{S}(i/t)\exists(j) &= \mathbf{S}(i/t)\mathbf{S}(j, k)\exists(k)\mathbf{S}(j, k)\\ &= \mathbf{S}(j, k)\mathbf{S}(i/t)\exists(k)\mathbf{S}(j, k) \ \ [\text{by } (12.5)]\\ &= \mathbf{S}(j, k)\exists(k)\mathbf{S}(i/t)\mathbf{S}(j, k) \ \ [\text{by } (T3)]\\ &= \mathbf{S}(j, k)\exists(k)\mathbf{S}(j, k)\mathbf{S}(i/t) \ \ [\text{by } (12.5)] = \exists(j)\mathbf{S}(i/t),\end{aligned}$$

and the proof of the theorem is complete.

(12.6) LEMMA. *If $j_1, \cdots, j_n$ are distinct elements of I and if t is a $\{j_1, \cdots, j_n\}$-term, then there exists a unique n-place operation T such that $T(j_1, \cdots, j_n) = t$.*

Proof. The operation T is defined by

$$T(i_1, \cdots, i_n) = (j_1, \cdots, j_n/i_1, \cdots, i_n)t.$$

To prove that the function T is indeed an operation, let τ be an arbitrary transformation on I and note that if $j \epsilon \{j_1, \cdots, j_n\}$, then

$$\tau\cdot(j_1, \cdots, j_n/i_1, \cdots, i_n)j = (j_1, \cdots, j_n/\tau i_1, \cdots, \tau i_n)j.$$

Since, by assumption, t is a $\{j_1, \cdots, j_n\}$-term, it follows (cf. (11.13)) that

$$\begin{aligned}\tau T(i_1, \cdots, i_n) &= \tau\cdot(j_1, \cdots, j_n/i_1, \cdots, i_n)t\\ &= (j_1, \cdots, j_n/\tau i_1, \cdots, \tau i_n)t = T(\tau i_1, \cdots, \tau i_n).\end{aligned}$$

The uniqueness of T is an immediate consequence of (12.3), with

$$(j_1, \cdots, j_n/i_1, \cdots, i_n)$$

in the role of τ.

(12.7) COROLLARY. *Every term of **A** is in the range of some operation of **A**.*

It is reasonable to ask whether the concept of an operation can be generalized the same way as the concept of a J-predicate generalizes the concept of a predicate. The answer is yes, but since neither the techniques nor the results of the generalized theory offer any points of novelty, we do not enter into the details.

We conclude this preliminary discussion of operations by observing that there is a natural way of combining predicates and operations to form predicates of higher degree. If P is a predicate of degree n, and if $T_1, \cdots, T_n$ are operations of degrees $m_1, \cdots, m_n$ respectively, we write $m = m_1 + \cdots + m_n$ and we define a predicate Q of degree m by writing $Q = P(T_1, \cdots, T_n)$. To be more precise, suppose that P is a binary predicate and that S and T are operations of degrees m and n respectively; in that case the predicate $P(S, T)$ $(=Q)$ is the $(n+m)$-place predicate defined by

$$(12.8) \qquad Q(i_1, \cdots, i_m, j_1, \cdots, j_n) = P(S(i_1, \cdots, i_m), T(j_1, \cdots, j_n)).$$

(The only essential difference between this special case and the most general case is that for the latter the number of indices that it is necessary to use makes the theory appear much more complicated than it really is.) If τ is a transformation on I, then

$$\begin{aligned} \mathbf{S}(\tau)Q(i_1, \cdots, i_m, j_1, \cdots, j_n) & \\ &= P(\tau S(i_1, \cdots, i_m), \tau T(j_1, \cdots, j_n)) \quad [\text{by } (11.17)] \\ &= P(S(\tau i_1, \cdots, \tau i_m), T(\tau j_1, \cdots, \tau j_n)) \quad [\text{by } (12.3)] \\ &= Q(\tau i_1, \cdots, \tau i_m, \tau j_1, \cdots, \tau j_n); \end{aligned}$$

this proves that the function Q is indeed a predicate.

13. **Terms in terms.** Since transformations act on elements p of $\boldsymbol{A}$ and, as we have seen, on terms t of $\boldsymbol{A}$, it makes sense to speak of replacing some variables by others in p or in t. It also makes sense to speak of replacing variables by terms in p; our next purpose is to discuss the analogous process of replacing variables by terms in a term t. Given two terms s and t and a subset K of I, we shall define a term, to be denoted by $(K/s)t$, that depends on s in just about the same way as the image of t under the transformation (K/j), i.e., $(K/j)t$, depends on the variable j. The definition of $(K/s)t$ is similar in spirit, and even in some details, to the definition (cf. §11) of the way a transformation acts on a term.

(13.1) LEMMA. *If s is an H-term, t is a J-term, K is a finite subset of I, and i is an element of $I-(H \cup J \cup K)$, then there exists a unique $(H \cup (J-K))$-term u such that*

$$(13.2) \qquad \mathbf{S}(i/u)p = \mathbf{S}(K/s)\mathbf{S}(i/t)p$$

whenever p is independent of K.

Proof. The plan of attack is similar to the one used in (11.5). If p is independent of K, we write

$$gp = \mathbf{S}(K/s)\mathbf{S}(i/t)p.$$

We construct u by applying (9.2); the set $\bar{J} = H \cup (J-K)$ will play the role of what was there denoted by J. Since $\bar{J} \cup K = H \cup J \cup K$, it follows from our assumption concerning i that $i \in I-(\bar{J} \cup K)$. It is clear that g is a Boolean homomorphism from the range of $\exists(K)$ into A; we proceed to verify that g satisfies the conditions (1)–(4) of (9.2) (with $\bar{J}$ in place of J).

We have

$$\begin{aligned}\exists(i)g\exists(K) &= \exists(i)\mathbf{S}(K/s)\mathbf{S}(i/t)\exists(K) = \mathbf{S}(K/s)\exists(i)\mathbf{S}(i/t)\exists(K) \ [\text{by } (\bar{\text{T}}4)]\\ &= \mathbf{S}(K/s)\mathbf{S}(i/t)\exists(K) \ [\text{by (T4)}] = g\exists(K)\end{aligned}$$

and

$$\begin{aligned}g\exists(K)\exists(i) &= \mathbf{S}(K/s)\mathbf{S}(i/t)\exists(K)\exists(i) = \mathbf{S}(K/s)\mathbf{S}(i/t)\exists(i)\exists(K)\\ &= \mathbf{S}(K/s)\exists(i)\exists(K) \ [\text{by (T3)}] = \mathbf{S}(K/s)\exists(K)\exists(i) = \exists(K)\exists(i).\end{aligned}$$

The justification of the last equation is that the image under $\exists(K)$ of every element of A has $I-K$ for a support, and consequently, by (7.6), $\mathbf{S}(K/s)\exists(K) = \mathbf{S}(\varnothing/s)\exists(K)$. This settles (9.2)(1) and (9.2)(2).

Suppose next that j is an element of $I-\bar{J}$ (i.e., that $j \in' H \cup (J-K)$) and that $j \neq i$. We split the discussion into two cases, according as $j \in K$ or $j \in' K$. If $j \in K$, and if p is an arbitrary element of A, then $I-K$ supports $\exists(K)p$, and therefore $J \cup (I-K)$ supports $\mathbf{S}(i/t)\exists(K)p$ (7.5). It follows (again by (7.5)) that $H \cup (I-K)$ supports $g\exists(K)p$. Since j does not belong to this set, $g\exists(K)p$ is independent of j; consequently

$$g\exists(K)\exists(j)p = g\exists(K)p \ [\text{since } j \in K] = \exists(j)g\exists(K)p.$$

If $j \in' K$, then $j \in' H \cup J$; it follows that

$$\begin{aligned}g\exists(K)\exists(j) &= \mathbf{S}(K/s)\mathbf{S}(i/t)\exists(K)\exists(j) = \mathbf{S}(K/s)\mathbf{S}(i/t)\exists(j)\exists(K)\\ &= \mathbf{S}(K/s)\exists(j)\mathbf{S}(i/t)\exists(K) \ [\text{by (T3)}] = \exists(j)\mathbf{S}(K/s)\mathbf{S}(i/t)\exists(K) \ [\text{by } (\bar{\text{T}}3)]\\ &= \exists(j)g\exists(K);\end{aligned}$$

this concludes the proof of (9.2)(3).

Suppose finally that τ is a finite transformation that lives on $I-(\{i\} \cup \bar{J} \cup K)$. It follows that

$$\begin{aligned}g\exists(K)\mathbf{S}(\tau) &= \mathbf{S}(K/s)\mathbf{S}(i/t)\exists(K)\mathbf{S}(\tau)\\ &= \mathbf{S}(K/s)\mathbf{S}(i/t)\mathbf{S}(\tau)\exists(K) \ [\text{since } \tau \text{ lives on } I-K]\\ &= \mathbf{S}(K/s)\mathbf{S}(\tau)\mathbf{S}(i/t)\exists(K) \ [\text{by (T5)}]\\ &= \mathbf{S}(\tau)\mathbf{S}(K/s)\mathbf{S}(i/t)\exists(K) \ [\text{by } (\bar{\text{T}}5)] = \mathbf{S}(\tau)g\exists(K).\end{aligned}$$

The existence of a $\bar{J}$-term u satisfying (13.2) is now an immediate consequence of (9.2); uniqueness is guaranteed by (9.8).

We prove next that the term u of (13.1) does not depend on the choice i.

(13.3) LEMMA. *Suppose that s is an H-term, t is a J-term, K is a finite subset of I, and i and j are elements of $I-(H \cup J \cup K)$. If u and v are $(H \cup (J-K))$-terms such that*

$$\mathbf{S}(i/u)p = \mathbf{S}(K/s)\mathbf{S}(i/t)p \quad \textit{and} \quad \mathbf{S}(j/v)p = \mathbf{S}(K/s)\mathbf{S}(j/t)p$$

whenever p is independent of K, then $u=v$.

Proof. Since the transformation (i, j) lives on $I-K$, it follows that if p is independent of K, then $\mathbf{S}(i, j)p=p$.[1] This implies that

$$\begin{aligned}\mathbf{S}(i/u)p &= \mathbf{S}(K/s)\mathbf{S}(i/t)p = \mathbf{S}(K/s)\mathbf{S}(i/t)\mathbf{S}(i, j)p \\ &= \mathbf{S}(K/s)\mathbf{S}(i, j)\mathbf{S}(j/t)p \text{ [by (T5)]} = \mathbf{S}(i, j)\mathbf{S}(K/s)\mathbf{S}(j/t)p \text{ [by }(\overline{\text{T5}})] \\ &= \mathbf{S}(i, j)\mathbf{S}(j/v)p = \mathbf{S}(j/v)\mathbf{S}(i, j)p \text{ [by (T5)]} = \mathbf{S}(j/v)p;\end{aligned}$$

the desired equality is now a consequence of (9.8).

We are now prepared to define $(K/s)t$ whenever s is an H-term, t is a J-term, and K is a finite set. Let i be an element of $I-(H \cup J \cup K)$, and define $(K/s)t$ to be the unique $(H \cup (J-K))$-term obtained by applying (13.1), so that

$$\mathbf{S}(i/(K/s)t)p = \mathbf{S}(K/s)\mathbf{S}(i/t)p \tag{13.4}$$

whenever p is independent of K. Lemma (13.3) guarantees that the definition is unambiguous (i.e., that as long as i satisfies the stated conditions, it does not matter how it is chosen). We proceed to derive some of the properties of the concept so defined.

(13.5) LEMMA. *If s is an H-term, t is a J-term, and K is a finite subset of I, then*

$$(K/s)t = (J \cap K/s)t.$$

Proof. Let i be an element of $I-(H \cup J \cup K)$ and let p be an element of A independent of K. If L is a finite support of p such that $L \cap K=\varnothing$, then (by (7.5)) $J \cup L$ supports $\mathbf{S}(i/t)p$, and therefore

$$\begin{aligned}&\mathbf{S}(i/(K/s)t)p = \mathbf{S}(K/s)\mathbf{S}(i/t)p \text{ [by (13.4)]} \\ &\quad = \mathbf{S}(K \cap (J \cup L)/s)\mathbf{S}(i/t)p \text{ [by (7.6)]} = \mathbf{S}(J \cap K/s)\mathbf{S}(i/t)p \text{ [since } L \cap K = \varnothing] \\ &\quad = \mathbf{S}(i/(J \cap K/s)t)p \text{ [by (13.4)]};\end{aligned}$$

the conclusion follows from (9.8).

Using (13.5), we can extend the definition of $(K/s)t$ to the case of a not necessarily finite K simply by writing

$$(K/s)t = (J \cap K/s)t.$$

[1] See p. 239.

If K happens to be finite, then the new definition of $(K/s)t$ is compatible with the old one. Observe that, since $J-K=J-(J \cap K)$, the term $(K/s)t$ is always an $(H \cup (J-K))$-term.

(13.6) THEOREM. *If j is a variable, t is a J-term, and K is a subset of I, then*

$$(K/t_j)t = (K/j)t.$$

Proof. By (13.5), $(K/t_j)t=(J \cap K/t_j)t$. Since the transformations (K/j) and $(J \cap K/j)$ agree on J, it follows from (11.10) that $(K/j)t=(J \cap K/j)t$. These two comments together show that there is no loss of generality in assuming that the set K is finite.

The transformation (K/j) lives on $\{j\} \cup K$. If $i \in I-(\{j\} \cup J \cup K)$ and if p is independent of $\{j\} \cup K$, then

$$\begin{aligned}\mathbf{S}(i/(K/j)t)p &= \mathbf{S}(K/j)\mathbf{S}(i/t)p \text{ [by (11.8)]}\\ &= \mathbf{S}(K/t_j)\mathbf{S}(i/t)p \text{ [by (10.9)]} = \mathbf{S}(i/(K/t_j)t)p \text{ [by (13.4)]}.\end{aligned}$$

The conclusion now follows from the uniqueness theorem (9.8).

(13.7) THEOREM. *If s is a term, if τ is a transformation, and if k is an element of I such that $\tau^{-1}\{k\}=\{k\}$, then*

$$(k/\tau s)\tau = \tau(k/s).$$

REMARK. The conclusion means that if t is a term, then $(k/\tau s)(\tau t) =\tau((k/s)t)$. Recall that if $j \in I$, and if k and τ are as in the theorem, then $(k/\tau j)\tau=\tau(k/j)$.

Proof. Suppose that s is an H-term, and let t be a J-term. If τ_0 is defined by writing $\tau_0=\tau$ on $H \cup J$ and $\tau_0=\delta$ outside $H \cup J$, then τ_0 is a finite transformation, $\tau_0^{-1}\{k\}=\{k\}$, and, in view of (11.13), the desired conclusion $(k/\tau s)\tau t=\tau(k/s)t$ is equivalent to the corresponding equation for τ_0. We may therefore assume that τ is finite in the first place.

Let K be a finite set containing k such that τ lives on K, and let i be an element of $I-(H \cup J \cup \tau H \cup \tau J \cup K)$. If p is independent of K, then

$$\begin{aligned}\mathbf{S}(i/(k/\tau s)\tau t)p &= \mathbf{S}(k/\tau s)\mathbf{S}(i/\tau t)p \text{ [by (13.4)]}\\ &= \mathbf{S}(k/\tau s)\mathbf{S}(\tau)\mathbf{S}(i/t)p \text{ [by (11.8)]} = \mathbf{S}(\tau)\mathbf{S}(k/s)\mathbf{S}(i/t)p \text{ [by (11.15)]}\\ &= \mathbf{S}(\tau)\mathbf{S}(i/(k/s)t)p \text{ [by (13.4)]} = \mathbf{S}(i/\tau(k/s)t)p \text{ [by (11.8)]}.\end{aligned}$$

The conclusion follows from the uniqueness theorem (9.8).

(13.8) COROLLARY. *If s is an H-term, if τ is a transformation such that $\tau=\delta$ on H, and if k is an element of I such that $\tau^{-1}\{k\}=\{k\}$, then*

$$(k/s)\tau = \tau(k/s).$$

REMARK. As in (13.7), the conclusion means that $(k/s)(\tau t)=\tau((k/s)t)$ for every term t.

Proof. Apply (11.13) and (13.8).

(13.9) THEOREM. *If s is an H-term, t is a J-term, K is a finite subset of I, and σ is a transformation of type $(K, H \cup J)$, then*

$$(K/s)t = (\sigma K/s)\sigma t.$$

Proof. Let i be an element of $I-(H \cup J \cup K \cup \sigma J \cup \sigma K)$. If p is an element of A such that p is independent of $K \cup \sigma K$, then

$$\begin{aligned} \mathbf{S}(i/(\sigma K/s)\sigma t)p &= \mathbf{S}(\sigma K/s)\mathbf{S}(i/\sigma t)p \text{ [by (13.4)]} \\ &= \mathbf{S}(\sigma K/s)\mathbf{S}(\sigma)\mathbf{S}(i/t)p \text{ [by (11.8)]} = \mathbf{S}(K/s)\mathbf{S}(i/t)p \text{ [by (4.1)]} \\ &= \mathbf{S}(i/(K/s)t)p \text{ [by (13.4)]}. \end{aligned}$$

The conclusion follows from the uniqueness theorem (9.8).

(13.10) COROLLARY. *If s is an H-term, t is a J-term, k is an element of I, and i is an element of $I-(H \cup J)$, then*

$$(k/s)t = (i/s)(k/i)t.$$

Proof. Apply (13.9) with $K=\{k\}$ and $\sigma=(k/i)$.

14. **Terms in operations.** Suppose that T is an n-place operation and that $t_1, \cdots, t_n$ are terms; the purpose of this section is to show how it is possible to assign a sensible meaning to the expression $T(t_1, \cdots, t_n)$. The theory is similar to the corresponding part of the theory of terms in predicates.

(14.1) LEMMA. *Suppose that $J_0, J_1, \cdots, J_n$ are finite subsets of I, and that t_0 is a J_0-term, t_1 is a J_1-term, $\cdots$, t_n is a J_n-term; write $J=J_1 \cup \cdots \cup J_n$. If $i_1, \cdots, i_n, k_1, \cdots, k_n$ are $2n$ distinct variables in $I-J$, and if $\{i_1, \cdots, i_n\} \subset I-J_0$, then*

$$(i_1/t_1) \cdots (i_n/t_n)(k_1/i_1) \cdots (k_n/i_n)t_0 = (k_1/t_1) \cdots (k_n/t_n)t_0. \tag{14.2}$$

Proof. If $n \neq 1$, then (13.8) (with $s=t_n$, $\tau=(k_1/i_1)$, and $k=i_n$) implies that $(i_n/t_n)(k_1/i_1)=(k_1/i_1)(i_n/t_n)$. After a repeated application of this reasoning, the desideratum reduces to

$$(i_1/t_1)(k_1/i_1) \cdots (i_n/t_n)(k_n/i_n)t_0 = (k_1/t_1) \cdots (k_n/t_n)t_0. \tag{14.3}$$

Corollary (13.10) implies that

$$(i_n/t_n)(k_n/i_n)t_0 = (k_n/t_n)t_0.$$

The term $(k_n/t_n)t_0$ is a $(J_n \cup J_0)$-term, and $\{i_1, \cdots, i_n\} \subset I-(J_n \cup J_0)$. Since this comment prepares the ground for the induction step, (14.3) follows from an n-fold application of (13.10).

(14.4) LEMMA. *Suppose that $J_1, \cdots, J_n$ are finite subsets of I, that t_1 is a J_1-term, $\cdots$, t_n is a J_n-term, and that T is an n-place operation; write $J=J_1 \cup \cdots \cup J_n$. If $i_1, \cdots, i_n$ are distinct variables in $I-J$, and $j_1, \cdots, j_n$*

are distinct variables in $I-J$, *then*

(14.5) $(i_1/t_1) \cdots (i_n/t_n)T(i_1, \cdots, i_n) = (j_1/t_1) \cdots (j_n/t_n)T(j_1, \cdots, j_n).$

Proof. Let $k_1, \cdots, k_n$ be distinct variables in $I-(J \cup \{i_1, \cdots, i_n\} \cup \{j_1, \cdots, j_n\})$ and write $T(k_1, \cdots, k_n)=t_0$. Since $(k_1/i_1) \cdots (k_n/i_n) = (k_1, \cdots, k_n/i_1, \cdots, i_n)$, it follows from (12.3) that

(14.6) $$T(i_1, \cdots, i_n) = (k_1/i_1) \cdots (k_n/i_n)t_0.$$

By (14.6) and (14.1), the left side of (14.5) is equal to $(k_1/t_1) \cdots (k_n/t_n)t_0$; the same argument with $j_1, \cdots, j_n$ in place of $i_1, \cdots, i_n$ shows that the right side of (14.5) is equal to the same thing.

If T is an n-place operation, and if t_1 is a J_1-term, $\cdots$, t_n is a J_n-term, we define $T(t_1, \cdots, t_n)$ by writing

(14.7) $$T(t_1, \cdots, t_n) = (i_1/t_1) \cdots (i_n/t_n)T(i_1, \cdots, i_n)$$

whenever $i_1, \cdots, i_n$ are distinct variables in $I-(J_1 \cup \cdots \cup J_n)$. Lemma (14.4) guarantees that this is an unambiguous definition.

(14.8) THEOREM. *If* T *is an* n*-place operation, if* $t_1, \cdots, t_n$ *are terms, and if* τ *is a transformation, then*

$$\tau T(t_1, \cdots, t_n) = T(\tau t_1, \cdots, \tau t_n).$$

Proof. Suppose that t_1 is a J_1-term, $\cdots$, t_n is a J_n-term, and write $J=J_1 \cup \cdots \cup J_n$. Let τ_0 be the transformation that agrees with τ on J and that is equal to δ outside J. Since the desired conclusion is the same for τ as for τ_0, we may henceforth assume that the transformation τ is finite to begin with.

If $i_1, \cdots, i_n$ are distinct elements of $I-(J \cup \tau J)$, then $\tau^{-1}\{i_1\}=\{i_1\}$, $\cdots$, $\tau^{-1}\{i_n\}=\{i_n\}$, and therefore

$$\begin{aligned}\tau T(t_1, \cdots, t_n) &= \tau(i_1/t_1) \cdots (i_n/t_n)T(i_1, \cdots, i_n) \quad [\text{by } (14.7)]\\ &= (i_1/\tau t_1) \cdots (i_n/\tau t_n)\tau T(i_1, \cdots, i_n) \quad [\text{by } (13.7)]\\ &= (i_1/\tau t_1) \cdots (i_n/\tau t_n)T(i_1, \cdots, i_n) \quad [\text{by } (12.3)]\\ &= T(\tau t_1, \cdots, \tau t_n).\end{aligned}$$

We conclude this discussion of terms in operations by observing that there is a natural way of combining operations with other operations to form operations of higher degree. If T is an operation of degree n, and if $T_1, \cdots, T_n$ are operations of degrees $m_1, \cdots, m_n$ respectively, we write $m=m_1+ \cdots +m_n$, and we define an operation $T(T_1, \cdots, T_n)$ of degree m by substituting $T_1, \cdots, T_n$ into T. The precise details can be obtained by an obvious modification of the corresponding discussion in § 12.

UNIVERSITY OF CHICAGO,
CHICAGO, ILL.

VII

EQUALITY

ALGEBRAIC LOGIC IV. EQUALITY IN POLYADIC ALGEBRAS[1]

Introduction. A standard way to begin the study of symbolic logic is to describe one after another the propositional calculus, the monadic functional calculus, the pure first-order functional calculus, and the functional calculus with equality. The algebraic aspects of these logical calculi belong to the theories of Boolean algebras, monadic algebras, polyadic algebras, and cylindric algebras respectively. The connection between the propositional calculus and Boolean algebras is well-known; for a recent exposition of it (and also of some aspects of the more advanced theories) see *The basic concepts of algebraic logic*, Amer. Math. Monthly vol. 63 (1956) pp. 363–387. Monadic algebras and polyadic algebras were studied in the first three papers of this sequence; [see *Algebraic logic* III, Trans. Amer. Math. Soc. vol. 83 (1956) pp. 430–470, and the references given there]. Cylindric algebras were introduced by Tarski and Thompson [*Some general properties of cylindric algebras*, Bull. Amer. Math. Soc. Abstract 58-1-85; see also Tarski, *A representation theorem for cylindric algebras*, Bull. Amer. Math. Soc. Abstract 58-1-86.]

Most of what was done for polyadic algebras in [II] and [III] was restricted to locally finite polyadic algebras of infinite degree. (The Roman numerals refer to the other parts of this sequence.) Since it is known that every locally finite cylindric algebra of infinite degree possesses a natural polyadic structure, the results of those papers apply to cylindric algebras without any change. This paper, on the other hand, is mostly pre-cylindric; its main purpose is to discuss (in algebraic language) the introduction of equality.

The paper is not self-contained. The notation introduced in §1 of [III] will be used without any further explicit reference, and some of the basic concepts studied in [III] (notably the concept of a predicate) will also be assumed known. (The most difficult part of [III], the theory of terms, is used in §9 only.) At one point (§6) the representation theorem for simple polyadic algebras is needed, and later (§7) we make use of the duality theory for monadic algebras. Most parts of the paper, however (and, in particular, all definitions and the statements of all the theorems) are accessible to anyone who has skimmed through [III], provided that, in addition, he is acquainted with the elementary theory of functional polyadic algebras. For the con-

Received by the editors January 5, 1956 and, in revised form, July 16, 1956.

(1) The work on this paper was sponsored in part by the National Science Foundation, NSF grant 2266.

venience of the reader, the necessary facts from that theory are summarized in §1 below.

§2 introduces the concept of equality and establishes the connection with cylindric algebras; §3 proves that equality is unique. §§4 and 6 study equality in functional algebras; the culmination of this work is the algebraic version of the completeness theorem for the functional calculus with equality. The most interesting negative results are in §5; it is shown there that an equality need not always exist (not even for the relatively well-behaved class of locally finite functional algebras). This fact makes it all the more pleasant to learn (§7) that an equality can always be adjoined. The last two sections are devoted to some results on the algebraic behavior of the description operator and touch on the algebraic meaning of the logical theorem that asserts the eliminability of that operator.

1. **Notation.** In addition to the Boolean notation described in [III], we shall sometimes use $p \to q$ to denote $p' \vee q$, and we shall use $\boldsymbol{O}$ to denote the simple Boolean algebra $\{0, 1\}$. The supremum and the infimum of a subset $\boldsymbol{E}$ of a Boolean algebra (if they exist) are denoted by $\bigvee \boldsymbol{E}$ and $\bigwedge \boldsymbol{E}$ respectively.

If I is a set and X is a nonempty set, we shall write X^I for the set of all functions from I into X; the value of a function x in X^I at an element i of I will always be denoted by x_i. If τ is a transformation on I, we denote by τ_* the mapping of X^I into itself such that $(\tau_* x)_i = x_{\tau i}$ for all i. If J is a subset of I, we denote by J_* the equivalence relation in X^I such that xJ_*y if and only if $x_i = y_i$ for all i in $I - J$.

If $\boldsymbol{B}$ is a Boolean algebra, then the set of all functions from X^I into $\boldsymbol{B}$ is a Boolean algebra with respect to the obvious pointwise operations. A function p from X^I into $\boldsymbol{B}$ is *independent* of a subset J of I if $p(x) = p(y)$ whenever xJ_*y; the set J is a *support* of p (or J *supports* p) if p is independent of $I - J$. A function p is *finite-dimensional* if it has a finite support. If p is a function from X^I into $\boldsymbol{B}$ (not necessarily finite-dimensional), we write $\mathbf{S}(\tau)p$ for the function defined by $\mathbf{S}(\tau)p(x) = p(\tau_* x)$. (A symbol such as $\mathbf{S}(\tau)p(x)$, here and below, means $(\mathbf{S}(\tau)p)(x)$.) If p is a function from X^I into $\boldsymbol{B}$ and if the set $\{p(y) : xJ_*y\}$ has a supremum in $\boldsymbol{B}$ for each x in X^I, we write $\exists(J)p$ for the function whose value at x is that supremum. A *functional polyadic algebra* is a Boolean subalgebra $\boldsymbol{A}$ of the algebra of all functions from X^I into $\boldsymbol{B}$, such that $\mathbf{S}(\tau)p \in \boldsymbol{A}$ whenever $p \in \boldsymbol{A}$ and τ is a transformation on I, and such that $\exists(J)p$ exists and belongs to $\boldsymbol{A}$ whenever $p \in \boldsymbol{A}$ and J is a subset of I. The set X is called the *domain* of this functional algebra; when we wish to indicate the dependence of the algebra on I, X, and $\boldsymbol{B}$, we call it a $\boldsymbol{B}$-valued I-algebra over X. A functional polyadic algebra is a polyadic algebra with respect to the operator mappings $\mathbf{S}$ and $\exists$ defined above, and for elements of a functional algebra the set-theoretic and the algebraic definitions of support are equivalent.

As in [III], we shall be working with a fixed (not necessarily functional) polyadic algebra $(A, I, \mathbf{S}, \exists)$; suitable warnings will indicate our few temporary deviations from this notation. The algebra $(A, I, \mathbf{S}, \exists)$ of [III] was assumed to be locally finite and of infinite degree. In this paper those conditions are needed only about half the time, and, consequently, they will not be assumed here; it will be clearer (and more elegant) to build them into the hypotheses of the theorems in which they are indispensable. When nothing is said to the contrary, the algebra A may have finite degree, or else, at the other extreme, it need not even be locally finite.

2. **Equality.** A binary predicate E is *reflexive* if $E(i, i) = 1$ for every variable i. Except in the trivial case when the set of variables is empty (i.e., when the algebra A is simply a Boolean algebra with no additional polyadic structure), a predicate E is reflexive if and only if there exists at least one variable i_0 such that $E(i_0, i_0) = 1$. Indeed, if this is the case, and if i is any variable, then

$$E(i, i) = \mathbf{S}(i_0/i)E(i_0, i_0) = \mathbf{S}(i_0/i)1 = 1.$$

A binary predicate E of A is *substitutive* if $p \wedge E(i, j) \leqq \mathbf{S}(i/j)p$ whenever i and j are variables and $p \epsilon A$. Except in the trivial case when the set of variables has fewer than two elements (i.e., when the algebra A is simply a Boolean algebra or else a monadic algebra), a predicate E is substitutive if and only if there exists at least one pair of distinct variables i_0 and j_0 such that $p \wedge E(i_0, j_0) \leqq \mathbf{S}(i_0/j_0)p$ whenever $p \epsilon A$. Suppose, indeed, that this is the case and let i and j be any two variables. If $i = j$, the assertion is that $p \wedge E(i, i) \leqq p$ for all p in A, and this is obvious. If $i \neq j$, then there exists a permutation τ such that $\tau i_0 = i$ and $\tau j_0 = j$, and, consequently, such that $\tau(i_0/j_0)\tau^{-1} = (i/j)$. Since, by assumption,

$$\mathbf{S}(\tau^{-1})p \wedge E(i_0, j_0) \leqq \mathbf{S}(i_0/j_0)\mathbf{S}(\tau^{-1})p$$

whenever $p \in A$, the desideratum follows by applying $\mathbf{S}(\tau)$ to both sides of this inequality.

The condition used to define substitutivity implies a superficially stronger version of itself; if E is substitutive, then

$$p \wedge E(i, j) = \mathbf{S}(i/j)p \wedge E(i, j) \tag{2.1}$$

whenever i and j are variables and $p \epsilon A$. Substitutivity implies that the left side of (2.1) is dominated by the right side; it is, therefore, sufficient to prove that

$$\mathbf{S}(i/j)p \wedge E(i, j) \leqq p.$$

The proof consists of the following straightforward computation:

$$\begin{aligned}(\mathbf{S}(i/j)p \wedge E(i, j)) \wedge p' &= \mathbf{S}(i/j)p \wedge (p' \wedge E(i, j)) \\ \leqq \mathbf{S}(i/j)p \wedge \mathbf{S}(i/j)p' &= \mathbf{S}(i/j)(p \wedge p') = 0.\end{aligned}$$

From (2.1) we can derive still another version of substitutivity; the new version treats the variables i and j symmetrically. The assertion is that if E is substitutive, then

$$p \wedge E(i, j) = \mathbf{S}(i, j)p \wedge E(i, j). \tag{2.2}$$

The proof of (2.2) is based on (2.1) and on the fact that $(i/j)(i, j) = (i/j)$; it runs as follows:

$$\begin{aligned} p \wedge E(i, j) &= \mathbf{S}(i/j)p \wedge E(i, j) \\ &= \mathbf{S}(i/j)\mathbf{S}(i, j)p \wedge E(i, j) = \mathbf{S}(i, j)p \wedge E(i, j). \end{aligned}$$

A binary predicate E is *symmetric* if $E(i, j) \leqq E(j, i)$ whenever i and j are variables; applying this condition with i and j interchanged, we conclude that E is symmetric if and only if

$$E(i, j) = E(j, i) \tag{2.3}$$

whenever i and j are variables. A binary predicate E is *transitive* if

$$E(i, j) \wedge E(j, k) \leqq E(i, k) \tag{2.4}$$

whenever i, j, and k are variables. Techniques similar to the ones used above (for reflexivity and for substitutivity) show that, except in the cases when the number of variables is too small, E is symmetric if and only if there exists at least one pair of distinct variables i_0 and j_0 such that $E(i_0, j_0) \leqq E(j_0, i_0)$, and similarly, E is transitive if and only if there exists at least one triple of distinct variables i_0, j_0, and k_0, such that $E(i_0, j_0) \wedge E(j_0, k_0) \leqq E(i_0, k_0)$.

An *equality* for $\mathbf{A}$ is a reflexive and substitute binary predicate of $\mathbf{A}$. In the remainder of this section we derive some elementary consequences of this definition.

(2.5) LEMMA. *Every equality is symmetric and transitive.*

Proof. To prove that an equality E is symmetric, apply substitutivity to $E'(i, j)$; to prove that it is transitive, apply substitutivity to $E(j, k)$.

(2.6) LEMMA. *If E is an equality, then $\exists(i)E(i, j) = 1$ for all i and j.*

Proof. $\exists(i)E(i, j) \geqq \mathbf{S}(i/j)E(i, j) = E(j, j) = 1$.

(2.7) LEMMA. *If E is an equality for $\mathbf{A}$ and if i and j are distinct variables, then $\exists(i)(p \wedge E(i, j)) = \mathbf{S}(i/j)p$ for every p in $\mathbf{A}$.*

Proof. Apply $\exists(i)$ to (2.1) and use (2.6).

(2.8) LEMMA. *If E is an equality for $\mathbf{A}$ and if p is an element of $\mathbf{A}$ independent of the variable i, then*

$$p = \exists(i)(\mathbf{S}(j/i)p \wedge E(i, j)).$$

Proof. If $i=j$, the assertion is merely that $p=\exists(i)p$, and this is true by assumption. If $i\neq j$, replace p in (2.7) by $\mathbf{S}(j/i)p$; the desired result follows from the fact that $(i/j)(j/i)=(i/j)$, together with the fact that $\mathbf{S}(i/j)p=p$ (since p is independent of i).

An alternative proof of (2.8) can be given directly (without reference to (2.7)) by another reference to (2.1).

(2.9) LEMMA. *If E is an equality for $\boldsymbol{A}$ and if i and j are distinct variables, then* $\exists(i)(p\wedge E(i,j))\wedge\exists(i)(p'\wedge E(i,j))=0$ *for every p in $\boldsymbol{A}$.*

Proof. By (2.7), the two elements whose infimum is asserted to be 0 are equal to $\mathbf{S}(i/j)p$ and $\mathbf{S}(i/j)p'$; the conclusion follows from the fact that $\mathbf{S}(i/j)$ is a Boolean homomorphism.

(2.10) LEMMA. *If E is an equality and if i, j, and k are variables such that $i\,\epsilon'\,\{j,k\}$, then*

$$\exists(i)(E(i,j)\wedge E(i,k)) = E(j,k).$$

Proof. Put $p=E(j,k)$ in (2.8); the assumption $i\,\epsilon'\,\{j,k\}$ guarantees that p is independent of i.

A cylindric algebra is a Boolean algebra $\boldsymbol{A}$, together with a set I, a mapping $\exists$ from I into quantifiers on $\boldsymbol{A}$, and a mapping E from I^2 into $\boldsymbol{A}$, such that (i) the values of $\exists$ commute with each other, (ii) E is reflexive, and (iii) Lemmas (2.9) and (2.10) hold for E. Thus the Lemmas (2.9) and (2.10), together with the more elementary properties of equalities, can be summed up by saying that a polyadic algebra with equality is a cylindric algebra.

3. **The uniqueness of equality.** How many equalities is a polyadic algebra likely to possess? The answer is: not more than one. We shall presently see some examples of polyadic algebras for which, for one reason or another, there is no equality at all. First, however, we prove (in two different ways) that there can never be more than one.

(3.1) LEMMA. *If E and F are equalities for the same polyadic algebra, then $E=F$ (i.e., $E(i,j)=F(i,j)$ for all i and j).*

Proof. Compute, as follows:

$$\begin{aligned} E(i,j) &= F(j,j)\wedge E(i,j) \ \text{[since } F \text{ is reflexive]}\\ &= \mathbf{S}(i/j)F(i,j)\wedge E(i,j) \ \text{[since } F \text{ is a predicate]}\\ &= F(i,j)\wedge E(i,j) \ \text{[since } E \text{ is substitutive]}\\ &= \mathbf{S}(i/j)E(i,j)\wedge F(i,j) \ \text{[since } F \text{ is substitutive]}\\ &= E(j,j)\wedge F(i,j) \ \text{[since } E \text{ is a predicate]}\\ &= F(i,j) \ \text{[since } E \text{ is reflexive]}. \end{aligned}$$

Using a different method, we can recapture this result, and more.

(3.2) THEOREM. *If E is an equality for* $\mathbf{A}$, *and if*

$$D(i,j) = \{p \in \mathbf{A} : \mathbf{S}(i/j)p = 1\}, \tag{3.3}$$

then the set $\mathbf{D}(i,j)$ *has an infimum in* $\mathbf{A}$ *for each i and j, and*

$$\bigwedge D(i,j) = E(i,j). \tag{3.4}$$

REMARK. In intuitive language, the theorem says that to assert $E(i, j)$ is the same as to assert, simultaneously, every proposition that becomes true when i is replaced by j.

Proof. If $\mathbf{S}(i/j)p = 1$, then, by (2.1), $p \wedge E(i,j) = E(i,j)$, so that $E(i,j) \leqq p$; in other words, $E(i,j)$ is a lower bound of $D(i,j)$. Since $\mathbf{S}(i/j)E(i,j) = E(j,j) = 1$, it follows that $E(i,j) \in D(i,j)$. The preceding two sentences imply that the set $D(i, j)$ contains a least element, namely $E(i, j)$; this completes the proof of the theorem.

Note that (3.2) does not assert that $D(i, j)$ always has an infimum in $\mathbf{A}$, nor that if $D(i,j)$ does have an infimum in $\mathbf{A}$, then $\mathbf{A}$ has an equality. There are polyadic algebras in which $D(i, j)$ does not have an infimum for any pair of distinct variables i and j, and there are polyadic algebras in which $D(i, j)$ has an infimum for every i and j, but if

$$F(i,j) = \bigwedge D(i,j), \tag{3.5}$$

then F is not an equality. The last point is worth a second glance. Suppose that $\mathbf{A}$ is such that $D(i, j)$ has an infimum for every i and j and define F by (3.5). Since $D(i, i) = \{1\}$, it follows that

$$F(i,i) = 1. \tag{3.6}$$

Since $p' \vee \mathbf{S}(i/j)p \in D(i,j)$ for every p in $\mathbf{A}$, it follows that

$$p \wedge F(i,j) \leqq p \wedge (p' \vee \mathbf{S}(i/j)p) = p \wedge \mathbf{S}(i/j)p \leqq \mathbf{S}(i/j)p. \tag{3.7}$$

These two comments show that if F were a predicate, then F would be reflexive and substitutive. The only way F can fail to be an equality is that it fails to be a predicate, i.e., that it fails to be suitably related to the transformation structure of $\mathbf{A}$. Our examples will show that this unpleasant phenomenon is realizable.

4. **Functional equality.** Suppose that I is a set, X is a nonempty set, and $\mathbf{B}$ is a Boolean algebra. If i and j are any two (not necessarily distinct) elements of I, let $E_0(i,j)$ be the function from X^I to $\mathbf{B}$ defined, for each x in X^I, by

$$E_0(i,j)(x) = \begin{cases} 1 \text{ if } x_i = x_j, \\ 0 \text{ if } x_i \neq x_j. \end{cases} \tag{4.1}$$

The function-valued function E_0 (whose domain is I^2 and whose values are functions on X^I) will be called the *functional equality* associated with I, X, and $\boldsymbol{B}$. Note that since $E_0(i, j)(x) \in \boldsymbol{O}$ for all i, j, and x, the algebra $\boldsymbol{B}$ does not play an important role in this definition.

If τ is a transformation on I, then

$$(4.2) \qquad \mathbf{S}(\tau)E_0(i, j) = E_0(\tau i, \tau j)$$

for all i and j. Indeed, for all x,

$$\mathbf{S}(\tau)E_0(i, j)(x) = E_0(i, j)(\tau_* x) = \begin{cases} 1 \text{ if } x_{\tau i} = x_{\tau j}, \\ 0 \text{ if } x_{\tau i} \neq x_{\tau j}. \end{cases}$$

This result implies that whenever $\boldsymbol{A}$ is a $\boldsymbol{B}$-valued functional I-algebra over X such that $E_0(i, j) \in \boldsymbol{A}$ for all i and j, then E_0 is a predicate of $\boldsymbol{A}$.

Since, trivially,

$$(4.3) \qquad E_0(i, i) = 1$$

for all i, it follows that whenever E_0 is a predicate of a functional algebra, it is automatically reflexive.

If p is an arbitrary function from X^I into $\boldsymbol{B}$, then

$$(4.4) \qquad p \wedge E_0(i, j) = \mathbf{S}(i/j)p \wedge E_0(i, j)$$

for all i and j. Indeed, if x is such that $x_i \neq x_j$, then both sides of (4.4) vanish at x; if, on the other hand, $x_i = x_j$, then both sides of (4.4) are equal, at x, to $p(x)$. This result implies that whenever E_0 is a predicate of a functional algebra, it is automatically substitutive.

From the preceding three paragraphs we conclude that whenever all the values of E_0 (i.e., all the functions $E_0(i, j)$) belong to a functional polyadic algebra $\boldsymbol{A}$, then E_0 is an equality for $\boldsymbol{A}$; this justifies the appellation "functional equality."

(4.5) LEMMA. *If $\boldsymbol{A}$ is a functional I-algebra, then, for all i and j in I,*

$$\boldsymbol{D}(i, j) = \{p \in \boldsymbol{A}: \mathbf{S}(i/j)p = 1\} = \{p \in \boldsymbol{A}: E_0(i, j) \leqq p\},$$

where E_0 is the associated functional equality.

REMARK. It is not claimed that $E_0(i, j)$ necessarily belongs to $\boldsymbol{A}$; the assertion is merely that a necessary and sufficient condition that $\mathbf{S}(i/j)p = 1$ is that $E_0(i, j)(x) \leqq p(x)$ for all x.

Proof. Suppose that $\mathbf{S}(i/j)p = 1$, i.e., that $p((i/j)_* x) = 1$ for all x. If x is such that $x_i = x_j$, then $(i/j)_* x = x$, so that $p(x) = 1$, and therefore $E_0(i, j)(x) \leqq p(x)$. If $x_i \neq x_j$, the same inequality holds, because in that case, $E_0(i, j)(x) = 0$. Suppose now, conversely, that $E_0(i, j) \leqq p$. It follows (cf. (4.2)) that $E_0(j, j) = \mathbf{S}(i/j)E_0(i, j) \leqq \mathbf{S}(i/j)p$. Since (cf. (4.3)) $E_0(j, j) = 1$, this implies that $\mathbf{S}(i/j)p = 1$.

(4.6) COROLLARY. *If E is an equality for a functional I-algebra, then, for all i and j in I,*

$$E_0(i, j) \leqq E(i, j),$$

where E_0 is the associated functional equality.

Proof. The result follows immediately from (4.5) and (3.2). It follows also, without the use of (3.2), by an application of (4.4) to $E'(i, j)$ in the role of p.

It is sometimes useful to know how a functional equality E_0 (associated with I, X, and $\mathbf{B}$, say) behaves with respect to quantification. Since the values of $E_0(i, j)$ are always in $\mathbf{O}$, the supremum that defines $\exists(J)E_0(i, j)$ always exists; it is, in fact, given by

$$(4.7) \qquad \exists(J)E_0(i, j) = \begin{cases} E_0(i, j) & \text{if } \{i, j\} \cap J = \varnothing, \\ 1 & \text{otherwise.} \end{cases}$$

Indeed, E_0 is an equality for the functional polyadic algebra $\mathbf{A}$ of all functions from X^I into $\mathbf{O}$; the desired conclusion follows from (2.6) together with the fact that (since E_0 is a predicate) $\{i, j\}$ supports $E_0(i, j)$.

5. **Examples.** All the examples to be constructed in this section will be locally finite, $\mathbf{O}$-valued functional algebras. Suppose that I is a set, X is a nonempty set, and $\mathbf{C}$ is a Boolean algebra of functions from X (not X^I) into $\mathbf{O}$; the sets I and X and the Boolean algebra $\mathbf{C}$ will be held fixed throughout the construction.

If J is a finite subset of I and if r_i is an element of $\mathbf{C}$ for each i in J, a function p from X^I into $\mathbf{O}$ is defined by

$$(5.1) \qquad p(x) = \bigwedge \{r_i(x_i): i \in J\}.$$

The function p may be called the *tensor product* of the functions r_i, for i in J. Let $\mathbf{A}$ be the Boolean algebra of functions from X^I into $\mathbf{O}$ generated by all such tensor products. It is easy to verify that every element of $\mathbf{A}$ is the supremum of a finite number of tensor products. It follows, in particular, that every function in $\mathbf{A}$ is finite-dimensional.

We prove that $\mathbf{A}$ is a functional polyadic I-algebra over X. Suppose, for this purpose, that $p \in \mathbf{A}$ and that τ is a transformation on I; we must show that $\mathbf{S}(\tau)p \in \mathbf{A}$. It is sufficient to treat the case in which p is given by (5.1). Since

$$\mathbf{S}(\tau)p(x) = p(\tau_* x) = \bigwedge \{r_i((\tau_* x)_i): i \in J\} = \bigwedge \{r_i(x_{\tau i}): i \in J\},$$

it follows that if, for each j in τJ,

$$s_j = \bigwedge \{r_i: i \in J \cap \tau^{-1}\{j\}\},$$

then

$$\mathbf{S}(\tau)p(x) = \bigwedge \{s_j(x_j): j \in \tau J\},$$

and hence that $\mathbf{S}(\tau)p$ is the tensor product of the s_j, for j in τJ. This settles transformations.

Suppose next that $p \in A$ and that K is a subset of I; we must show that $\exists(K)p \in A$. Again it is sufficient to treat the case in which p is given by (5.1). Since, in that case,

$$\begin{aligned}\exists(K)p(x) &= \bigvee\{p(y): xK_*y\} = \bigvee\{\bigwedge\{r_i(y_i): i \in J\}: xK_*y\} \\ &= \bigvee\{\bigwedge\{r_i(y_i): i \in J - K\} \wedge \bigwedge\{r_i(y_i): i \in J \cap K\}: xK_*y\} \\ &= \bigvee\{\bigwedge\{r_i(x_i): i \in J - K\} \wedge \bigwedge\{r_i(y_i): i \in J \cap K\}: xK_*y\} \\ &= \bigwedge\{r_i(x_i): i \in J - K\},\end{aligned}$$

it follows that $\exists(K)p$ is the tensor product of the r_i, for i in $J-K$, and hence that A is closed under the application of $\exists(K)$.

The time has come to specialize: from now on we shall assume that X is the set of all integers. Let m be a fixed positive integer, and let $\mathbf{C}_m$ be the set of all those functions from X into $\mathbf{O}$ that are periodic of period m; in other words, $r \in \mathbf{C}_m$ if and only if $r(x+m)=r(x)$ for all x in X. It is clear that $\mathbf{C}_m$ is a Boolean algebra; we denote by $\mathbf{A}_m$ the functional I-algebra obtained from $\mathbf{C}_m$ by the method of tensor products discussed above. We shall say that a function p from X^I into $\mathbf{O}$ is periodic of period m if it has that property in each coordinate separately, or, in other words, if, for each i in I, $p(x)=p(y)$ whenever xi_*y and $x_i=y_i+m$.

(5.2) Lemma. *A necessary and sufficient condition that a function p from X^I into $\mathbf{O}$ belong to $\mathbf{A}_m$ is that it be finite-dimensional and periodic of period m.*

Proof. The necessity of the condition is easy; we omit the proof. To prove sufficiency, suppose that p is a function from X^I into $\mathbf{O}$ with finite support J and period m. Let z be an element of X^I such that $0 \leqq z_i \leqq m-1$ when $i \in J$ and $z_i=0$ when $i \in I-J$. The set of all such z's is finite, and, since p is independent of $I-J$ and p is periodic of period m, the function p is uniquely determined by its restriction to that finite set. Suppose that $p(z)=1$ at one such z. For each i in J, let r_i be the function from X into $\mathbf{O}$ such that r_i is equal to 1 at every integer that is congruent to z_i modulo m and r_i is equal to 0 otherwise; it follows that $r_i \in \mathbf{C}_m$ and that the tensor product of the r_i, for $i \in J$, is dominated by p. Since the set of all z's is finite, the set of all such tensor products is also finite. The supremum of all such tensor products is exactly p, and therefore $p \in \mathbf{A}_m$, as asserted.

If i and j are in I, we define a function $E_m(i, j)$ from X^I into $\mathbf{O}$ by

$$E_m(i,j)(x) = \begin{cases} 1 \text{ if } x_i \equiv x_j \pmod{m}, \\ 0 \text{ if } x_i \not\equiv x_j \pmod{m}. \end{cases} \tag{5.3}$$

The definition implies that $E_m(i, j)$ is finite-dimensional (in fact, $\{i, j\}$ supports $E_m(i, j)$) and that $E_m(i, j)$ is periodic of period m. It follows from (5.2) that $E_m(i, j) \in A_m$ for all i and j. An obvious modification of the arguments used to show that a functional equality has the properties of an equality (cf. (4.2), (4.3), and (4.4)) serves to show that E_m is an equality for A_m.

If A_0 is the set of all functions from X^I into O, then A_0 is a functional polyadic algebra and A_m is a polyadic subalgebra of A_0. Since E_m is not equal to the functional equality E_0, this example shows that a subalgebra of an algebra with equality may possess an equality different from the one that works for the large algebra. Since the functional algebras A_m and A_0 have the same domain, their elements, regarded as propositional functions, have the same subject matter. Accordingly we may say, in a very crude but nevertheless suggestive phrase, that the meaning of equality depends not so much on what we are talking about as on how much we are allowed to say about it. If we are allowed to say everything, then equality is the same as identity; if we are allowed to talk only in terms of a fixed modulus m, then equality is congruence modulo m.

We now construct a new example by varying the auxiliary algebra C (but holding I and X fixed). We say that a function r from X into O is periodic (with unspecified period) if there exists a positive integer m such that r is periodic of period m. (Here it is essential that m be strictly positive.) Let C_* be the set of all periodic functions from X into O. It is clear that C_* is a Boolean algebra; we denote by A_* the functional I-algebra obtained from C_* by the method of tensor products.

There is a useful relation connecting the "periodic" algebras defined so far (i.e., A_* and A_m, for $m=1, 2, \cdots$). Since C_m is a Boolean subalgebra of C_*, it follows that A_m is a polyadic subalgebra of A_*; since, moreover, $C_* = \bigcup_{m=1}^{\infty} C_m$, it follows that

$$A_* = \bigcup_{m=1}^{\infty} A_m. \tag{5.4}$$

All that needs proof is that if $p \in A_*$, then $p \in A_m$ for some positive m. It is sufficient to treat the case in which p is given by (5.1), with $r_i \in C_*$ for each i in J. If r_i is periodic of period m_i, let m be the least common multiple of all the m_i $(i \in J)$; it follows that r_i is periodic of period m, and hence that $p \in A_m$. This result implies an analogue of (5.2) for A_*. We shall say that a function p from X^I into O is periodic (with unspecified period) if it has that property in each coordinate separately, or, in other words, if, for each i in I, there exists a (strictly) positive integer m such that $p(x)=p(y)$ whenever xi_*y and $x_i=y_i+m$. It follows from (5.2) and from (5.4) that a necessary and sufficient condition that a function p from X^I into O belong to A_* is that it be finite-dimensional and periodic.

(5.5) LEMMA. *If* $D_*(i, j)=\{p \in A_*: \mathsf{S}(i/j)p=1\}$, *then the set* $D_*(i, j)$ *has an infimum in* A_* *for each* i *and* j, *and*

$$\bigwedge D_*(i, j) = \begin{cases} 1 & \textit{if } i = j, \\ 0 & \textit{if } i \neq j. \end{cases}$$

Proof. Since $D_*(i, i) = \{1\}$, the case $i = j$ is trivial. Suppose now that $i \neq j$. By (4.5)

$$D_*(i, j) = \{p \in A_*: E_0(i, j) \neq p\}.$$

Since (cf. (5.4)) $E_m(i, j) \in A_*$ for every m, and since $E_0(i, j) \leqq E_m(i, j)$ (by the definitions of E_0 and E_m), it follows that if q is a lower bound of $D_*(i, j)$ in A_*, then $q \leqq E_m(i, j)$ for every m. We shall show that this implies that $q = 0$. Indeed, if $q(x) = 1$, then $E_m(i, j)(x) = 1$ for every m, and therefore $x_i \equiv x_j \pmod{m}$ for every m. If, in other words, $q(x) = 1$, then $x_i = x_j$. By (5.4), $q \in A_m$ for some positive m. It follows from (5.2) that q is periodic of period m, and hence that $q(y) = 1$ whenever each coordinate of y is congruent modulo m to the corresponding coordinate of x. This is a contradiction: even if $x_i = x_j$ (as it must be if $q(x) = 1$), it does not follow that $y_i = y_j$ (as it should be if $q(y) = 1$). The conclusion is that $q(x) \neq 1$ for all x, and hence that $q = 0$.

Lemma (5.5) implies that the algebra A_* has no equality. Indeed, if E_* were an equality for A_*, then, by (3.3) and (5.5), we should have $E_*(i, j) = 1$ if $i = j$ and $E_*(i, j) = 0$ if $i \neq j$. Since this particular function E_* from I^2 into A_* is not a predicate ($\mathsf{S}(i/j)E_*(i, j) \neq E_*(j, j)$), the algebra A_* can have no equality at all. Since $A_m \subset A_*$, it follows that an extension of an algebra with equality need not have an equality and, since $A_* \subset A_0$, it follows that a subalgebra of an algebra with equality need not have one either.

For our final example (still with the same I and X) we let $\mathbf{C}^*$ be the Boolean algebra of all functions from X into $\mathbf{O}$, and we denote by A^* the functional I-algebra obtained from $\mathbf{C}^*$ by the method of tensor products. There is a superficial reason for thinking that $A^* = A_0$, but it is not so. In crude language the difference between A^* and A_0 is that in A^* we are allowed to say everything about the integers, but, so to speak, only a finite number of times, whereas in A_0 there are no restrictions.

We already have an example (namely A_*) of a polyadic algebra that has no equality because the pertinent infima, although they exist, do not define a predicate. We shall prove that the algebra A^* does not have an equality either, but this time the reason is that the infima do not even exist. (This implies, in particular, that $A^* \neq A_0$.) We assert, in other words, that if

$$\begin{aligned} D^*(i, j) &= \{p \in A^*: \mathsf{S}(i/j)p = 1\} \\ &= \{p \in A^*: E_0(i, j) \leqq p\}, \end{aligned}$$

then the set $D^*(i, j)$ has no infimum in A^* (unless $i = j$).

We assert, to begin with, that if q is a lower bound of $D^*(i, j)$ in A^* (i.e., if $q \leqq p$ whenever $E_0(i, j) \leqq p$), then $q \leqq E_0(i, j)$. Suppose, indeed, that $E_0(i, j)(z) = 0$; we are to show that $q(z) = 0$. Let r be the characteristic function of the singleton $\{z_i\}$ in X and let s be the characteristic function of $\{z_j\}$. If, for every x in X^I,

$$p(x) = (r(x_i) \wedge s'(x_j)) \vee (r'(x_i) \wedge s(x_j)) \vee (r'(x_i) \wedge s'(x_j)),$$

then $p \in \mathbf{A}^*$ and $p(z) = 0$. Since a direct verification shows that $E_0(i, j) \leqq p$, it follows that $q \leqq p$, and hence that $q(z) = 0$, as asserted.

In view of the preceding paragraph, the assertion about $\mathbf{D}^*(i, j)$ reduces to this: among the q's in $\mathbf{D}^*$ such that $q \leqq E_0(i, j)$ (i.e., among the lower bounds of $E_0(i, j)$ in $\mathbf{A}^*$) there is no greatest. For each c in X, let r^c be the characteristic function of $\{c\}$ in X, and write

$$q^c(x) = r^c(x_i) \wedge r^c(x_j).$$

If q is the supremum of a finite number of functions such as q^c, then $q \leqq E_0(i, j)$ but q can certainly not be the greatest lower bound of $E_0(i, j)$ in $\mathbf{A}^*$, because the adjunction of an extra c makes it strictly greater. The proof will be completed by showing that if q is any lower bound of $E_0(i, j)$ in $\mathbf{A}^*$, then q is dominated by the supremum of a finite number of functions such as q^c. In this argument we may and do assume that $q \neq 0$.

Since $q \in \mathbf{A}^*$, the function q is the supremum of a finite number of nonzero tensor products such as q_0, where

$$q_0(x) = \bigwedge \{r_k(x_k) : k \in J\}.$$

Here J is a finite subset of I and $r_k \in \mathbf{C}^*$ for all k in J. We prove that i and j must belong to J. Indeed if, say, $i \in' J$, let x be an element of X^I such that $r_k(x_k) = 1$ for all k in J (recall that $q_0 \neq 0$) and such that $x_i \neq x_j$. It follows that $q_0(x) = 1$, and therefore $q(x) = 1$, whereas $E_0(i, j)(x) = 0$; this contradiction implies that $i \in J$, as asserted. It remains only to prove now that both r_i and r_j have the form r^c for some c in X. We know that neither r_i nor r_j vanishes identically. If either one took the value 1 at two distinct elements of X, then we could find an element x of X^I such that $r_k(x_k) = 1$ for all k in J and such that $x_i \neq x_j$; since we just saw that this leads to a contradiction, the proof is complete.

6. **Equalities for functional algebras.** We have seen that an equality for a locally finite functional algebra need not be a functional equality. Our next purpose is to show that, at least in the $\mathbf{O}$-valued case, every example of such a situation is similar to the particular example (congruence with respect to a fixed modulus) that we constructed above, and that, by a suitable modification, every such example can be converted into a functional algebra with a functional equality. Throughout this section we shall assume that E is an equality (not necessarily the same as the functional equality E_0) for a locally finite, $\mathbf{B}$-valued functional polyadic I-algebra $\mathbf{A}$ over a domain X; to exclude trivial cases we shall assume also that the cardinal number of I is greater than or equal to 3.

Suppose that a and b are elements of X with the following property: there exists an element z of X^I and there exist distinct variables i and j such that $z_i = a$, $z_j = b$, and $E(i, j)(z) = 1$. If that is the case, we shall write $a \sim b$; we proceed to investigate the binary relation $\sim$ in X.

(6.1) LEMMA. *If $a \sim b$, then $E(i, j)(x) = 1$ for all i and j and x such that $x_i = a$ and $x_j = b$.*

Proof. By assumption, there exists an element z of X^I and there exist distinct variables i_0 and j_0 such that $z_{i_0} = a$, $z_{j_0} = b$, and $E(i_0, j_0)(z) = 1$. If $\tau = (i_0, j_0/i, j)$, then

$$E(i, j)(x) = E(\tau i_0, \tau j_0)(x) = \mathbf{S}(\tau)E(i_0, j_0)(x) = E(i_0, j_0)(\tau_* x)$$

for all i, j, and x. If, in particular, $x_i = a$ and $x_j = b$, then $(\tau_* x)_{i_0} = a$ and $(\tau_* x)_{j_0} = b$. It follows that $\tau_* x(I - \{i_0, j_0\})_* z$; since $E(i_0, j_0)$ is independent of $I - \{i_0, j_0\}$, this implies that $E(i, j)(x) = E(i_0, j_0)(z) = 1$.

(6.2) LEMMA. *The relation $\sim$ is an equivalence.*

Proof. Suppose that $a \in X$. If $z_i = a$ for all i, then $E_0(i, j)(z) = 1$ for all i and j; it follows from (4.6) that $E(i, j)(z) = 1$, and hence that $a \sim a$.

Suppose that $a \sim b$. If $z \in X^I$ and if i and j are distinct variables such that $z_i = a$, $z_j = b$, and $E(i, j)(z) = 1$, then it follows from the symmetry of E that $E(j, i)(z) = 1$, and hence that $b \sim a$.

Suppose, finally, that $a \sim b$ and $b \sim c$. Let u and v be elements of X^I and let i, j, and k be distinct variables such that $u_i = a$, $u_j = v_j = b$, $v_k = c$ and $E(i, j)(u) = E(j, k)(v) = 1$. (The existence of these objects follows from (6.1).) Let z be an element of X^I such that $z_i = a$, $z_j = b$, and $z_k = c$. It follows that $E(i, j)(z) = E(i, j)(u) = 1$ and $E(j, k)(z) = E(j, k)(v) = 1$, so that, by the transitivity of E, $E(i, k)(z) = 1$, and hence $a \sim c$.

REMARK. Lemma (6.2) is false for dyadic algebras. Suppose that $I = \{i, j\}$ (with $i \neq j$); write $X = \{0, 1, 2, 3\}$ and $\boldsymbol{B} = \boldsymbol{O}$. If p is the characteristic function of $\{z \in X^I : |z_i - z_j| \leq 1\}$, then $\boldsymbol{A} = \{0, p, p', 1\}$ is a functional I-algebra over X. If $E(i, i) = E(j, j) = 1$ and $E(i, j) = E(j, i) = p$, then E is an equality for $\boldsymbol{A}$. In this situation the assertions $0 \sim 1$, $1 \sim 2$, and $2 \sim 3$ are true, but the assertions $0 \sim 2$, $0 \sim 3$, and $1 \sim 3$ are false.

(6.3) LEMMA. *If p is an element of $\boldsymbol{A}$ with support J and if x and y are elements of X^I such that $x_i \sim y_i$ whenever $i \in J$, then $p(x) = p(y)$.*

Proof. Since p has a finite support, there is no loss of generality in assuming that J is finite. Let i be an element of J and let j be an element of $I - J$. If u is defined by

$$u_j = y_i \quad \text{and} \quad u_k = x_k \text{ whenever } k \neq j,$$

then $xj_* u$ and, by (6.1), $E(i, j)(u) = 1$. Since p is independent of j, it follows that $p(x)=p(u)$, and, since $E(i, j)(u)=1$, it follows that

$$p(u) = p(u) \wedge E(i, j)(u) = \mathbf{S}(i/j)p(u) \wedge E(i, j)(u) = p((i/j)_* u).$$

If $v=(i/j)_* u$, then $v_i=u_j=y_i$, $v_j=u_j=y_i$, and $v_k=u_k=x_k$ whenever $k \epsilon' \{i, j\}$. We know that $p(x)=p(v)$. If

$$w_i = v_i \; (= y_i) \quad \text{and} \quad w_k = x_k \text{ whenever } k \neq i,$$

then $vj_* w$; it follows that $p(v)=p(w)$ and hence that $p(x)=p(w)$. We have proved thus that if we replace x_i by y_i for some i in J, then $p(x)$ remains unchanged. An inductive repetition of this result implies that if $z_i=y_i$ whenever $i \in J$ and $z_k=x_k$ whenever $k \epsilon' J$, then $p(x)=p(z)$. Since $zJ_* y$, it follows that $p(z)=p(y)$ and hence that $p(x)=p(y)$.

Our next result has nothing directly to do with equality; it is a general statement (of some independent interest) about functional polyadic algebras. (The result, by the way, is true without any restrictions on the cardinal number of I.) Suppose that π is a mapping from the set X onto a set $X^\#$. The mapping π induces a mapping π_* from X^I onto $(X^\#)^I$; by definition,

$$(\pi_* x)_i = \pi x_i$$

for all i in I. The mapping π_*, in turn, induces a mapping f that sends $\boldsymbol{B}$-valued functions on $(X^\#)^I$ onto $\boldsymbol{B}$-valued functions on X^I; by definition,

$$fq(x) = q(\pi_* x).$$

(6.4) Lemma. *If π is a mapping from X onto a set $X^\#$, if f is the functional mapping induced by π, and if $\boldsymbol{A}^\#$ is the set of all those $\boldsymbol{B}$-valued functions q on $(X^\#)^I$ for which $fq \in \boldsymbol{A}$, then $\boldsymbol{A}^\#$ is a $\boldsymbol{B}$-valued functional I-algebra over $X^\#$ and the mapping f is a polyadic monomorphism from $\boldsymbol{A}_\#$ into $\boldsymbol{A}$.*

Proof. The verification that f is a Boolean homomorphism (and that, therefore, $\boldsymbol{A}^\#$ is a Boolean algebra) is routine. If $q \in \boldsymbol{A}^\#$ and $fq=0$, then $q(\pi_* x)=0$ for all x in X^I; since π_* maps X^I onto $(X^\#)^I$, it follows that $q=0$, and hence that f is a monomorphism.

If τ is a transformation on I and if $x \in X^I$, then

$$(\pi_* \tau_* x)_i = \pi(\tau_* x)_i = \pi x_{\tau i} = (\pi_* x)_{\tau i} = (\tau_* \pi_* x)_i.$$

From this it follows that if $q \in \boldsymbol{A}^\#$, then

$$\mathbf{S}(\tau)fq(x) = f\mathbf{S}(\tau)q(x).$$

This, in turn, implies that $\boldsymbol{A}^\#$ contains $\mathbf{S}(\tau)q$ along with q, and that f preserves the transformation structure of $\boldsymbol{A}^\#$.

Suppose, finally, that J is a subset of I and that $x \in X^I$. If $x^\#$ is an element of $(X^\#)^I$ such that

$$\pi_* x \; J_* \; x^\#, \tag{6.5}$$

write $y_i = x_i$ if $i \epsilon' J$, and, if $i \epsilon J$, let y_i be an element of X such that $\pi y_i = x_i^\#$. It follows that

$$(6.6) \qquad x^\# = \pi_* y \quad \text{and} \quad xJ_*y.$$

If, conversely, x and y are elements of X^I and $x^\#$ is an element of $(X^\#)^I$ such that (6.6) holds, then $x_i = y_i$ (and therefore $\pi x_i = \pi y_i$) whenever $i \epsilon' J$; it follows that $(\pi_* x)_i = x_i^\#$ whenever $i \epsilon' J$, and hence that (6.5) holds. From these considerations we deduce that if $q \epsilon \mathbf{A}^\#$, then $\exists(J)q(x^\#)$ exists for all $x^\#$ in $(X^\#)^I$, the function $\exists(J)q$ belongs to $\mathbf{A}^\#$, and $f\exists(J)q = \exists(J)fq$; the relevant computations run as follows:

$$\begin{aligned} \exists(J)fq(x) &= \bigvee \{fq(y) : xJ_*y\} = \bigvee \{q(\pi_* y) : xJ_*y\} \\ &= \bigvee \{q(x^\#) : \text{for some } y,\ x^\# = \pi_* y \text{ and } xJ_*y\} \\ &= \bigvee \{q(x^\#) : \pi_* x J_* x^\#\} = \exists(J)q(\pi_* x) = f\exists(J)q(x). \end{aligned}$$

This completes the proof of Lemma (6.4).

We shall say that an equality (such as E) on a functional algebra (such as $\mathbf{A}$) is *reduced* if $E(i, j)(x) = 1$ implies that $x_i = x_j$ whenever i and j are in I and x is in X^I.

(6.7) THEOREM. *If $\mathbf{A}$ is a locally finite, $\mathbf{B}$-valued functional I-algebra of degree greater than or equal to 3, and if E is an equality for $\mathbf{A}$, then $\mathbf{A}$ is isomorphic to a $\mathbf{B}$-valued functional I-algebra $\mathbf{A}^\#$ with a reduced equality $E^\#$.*

Proof. Let X be the domain of $\mathbf{A}$ and let $\sim$ be the equivalence relation induced by E in X. Let $X^\#$ be the set of all equivalence classes and let π be the canonical mapping from X onto $X^\#$, so that $\pi a = \pi b$ if and only if $a \sim b$. Let f and $\mathbf{A}^\#$ be the functional mapping and the functional algebra described in (6.4). We shall prove that f is an isomorphism; all we need to do for this purpose is to prove that f maps $\mathbf{A}^\#$ onto $\mathbf{A}$. Given an element p of $\mathbf{A}$ and an element $x^\#$ of $(X^\#)^I$, find an x in X^I such that $\pi_* x = x^\#$ and write $q(x^\#) = p(x)$. This definition of q is unambiguous. Indeed, if $\pi_* x = \pi_* y$, then $\pi x_i = \pi y_i$ for all i, so that $x_i \sim y_i$ for all i; Lemma (6.3) implies that $p(x) = p(y)$. The function q so defined belongs to $\mathbf{A}^\#$ and is such that $fq = p$; indeed

$$fq(x) = q(\pi_* x) = q(x^\#) = p(x).$$

Let $E^\#$ be the equality for $\mathbf{A}^\#$ corresponding to E, i.e., $E^\# = f^{-1}E$, or, equivalently, $fE^\# = E$. If $E^\#(i, j)(x^\#) = 1$, find x so that $\pi_* x = x^\#$. Then $E(i, j)(x) = E^\#(i, j)(x^\#) = 1$, so that (by the definition of $\sim$) $x_i \sim x_j$; this implies that $\pi x_i = \pi x_j$ and hence that $x_i^\# = x_j^\#$.

(6.8) COROLLARY. *If $\mathbf{A}$ is a locally finite, $\mathbf{O}$-valued functional I-algebra of degree greater than or equal to 3, and if E is an equality for $\mathbf{A}$, then $\mathbf{A}$ is isomorphic to an $\mathbf{O}$-valued functional I-algebra $\mathbf{A}_0$ such that the functional equality E_0 is an equality for $\mathbf{A}_0$.*

Proof. It follows from (4.6) and the definition of a reduced equality that a reduced equality for an $\mathbf{O}$-valued functional algebra is the same as the associated functional equality.

The results of this section furnish a proof of the algebraic version of the completeness theorem for first-order functional calculi with equality. Since all the details of this theory are similar to the corresponding details for pure functional calculi, we shall be satisfied with a highly condensed discussion.

The relevant algebraic object is an *equality algebra*, i.e., a pair (A, E), where A is a polyadic algebra (an I-algebra, say), and E is an equality for A. If (A_1, E_1) and (A_2, E_2) are two such objects, then an *equality homomorphism* is defined to be a polyadic homomorphism f from A_1 into A_2 such that $fE_1 = E_2$. (This means, of course, that $fE_1(i, j) = E_2(i, j)$ for all i and j in I.) Not every polyadic homomorphism is an equality homomorphism (witness the embedding of $\mathbf{O}$ into some nontrivial equality algebra), but if f is a polyadic homomorphism from A_1 *onto* A_2, then it is necessarily an equality homomorphism. The reason is that, in that case, fE_1 is an equality for A_2, and hence, by the uniqueness of equality, $fE_1 = E_2$.

It follows from the preceding discussion of equality homomorphisms that concepts such as ideal, maximal ideal, simplicity, semisimplicity, and quotient algebra are exactly the same for equality algebras as they are for just plain polyadic algebras. The only new concept is that of an *equality model*; this, by definition, is an algebra A with equality E such that A is a model in the polyadic sense (i.e., an $\mathbf{O}$-valued functional algebra) and such that E is the associated functional equality. (The concept of an equality model is the algebraic version of what in logical terms is called a *standard model* for a first-order functional calculus with equality. The terminology is Henkin's; see *Completeness in the theory of types*, J. Symbolic Logic vol. 15 (1950) pp. 81–91.) The only difficult part of the completeness theorem for equality algebras follows immediately from the corresponding theorem for polyadic algebras [II, (17.3)] together with (6.8); it is the following characterization of simple equality algebras.

(6.9) THEOREM. *Every locally finite simple equality algebra of infinite degree is isomorphic to an equality model.*

7. **Adjunction of equality.** We have seen that the existence of an equality for an I-algebra A (not necessarily functional, and with no restrictions on the size of I) depends on the existence and properties of certain infima. The sets whose infima are relevant are defined by

$$D(i, j) = \{p \in A : \mathbf{S}(i/j)p = 1\}$$

for each pair of variables i and j. Even if $\bigwedge D(i, j)$ always exists, the function (from I^2 into A) that it defines need not be a predicate. There are two difficulties encountered in trying to prove that it is one. If τ is a transformation on I, then we could try to prove that $\mathbf{S}(\tau)(\bigwedge D(i, j)) = \bigwedge D(\tau i, \tau j)$ by proving first that

$$(7.1) \qquad \mathbf{S}(\tau)D(i, j) = D(\tau i, \tau j)$$

and by proving next that

$$(7.2) \qquad \mathbf{S}(\tau)(\bigwedge D(i, j)) = \bigwedge \mathbf{S}(\tau)D(i, j).$$

Neither of these equations is true in general. (A counter example to (7.2) is furnished by (5.5); take $\tau = (i/j)$.) It turns out, however, that (7.1) is nearly true and that (7.2) is often true. The purpose of this section is to make these assertions precise, and to infer from them that in a certain sense every polyadic algebra has an equality after all.

(7.3) LEMMA. *If i and j are in I and τ is a transformation on I, then $\mathbf{S}(\tau)D(i, j) \subset D(\tau i, \tau j)$.*

Proof. We are to prove that if $\mathbf{S}(i/j)p = 1$, then $\mathbf{S}(\tau i/\tau j)\mathbf{S}(\tau)p = 1$. The result follows immediately from the observation that $(\tau i/\tau j)\tau = (\tau i/\tau j)\tau(i/j)$.

Lemma (7.3) asserts that (7.1) is half true. This can be improved, as follows. Let $E(i, j)$ be the set of all those elements of $D(i, j)$ that depend on i and j only; more precisely

$$(7.4) \qquad E(i, j) = \{p \in A : \mathbf{S}(i/j)p = 1 \text{ and } \exists(I - \{i, j\})p = p\}.$$

The following lemma asserts that $E(i, j)$ is not much smaller than $D(i, j)$; combined with its successor it shows that (7.1) is nearly true, as asserted.

(7.5) LEMMA. *If $p \in D(i, j)$ then there exists an element q of $E(i, j)$ such that $q \leqq p$. If either $\bigwedge D(i, j)$ or $\bigwedge E(i, j)$ exists, then so does the other and the two are equal.*

Proof. To prove the first assertion, write $q = \forall(I - \{i, j\})p$ $(= (\exists(I - \{i, j\})p')')$. Then $q \leqq p$ and $\{i, j\}$ supports q; since (i, j) lives outside $I - \{i, j\}$, it follows that

$$\mathbf{S}(i/j)q = \mathbf{S}(i/j)\forall(I - \{i, j\})p = \forall(I - \{i, j\})\mathbf{S}(i/j)p = 1,$$

and hence that $q \in E(i, j)$. The second assertion is a consequence of the first assertion and of the fact that $E(i, j) \subset D(i, j)$.

(7.6) LEMMA. *If i and j are in I and τ is a transformation on I, then $\mathbf{S}(\tau)E(i, j) = E(\tau i, \tau j)$.*

Proof. If $p \in D(i, j)$, then $\mathbf{S}(\tau)p \in D(\tau i, \tau j)$ (by (7.3)), and if $\{i, j\}$ supports p, then $\{\tau i, \tau j\}$ supports $\mathbf{S}(\tau)p$. This proves that $\mathbf{S}(\tau)E(i, j) \subset E(\tau i, \tau j)$.

We must show next that if $q \in E(\tau i, \tau j)$, then $q = \mathbf{S}(\tau)p$ for some p in $E(i, j)$. If $\tau i = \tau j$, then $q = 1$, and we may write $p = 1$. If $\tau i \neq \tau j$, then there exists a permutation π on I such that $\pi\tau i = i$ and $\pi\tau j = j$; we write $p = \mathbf{S}(\pi)q$. Since $\{\tau i, \tau j\}$ supports q and since $\pi\{\tau i, \tau j\} = \{i, j\}$, it follows that $\{i, j\}$ supports

p. Since $(i/j)\pi = \pi(\tau i/\tau j)$ (both sides map τi onto j and every k distinct from τi onto τk), it follows that

$$\mathsf{S}(i/j)p = \mathsf{S}(i/j)\mathsf{S}(\pi)q = \mathsf{S}(\pi)\mathsf{S}(\tau i/\tau j)q = 1,$$

and hence that $p \in E(i, j)$. Since, finally, $\tau = \pi^{-1}$ on $\{i, j\}$, it follows that

$$\mathsf{S}(\tau)p = \mathsf{S}(\pi^{-1})p = \mathsf{S}(\pi^{-1})\mathsf{S}(\pi)q = q,$$

and the proof of the lemma is complete.

We go on now to show that (7.2) is often true, in the sense that every algebra can be embedded into one in which it is true. The idea is to use one of the standard methods of embedding a Boolean algebra into a complete Boolean algebra. The method that it is convenient to use is the one that goes via the representation theory; all we have to do is to pay attention to some polyadic details in the course of the embedding(2).

If A is an arbitrary Boolean algebra, then there exists a Boolean space X such that A is isomorphic to the set of all continuous functions from X into O; there is, therefore, no loss of generality in assuming that A *is* the set of all such functions. If A^+ is the Boolean algebra of all (not necessarily continuous) functions from X into O, then A is a Boolean subalgebra of A^+. (For a detailed discussion of the concepts and results used in the remainder of this section see [I, Part 2].)

In what follows we apply the duality theory of hemimorphisms and Boolean relations. If f is a hemimorphism on A, we write f^* for its dual (so that f^* is a Boolean relation in X). The hemimorphism f can be extended to a hemimorphism f^+ on A^+; we write

$$f^+r(x) = \bigvee \{r(y) : xf^*y\} \tag{7.7}$$

whenever $r \in A^+$ and $x \in X$. It is trivial that f^+ is indeed a hemimorphism on A^+; the fact that f^+ is an extension of f is expressed by

$$f^+p = fp \text{ whenever } p \in A. \tag{7.8}$$

We proceed to show that all the essential properties of f are reflected by f^+.

(7.9) LEMMA. *If f is the identity mapping on A, then f^+ is the identity mapping on A^+; if f and g are hemimorphisms on A, then*

$$(fg)^+ = f^+g^+. \tag{7.10}$$

Proof. The first assertion follows from the fact that if f is the identity mapping on A, then f^* is the identity mapping on X. The second assertion is proved by the computation:

(2) The method was used for essentially the same purpose by Jonsson and Tarski, *Boolean algebras with operators*, Amer. J. Math. vol. 73 (1951) pp. 891–939.

$$f^+g^+r(x) = \bigvee \{g^+r(y): xf^*y\} = \bigvee \{ \bigvee \{r(z): yg^*z\} : xf^*y\}$$
$$= \bigvee \{r(z): \text{for some } y,\ xf^*y \text{ and } yg^*z\} = \bigvee \{r(z): (g^*f^*)z\}$$
$$= \bigvee \{r(z): x(fg)^*z\} = (fg)^+r(x).$$

(7.11) LEMMA. *If f is a Boolean endomorphism (or a quantifier) on $\boldsymbol{A}$, then f^+ is a Boolean endomorphism (or a quantifier) on $\boldsymbol{A}^+$.*

Proof. If f is an endomorphism, then f^* is a function (from X to X), so that xf^*y means $y=f^*x$; the assertion about endomorphisms now follows easily from the definition (7.7). If f is a quantifier, then f^* is an equivalence relation (in X); in that case

$$f^+(r \wedge f^+s)(x) = \bigvee \{r(y) \wedge f^+s(y): xf^*y\}$$
$$= \bigvee \{r(y) \wedge \bigvee \{s(z): yf^*z\} : xf^*y\} = \bigvee \{r(y) \wedge \bigvee \{s(z): xf^*z\} : xf^*y\}$$
$$= \bigvee \{r(y) \wedge f^+s(x): xf^*y\} = \bigvee \{r(y): xf^*y\} \wedge f^+s(x)$$
$$= f^+r(x) \wedge f^+s(x),$$

and therefore f^+ is a quantifier.

The algebra $\boldsymbol{A}^+$ is, of course, a complete Boolean algebra; the supremum and the infimum of any family $\{r_\alpha\}$ of elements of $\boldsymbol{A}^+$ are given by

$$(\bigvee_\alpha r_\alpha)(x) = \bigvee_\alpha r_\alpha(x) \quad \text{and} \quad (\bigwedge_\alpha r_\alpha)(x) = \bigwedge_\alpha r_\alpha(x).$$

If f is a hemimorphism on $\boldsymbol{A}$, then

$$(f^+(\bigvee_\alpha r_\alpha))(x) = \bigvee \{(\bigvee_\alpha r_\alpha)(y): xf^*y\}$$
$$= \bigvee \{\bigvee_\alpha r_\alpha(y): xf^*y\} = \bigvee_\alpha(\bigvee \{r_\alpha(y): xf^*y\})$$
$$= \bigvee_\alpha(f^+r_\alpha)(x) = (\bigvee_\alpha f^+r_\alpha)(x).$$

It follows that if f is an endomorphism, then

$$f^+(\bigwedge_\alpha r_\alpha) = \bigwedge_\alpha f^+r_\alpha. \tag{7.12}$$

Suppose now that $\boldsymbol{A}$ has a polyadic structure, so that, say, $(\boldsymbol{A}, I, \mathbf{S}, \exists)$ is a polyadic algebra. If τ is a transformation on I, we write $\mathbf{S}^+(\tau) = (\mathbf{S}(\tau))^+$, and, similarly, if J is a subset of I, we write $\exists^+(J) = (\exists(J))^+$. It follows from (7.9) and (7.11) that $(\boldsymbol{A}^+, I, \mathbf{S}^+, \exists^+)$ is a polyadic algebra; (7.8) implies that this algebra has $(\boldsymbol{A}, I, \mathbf{S}, \exists)$ as a polyadic subalgebra. Caution: the algebra $\boldsymbol{A}^+$ need not be locally finite, even if $\boldsymbol{A}$ is such.

Let $\boldsymbol{D}^+(i, j)$ and $\boldsymbol{E}^+(i, j)$ be the sets defined for $\boldsymbol{A}^+$ as $\boldsymbol{D}(i, j)$ and $\boldsymbol{E}(i, j)$ were defined for $\boldsymbol{A}$, and write

$$E^+(i, j) = \bigwedge \boldsymbol{E}^+(i, j). \tag{7.13}$$

If τ is a transformation on I, then, by (7.12),

$$\mathbf{S}^+(\tau)E^+(i, j) = \bigwedge \mathbf{S}^+(\tau)\boldsymbol{E}^+(i, j),$$

and therefore, by (7.6),

$$\mathbf{S}^{+}(\tau)E^{+}(i, j) = E^{+}(\tau i, \tau j).$$

We see thus that E^{+} is a predicate of $\boldsymbol{A}^{+}$. Since (7.5) implies that

$$E^{+}(i, j) = \bigwedge D^{+}(i, j),$$

it follows from (3.6) and (3.7) that E^{+} is an equality for $\boldsymbol{A}^{+}$.

The set of all those elements of $\boldsymbol{A}^{+}$ that have a finite support is a polyadic subalgebra of $\boldsymbol{A}^{+}$; if $\boldsymbol{A}$ is locally finite, then that subalgebra includes $\boldsymbol{A}$ and contains every $E^{+}(i, j)$. It follows that E^{+} is an equality for the subalgebra; note that the subalgebra is locally finite if $\boldsymbol{A}$ is such.

The important part of our results can be summed up as follows.

(7.14) THEOREM. *Every (locally finite) polyadic algebra is a polyadic subalgebra of a (locally finite) algebra with equality.*

It follows from (7.14) that every polyadic algebra can be embedded into a cylindric algebra. (Note that (7.14) does not say that if the given algebra already has an equality, then the embedding algebra must have the same equality.) The logical counterpart of the theorem asserts the consistency of equality. One way to put it is this: a consistent first-order functional calculus remains consistent when a (new) sign of equality is adjoined to its symbols and the usual requirements on an equality are adjoined to its axioms.

8. **Unique existence.** We have studied the basic properties of equality and we have seen some situations in which equalities exist and others in which they do not. The question is: what can we do with an equality when there is one? Since equality is a fundamental logical concept of universal importance, and since algebraic logic is a faithful mirror of ordinary logic, the answer to the question is not a theorem but an extensive theory. The methods of developing that theory are algebraic adaptations of known logical methods. In what follows we shall illustrate the relevant techniques by studying the algebraic counterparts of propositions that assert the existence of a unique object satisfying certain conditions. From now on we shall always assume that the polyadic algebra we are given (denoted, as always, by $(\boldsymbol{A}, I, \mathbf{S}, \exists)$) is locally finite and of infinite degree, and that it comes equipped with an equality (denoted, as before, by E).

We begin with a theorem that asserts, essentially, that for constants equality is the same as identity.

(8.1) THEOREM. *If b and c are constants of $\boldsymbol{A}$, then a necessary and sufficient condition that $b = c$ is that $E(b, c) = 1$.*

REMARK. The theory of expressions such as $E(b, c)$ is treated in [III, §10]. For present purposes it is sufficient to know that if P is a unary predicate, then $P(c) = \mathbf{S}(i/c)P(i)$ whenever $i \in I$, and if Q is a binary predicate, then

$Q(b, c) = \mathbf{S}(i/b)\mathbf{S}(j/c)Q(i, j)$ whenever i and j are distinct elements of I. Observe that if Q is a binary predicate, and if $P(i) = Q(i, i)$ for every i in I, then P is a unary predicate.

Proof. The necessity of the condition is trivial; since $E(i, i) = 1$, it follows that

$$E(c, c) = \mathbf{S}(i/c)E(i, i) = 1.$$

Suppose now that the condition is satisfied. We shall prove that $b = c$ by proving that

$$\mathbf{S}(i/b)p = \mathbf{S}(i/c)p$$

whenever $i \in I$ and $p \in \mathbf{A}$; cf. the uniqueness assertion of the constant-construction theorem [II, (14.1)]. For this purpose, let j be a variable distinct from i and such that p is independent of j. Since (cf. (2.1))

$$p \wedge E(i, j) = \mathbf{S}(i/j)p \wedge E(i, j),$$

it follows that

$$\mathbf{S}(i/b)\mathbf{S}(j/c)(p \wedge E(i, j)) = \mathbf{S}(i/b)\mathbf{S}(j/c)(\mathbf{S}(i/j)p \wedge E(i, j)). \tag{8.2}$$

Since p is independent of j, it is clear that

$$\mathbf{S}(i/b)\mathbf{S}(j/c)p = \mathbf{S}(i/b)p. \tag{8.3}$$

On the other hand, since $\mathbf{S}(i/c)p$ is independent of both i and j, it follows that

$$\begin{aligned} \mathbf{S}(i/b)\mathbf{S}(j/c)\mathbf{S}(i/j)p &= \mathbf{S}(i/b)\mathbf{S}(i/j)\mathbf{S}((i/j)^{-1}\{j\}/c)p \\ &= \mathbf{S}(i/b)\mathbf{S}(i/j)\mathbf{S}(j/c)\mathbf{S}(i/c)p = \mathbf{S}(i/c)p. \end{aligned} \tag{8.4}$$

Putting these facts together with (8.2), and using the fact that $E(b, c) = 1$, we conclude that

$$\mathbf{S}(i/b)p = \mathbf{S}(i/c)p,$$

as asserted.

Our next result is an auxiliary statement (of some independent interest) about monadic algebras. If $\mathbf{A}$ is a monadic algebra with quantifier $\exists$, and if c is a constant of $\mathbf{A}$ [I, §13], then $\mathbf{A}$ might contain an element q whose intuitive interpretation is the proposition "the (only) variable of $\mathbf{A}$ is equal to c." The phrase "is equal to" is, to be sure, not expressible in a monadic algebra. The possibility just mentioned is realizable nevertheless. If, for instance, $\mathbf{A}$ is a functional monadic algebra over some domain X, if $x_0 \in X$, and if the characteristic function q of the singleton $\{x_0\}$ belongs to $\mathbf{A}$, then, intuitively, $q(x)$ can be thought of as the proposition "$x = x_0$." In this situation $\exists q$ is "true" (i.e., it is equal to 1), and, if $p \in \mathbf{A}$, then $\exists(p \wedge q) \wedge \exists(p' \wedge q)$ is "false" (i.e., it is equal to 0). Conversely, and this is the main point, these two conditions guarantee the existence of a unique constant c such that "what q says is true about c and about nothing else."

(8.5) THEOREM. *If* A *is a monadic algebra with quantifier* $\exists$, *and if* q *is an element of* A *such that*

(8.6) $$\exists q = 1$$

and

(8.7) $$\exists(p \wedge q) \wedge \exists(p' \wedge q) = 0 \textit{ for all } p \textit{ in } A,$$

then there exists a unique constant c *of* A *such that* $cq = 1$. *The constant* c *is defined by*

(8.8) $$cp = \exists(p \wedge q)$$

for all p *in* A.

Proof. If we define an operator c on A by (8.8), then, clearly, $c1 = 1$, and $c(p_1 \vee p_2) = cp_1 \vee cp_2$ whenever p_1 and p_2 are in A. Since, by (8.7), $cp \wedge cp' = 0$ for all p, and since

$$cp \vee cp' = c(p \vee p') = c1 = 1,$$

it follows that $cp' = (cp)'$, so that c is a Boolean endomorphism of A. Clearly

$$c\exists p = \exists(\exists p \wedge q) = \exists p \wedge \exists q = \exists p \text{ [by (8.6)]}$$

and

$$\exists cp = \exists\exists(p \wedge q) = \exists(p \wedge q) = cp,$$

so that c is a constant of A; we note that

$$cq = \exists(q \wedge q) = \exists q = 1.$$

To prove uniqueness, suppose that b is a constant of A such that $bq = 1$. Since

$$bp = bp \wedge bq = b(p \wedge q) \leq \exists(p \wedge q) = cp,$$

and since both b and c are Boolean homomorphisms, it follows that $b = c$, and the proof of the theorem is complete.

We return now to the polyadic algebra $(A, I, \mathbf{S}, \exists)$ with equality E. An existential assertion ("there is at least one i such that q") corresponds, in algebraic terms, to an element of the form $\exists(i)q$. What is the algebraic correspondent of an assertion of uniqueness ("there is at most one i such that q")? The answer to this question will be denoted by $!(i)q$. To define $!(i)q$, select a variable j distinct from i and such that q is independent of j, and write

(8.9) $$!(i)q = \forall(i)\forall(j)(q \wedge \mathbf{S}(i/j)q \rightarrow E(i, j)).$$

(Recall that if J is a subset of I, the universal quantifier $\forall(J)$ is defined by

$\forall(J)p = (\exists(J)p')'$ for all p in A.) It is, of course, important to know that the definition (8.9) is unambiguous, i.e., that if $k \neq i$ and q is independent of k, then the right side of (8.9) remains unchanged when we replace j by k. The proof is a simple computation; since q is independent of $\{j, k\}$, we have

$$\begin{aligned} &\forall(i)\forall(j)(q \wedge \mathbf{S}(i/j)q \to E(i, j)) \\ &= \forall(i)\mathbf{S}(j, k)\forall(k)\mathbf{S}(j, k)(q \wedge \mathbf{S}(i/j)q \to E(i, j)) \\ &= \forall(i)\mathbf{S}(j, k)\forall(k)(q \wedge \mathbf{S}(i/k)q \to E(i, k)) \\ &= \forall(i)\forall(k)(q \wedge \mathbf{S}(i/k)q \to E(i, k)). \end{aligned}$$

The algebraic correspondent of an assertion of unique existence ("there is exactly one i such that q") is denoted by $\exists!(i)q$; it is defined by

$$\exists!(i)q = \exists(i)q \wedge !(i)q. \tag{8.10}$$

The principal result concerning unique existence is that if $\exists!(i)q$ is true, then there does indeed exist a unique constant for which q is true.

(8.11) THEOREM. *If* $\{i\}$ *supports* q *and if* $\exists!(i)q = 1$, *then there exists a unique constant* c *of* A *such that* $\mathbf{S}(i/c)q = 1$.

Proof. The assumption $\exists!(i)q = 1$, together with the definition (8.10), implies that

$$\exists(i)q = 1 \tag{8.12}$$

and

$$q \wedge \mathbf{S}(i/j)q \leqq E(i, j) \tag{8.13}$$

whenever $j \neq i$. (Since $\{i\}$ supports q, the element q is automatically independent of every j distinct from i.) To construct c, we shall apply (8.5) to A regarded as a monadic algebra with quantifier $\exists(i)$.

From (8.12) we know that (8.6) is satisfied in the present situation. To prove (8.7), it is convenient to know that if $p \in A$ and if j is a variable distinct from i such that p is independent of j, then

$$\exists(i)(p \wedge q) = \exists(j)(\mathbf{S}(i/j)p \wedge \mathbf{S}(i/j)q). \tag{8.14}$$

(To prove (8.14), replace $\exists(i)$ by $\mathbf{S}(i, j)\exists(j)\mathbf{S}(i, j)$, observe that since $p \wedge q$ is independent of j, it follows that $\mathbf{S}(i, j)(p \wedge q) = \mathbf{S}(i/j)(p \wedge q)$, and use the fact that the right side of (8.14) is independent of $\{i, j\}$.) We are now ready to prove that

$$\exists(i)(p \wedge q) \wedge \exists(i)(p' \wedge q) = 0 \text{ for all } p \text{ in } A. \tag{8.15}$$

For this purpose, we select j distinct from i and such that p is independent of j; then

$$
\begin{aligned}
&\exists(i)(p \wedge q) \wedge \exists(i)(p' \wedge q)\\
&\quad = \exists(j)(\mathbf{S}(i/j)p \wedge \mathbf{S}(i/j)q) \wedge \exists(i)(p' \wedge q) \text{ [by (8.14)]}\\
&\quad = \exists(i)\exists(j)(\mathbf{S}(i/j)p \wedge \mathbf{S}(i/j)q \wedge p' \wedge q)\\
&\quad = \exists(i)\exists(j)((\mathbf{S}(i/j)p \wedge p') \wedge (q \wedge \mathbf{S}(i/j)q))\\
&\quad = \exists(i)\exists(j)((\mathbf{S}(i/j)p \wedge p') \wedge E(i, j)) \text{ [by (8.13)]}\\
&\quad = \exists(i)\exists(j)((\mathbf{S}(i/j)p \wedge E(i, j)) \wedge p')\\
&\quad = \exists(i)\exists(j)(p \wedge p') \text{ [by 2.1)]}.
\end{aligned}
$$

It follows from (8.5) that if $fp = \exists(i)(p \wedge q)$, then f is a Boolean endomorphism of A such that $\exists(i)f = f$, $f\exists(i) = \exists(i)$, and $fq = 1$. If $p \in A$ and if $j \neq i$, then

$$
\begin{aligned}
f\exists(j)p &= \exists(i)(\exists(j)p \wedge q) = \exists(i)\exists(j)(p \wedge q) \text{ [since } q = \exists(j)q]\\
&= \exists(j)\exists(i)(p \wedge q) = \exists(j)fp.
\end{aligned}
$$

It follows from the constant-construction theorem [II, (14.1)] that there exists a constant c of A such that $\mathbf{S}(i/c) = f$; this proves the existential assertion of the theorem.

To prove uniqueness, suppose that b and c are constants of A such that

(8.16) $$\mathbf{S}(i/b)q = \mathbf{S}(i/c)q = 1.$$

It follows from (8.13) that if $j \neq i$, then

$$\mathbf{S}(i/b)\mathbf{S}(j/c)(q \wedge \mathbf{S}(i/j)q) \leqq E(b, c).$$

Since (cf. (8.3) and (8.4)) this implies that

$$\mathbf{S}(i/b)q \wedge \mathbf{S}(i/c)q \leqq E(b, c),$$

it follows from (8.16) that $E(b, c) = 1$. The proof of the theorem is completed by an application of (8.1).

The constant c constructed in (8.11) ("the i such that q") will be denoted by $\perp(i)q$. (The traditional symbol is an inverted iota instead of an inverted T. The symbol here proposed is easier to read and more in harmony with the standard symbols $\forall$ and $\exists$; it might also serve as mnemonic for the initial letter of the article "the". The standard discussion of the description operator is given by Hilbert and Bernays, *Grundlagen der Mathematik*, Berlin, 1934; vol. I, §8.) The characteristic property of $\perp$ may be expressed as follows: if i supports q and if $\exists!(i)q = 1$, then

(8.17) $$\mathbf{S}(i/\perp(i)q)q = 1.$$

The construction of $\perp(i)q$ sheds light on the once controversial problem of contextual definitions. It might be said that the intuitively most satisfactory definition of some particular constant is an explicit definition, i.e., a definition that constructs the new constant from other constants, terms, and operations that are already available. It might also be said that a contextual definition of a constant c does not really define c, but, instead, describes in detail the effect of substituting c for a variable. According to the algebraic point of view, however, the latter kind of description *is* a constant, and no more primitive kind of definition can be hoped for. *Assuming* the existence of certain constants and manufacturing others from them is relatively easy and does not get at the root of the matter. Assuming merely a proposition of unique existence and *constructing* the unique constant whose existence is thereby implied is much harder, and, from the algebraic point of view much more explicit. We see thus that in algebraic logic the traditional terminology gets turned around; the so-called explicit definitions can be formulated only within a previously assumed context, whereas the so-called contextual definitions can be explicitly written down.

9. **Operations and predicates.** We conclude our present study of equality by establishing its connection with the theory of terms, operations, and predicates. These concepts are studied in [III], and, in this section, we shall make full use of the definitions and theorems of that paper.

As for terms, the fact is that everything that can be said about constants can be generalized to terms. We illustrate this assertion by proving the generalization of (8.11).

(9.1) THEOREM. *If J is a finite set not containing i, if $J \cup \{i\}$ supports q, and if $\exists!(i)q=1$, then there exists a unique J-term t of A such that $\mathbf{S}(i/t)q=1$.*

Proof. Write $I^-=I-J$ and let $(A^-, I^-, \mathbf{S}^-, \exists^-)$ be the algebra obtained from A by fixing the variables of J. If $j\neq i$ and $j \in' J$, then $\exists(j)q=q$ by assumption, and, therefore, $\exists^-(j)q=q$; this implies that $\{i\}$ supports q in A^-. The fact that $\exists!(i)q=1$ in A implies that $\exists^-!(i)q=1$ in A^-. It follows from (8.11) that there exists a unique J-constant c of A such that $\mathbf{S}(i/c)q=1$. From this and from the theorem on extending J-constants to J-terms [III, (6.7)] we conclude that there exists a J-term t such that $\mathbf{S}(i/t)q=1$. The uniqueness of t is implied by the uniqueness assertions of (8.11) and of the extension theorem just referred to.

The term t constructed in (9.1) will be denoted by $\perp(i)q$.

The treatment of predicates in [III] is similar to the treatment of operations; the similarity is so great that it is natural to suspect the existence of a close connection between the two concepts. The connection is most clearly visible in the presence of an equality. There is a one-to-one correspondence between all operations and some predicates (namely, the ones that are single-valued); this is still another example of a well-known logical phenomenon that can be discussed within the framework of polyadic algebras with equality. The relevant definition is obvious: an $(n+1)$-place predicate P is *single-valued* (with respect to its first n arguments) if $\exists!(j)P(i_1, \cdots, i_n, j)=1$

whenever $(i_1, \cdots, i_n) \in I^n$ and $j \in I - \{i_1, \cdots, i_n\}$. The connection between single-valued predicates and operations can be stated as follows.

(9.2) THEOREM. *If T is an n-place operation and if*

$$P(i_1, \cdots, i_n, j) = E(T(i_1, \cdots, i_n), j) \tag{9.3}$$

whenever $(i_1, \cdots, i_n, j) \in I^{n+1}$, then P is a predicate that is single-valued with respect to its first n arguments. If, conversely, P is an $(n+1)$-place predicate that is single-valued with respect to its first n arguments, then there exists a unique n-place operation T such that (9.3) holds; the operation T is such that

$$T(i_1, \cdots, i_n) = \bot(j)P(i_1, \cdots, i_n, j) \tag{9.4}$$

whenever $(i_1, \cdots, i_n) \in I^n$ and $j \in I - \{i_1, \cdots, i_n\}$.

Proof. For simplicity we shall discuss the case $n=1$; the general case differs from this special case in notation only. The first assertion of the theorem (in the special case) starts with a unary operation T and the equation

$$P(i, j) = E(T(i), j). \tag{9.5}$$

If τ is a transformation on I, then

$$\begin{aligned} \mathbf{S}(\tau)P(i, j) &= \mathbf{S}(\tau)E(T(i), j) = E(\tau T(i), \tau j) \\ &= E(T(\tau i), \tau j) = P(\tau i, \tau j), \end{aligned}$$

so that P is a binary predicate. If i, j, and k are distinct variables, then

$$\begin{aligned} \exists(j)P(i, j) &= \exists(j)E(T(i), j) = \exists(j)\mathbf{S}(k/T(i))E(k, j) \\ &= \mathbf{S}(k/T(i))\exists(j)E(k, j) = \mathbf{S}(k/T(i))1 = 1. \end{aligned}$$

In order to complete the proof that P is single-valued, suppose now that h, i, j, and k are distinct variables, and recall that

$$E(h, j) \wedge E(h, k) \leqq E(j, k).$$

Applying $\mathbf{S}(h/T(i))$ to this inequality and using (9.5), we obtain

$$P(i, j) \wedge P(i, k) \leqq E(j, k);$$

since this is equivalent to $!(j)P(i, j) = 1$, the proof of the first assertion of the theorem is complete.

The second assertion (case $n=1$ only) starts with a binary predicate P such that

$$\exists!(j)P(i, j) = 1 \tag{9.6}$$

whenever $i \neq j$. It follows from (9.1) that for each i in I there exists a unique $\{i\}$-term, say $T(i)$, such that

$$P(i, T(i)) = 1; \tag{9.7}$$

in fact $T(i)$ is $\bot(j)P(i, j)$ whenever $i \neq j$. Suppose now that i, j, and k are distinct variables, and observe that, by (9.6),

$$P(i, j) \wedge P(i, k) \leqq E(k, j).$$

Applying $\mathbf{S}(k/T(i))$ to this inequality and using (9.7) we obtain

$$P(i, j) \leqq E(T(i), j).$$

On the other hand

$$P(i, k) \wedge E(k, j) = P(i, j) \wedge E(k, j);$$

applying $\mathbf{S}(k/T(i))$ to this equation and using (9.7), we obtain

$$E(T(i), j) \leqq P(i, j).$$

We have proved thus that $P(i, j) = E(T(i), j)$; it remains only to prove that T is an operation.

We must show that if τ is a transformation on I, then

$$\tau T(i) = T(\tau i).$$

Since $P(i, T(i)) = 1$ and since P is a predicate, we know that

$$P(\tau i, \tau T(i)) = \mathbf{S}(\tau)P(i, T(i)) = 1. \tag{9.8}$$

On the other hand, replacing i by τi in (9.7), we obtain

$$P(\tau i, T(\tau i)) = 1. \tag{9.9}$$

The desired conclusion follows from (9.8) together with the fact that (9.9) characterizes $T(\tau i)$, i.e., if t is a $\{\tau i\}$-term such that $P(\tau i, t) = 1$, then $t = T(\tau i)$. This completes the proof of the theorem.

10. **Errata** (*Added in proof*, June 4, 1957.) In the proofs of [III, (11.7)] and [III, (13.3)] it is asserted that $\mathbf{S}(i, j)p = p$. This is false; what is true is that $\mathbf{S}(i, j)p$ is independent of $H \cup K$ (for (11.7)) and of K (for (13.3)). In view of this, the computations in the proofs have to be changed after the first step. The change is the same in both cases: replace $\mathbf{S}(i/t)$ by $\mathbf{S}(i, j)\mathbf{S}(j/t)\mathbf{S}(i, j)$. The remaining steps in both computations are almost automatic; use the obvious commutation relations to pull the left $\mathbf{S}(i, j)$ to the extreme left, use the fact that the assumptions apply to $\mathbf{S}(i, j)p$ as well as to p, and, finally, replace $\mathbf{S}(i, j)\mathbf{S}(j/v)\mathbf{S}(i, j)$ by $\mathbf{S}(i/v)$. I am grateful to Mr. Leon LeBlanc for calling these errors to my attention.

University of Chicago,
Chicago, Ill.

VIII

GENERAL THEORY

POLYADIC BOOLEAN ALGEBRAS

1. *Introduction.*—It is well known that when a logician speaks of the propositional calculus and when a mathematician speaks of Boolean algebras there is a sense in which they are talking about the same thing. What, in that sense, is the mathematical version of the so-called first-order functional calculus? Several answers to this question have been proposed in recent years;[1] the most satisfying of these answers is the concept of cylindric algebra introduced by Tarski and Thompson. The answer offered in this note (namely, polyadic Boolean algebra) differs from previous proposals mainly in being able to imitate the substitution processes that give the functional calculi their characteristic flavor. The imitation appears to be successful. It turns out, for example, just as it does in the propositional case, that the customary approach to the functional calculus can be viewed as nothing but a detailed description of the symbolism of a *free* polyadic algebra.

It should be emphasized that the theory of polyadic algebras is discussed here not as a possible tool for solving problems about the foundations of mathematics, but as an independently interesting part of modern algebra. The possibility of making a dictionary that translates logical terms (such as *interpretation* and *semantic completeness*) into algebraic terms (such as *homomorphism* and *semisimplicity*) is to be regarded as merely an amusing by-product of the discussion.

2. *Monadic Algebras.*—If a *proposition* is taken to mean an element of a Boolean algebra $\boldsymbol{B}$, then it is natural to interpret a *propositional function* as a function from some set X into $\boldsymbol{B}$. The set $\boldsymbol{A}^*$ of all such functions is a Boolean algebra with respect to the pointwise operations. A *functional monadic algebra* is a Boolean subalgebra $\boldsymbol{A}$ of $\boldsymbol{A}^*$ such that, for each p in $\boldsymbol{A}$, the range of p has a supremum in $\boldsymbol{B}$, and such that the (constant) function $\exists p$, whose value at every point of X is that supremum, belongs to $\boldsymbol{A}$. The mapping $\exists$ of $\boldsymbol{A}$ into itself is called a *functional quantifier;* it is easy to verify that it satisfies the conditions

$$(Q_1)\quad \exists 0 = 0,$$
$$(Q_2)\quad p \leqq \exists p,$$
$$(Q_3)\quad \exists(p \wedge \exists q) = \exists p \wedge \exists q,$$

for all p and q in $\boldsymbol{A}$. (Properly speaking, $\exists$ should be called a functional *existential* quantifier; *universal* quantifiers are defined by an obvious dualization.) A general concept that applies to any Boolean algebra is obtained by abstraction from the functional case: a *quantifier* is a mapping $\exists$ of a Boolean algebra into itself satisfying (Q_1)–(Q_3). (This concept occurs also in the announcement of the results of Tarski and Thompson.) A *monadic algebra* is a Boolean algebra with a quantifier.

The entire theory of Aristotelian (syllogistic) logic is easily subsumed under the elementary algebraic theory of monadic algebras.

3. *Polyadic Algebras.*—Propositional functions of several variables are treated by considering functions from a Cartesian power, say X^I, into a Boolean algebra $\boldsymbol{B}$. Even if the index set I is infinite, the interesting functions are those that depend on finitely many co-ordinates only. The set $\boldsymbol{A}^*$ of all such functions is a Boolean algebra, as before.

Let T^* be the semigroup of all transformations of I into itself (not necessarily one-to-one and not necessarily onto). Every element τ of T^* induces a Boolean endomorphism $p \to \tau p$ on $\boldsymbol{A}^*$, defined by allowing τ to act on the indices of the co-ordinates of the argument of p. [Example: if $I = \{1, 2, 3\}$ and if τ is the cyclic permutation (1, 2, 3), then $\tau p(x_1, x_2, x_3) = p(x_2, x_3, x_1)$.]

If $x \in X^I$ and $J \subset I$, it is natural to consider the set of points in X^I that can be obtained by varying arbitrarily those co-ordinates of x whose index is in J. [Example: if $I = \{1, 2, 3\}$, $x = (x_1, x_2, x_3)$, and $J = \{1, 3\}$, then the set so obtained consists of all points whose second co-ordinate is x_2.] If $p \in \boldsymbol{A}^*$ and if the image of that set under p has a supremum in $\boldsymbol{B}$, that supremum is denoted by $\exists_J p(x)$; if J contains only one element, say i, it is convenient to write $\exists_i$ for $\exists_J$.

A *functional polyadic algebra* (in more detail, a $\boldsymbol{B}$-valued functional polyadic algebra with domain X) is a Boolean subalgebra $\boldsymbol{A}$ of $\boldsymbol{A}^*$ such that, for each p in $\boldsymbol{A}$, (1) $\exists_J p(x)$ exists whenever $J \subset I$ and $x \in X^I$, (2) the function $\exists_J p$ belongs to $\boldsymbol{A}$ whenever $J \subset I$, and (3) $\tau p \in A$ whenever $\tau \in T^*$. Because of the finiteness condition imposed on the elements of $\boldsymbol{A}^*$, it turns out that it is sufficient to assume (1) and (2) for finite subsets J and to assume (3) for finite transformations τ; the infinite ones can then be recaptured. (A transformation in T^* is *finite* if it coincides with the identity transformation outside some finite subset of I. The set T of all finite transformations is a subsemigroup of T^*.)

The operators $\exists_J$ are quantifiers on $\boldsymbol{A}$; the quantifiers $\exists_J$ (J finite) and the endomorphisms τ(τ finite) have the following properties.

(P_1) *If J is empty, then* $\exists_J p = p$ *for all* p.
(P_2) $\exists_J \exists_K = \exists_{J \cup K}$.
(P_3) *If τ is the identity, then* $\tau p = p$ *for all* p.
(P_4) $(\sigma\tau)p = \sigma(\tau p)$.
(P_5) *If* $\sigma = \tau$ *outside J, then* $\sigma \exists_J = \tau \exists_J$.
(P_6) *If τ is one-to-one on* $\tau^{-1}J$, *then* $\exists_J \tau = \tau \exists_{\tau^{-1}J}$.
(P_7) *For every p in $\boldsymbol{A}$ there exists a finite subset J of I such that* $\exists_i p = p$ *whenever i is not in J.*

(The transformation τ is said to be one-to-one on $\tau^{-1}J$ if it never maps two distinct elements of I on the same element of J.)

The superficially complicated results (P_5) and (P_6) are merely a telegraphic summary of the usual intuitively obvious relations between quantification and substitution. A couple of examples will make them clear. Suppose that i and j are distinct elements of I, and let τ be the transformation that maps i on j and maps everything else (including j) on itself. Since τ agrees with the identity outside $\{i\}$, it follows from (P_5) that $\tau\exists_i = \exists_i$. This relation corresponds to the familiar fact that once a variable has been quantified, the replacement of that variable by another one has no further effect. To get another example, note, for the same τ, that $\tau^{-1}\{i\}$ is empty; it follows from (P_6) and (P_1) that $\exists_i\tau = \tau$. This relation corresponds to the familiar fact that once a variable has been replaced by another one, a quantification on the replaced variable has no further effect.

Once again a general concept is obtained by abstraction. A *polyadic algebra* is a Boolean algebra A, an index set I (whose elements are called *variables*), a correspondence from finite subsets of I to quantifiers on A, and a correspondence from finite transformations on I to Boolean endomorphisms on A, such that (P_1)–(P_7) are satisfied. It is useful to know that, just as in the functional case, the infinite quantifiers and endomorphisms can be defined in terms of the finite ones, and these extended operators satisfy the same conditions as their finite counterparts.

4. *Polyadic Logics.*—The usual machinery of universal algebra is immediately applicable to polyadic Boolean algebras; terms such as *subalgebra, homomorphism, kernel, ideal, maximal ideal, simple algebra,* and *free algebra* are defined as always. The essential point is to insist, in every case, that the corresponding Boolean concepts be invariant under the operators (quantifiers and endomorphisms) that define the polyadic structure. (Caution: polyadic algebras with different sets of variables are distinct types of algebraic systems. A homomorphism, for instance, is defined only between algebras with the same sets of variables. The situation is analogous to the theory of groups with operators.) A minor difficulty occurs in the definition of a free algebra: because of the assumed finiteness condition (P_7), a polyadic algebra can never be really free. The difficulty is easily conquered by an appropriate relativization of the concept of freedom.

With the algebraic nomenclature at hand, many logical terms are easily translatable into algebraic language. A *polyadic logic*, to begin with, is defined as a pair (A, I), where A is a polyadic algebra and I is a polyadic ideal in A. The motivation for this definition is the logical concept of refutability. In the customary treatment of logic, certain elements of an appropriate algebraic system (usually, though only implicitly, a polyadic algebra) are called refutable. From the algebraic point of view the definition of refutability in any particular case is inessential. What is important is the structure of the set I of all refutable elements; it turns out (and intuitive considerations demand that it should turn out so) that I is a polyadic ideal. (This discussion is, in fact, the dual of the usual one; reasons of algebraic convenience make refutability a more desirable concept than provability.)

A polyadic logic (A, I) is *simply consistent* if I is a proper ideal in A (not everything is refutable). Simple consistency is equivalent to the requirement that for no element p of A do both p and p' belong to I (where p' is the Boolean complement, or negation, of p). A polyadic logic (A, I) is *simply complete* if either $I = A$ or else I is a maximal ideal in A. Simple completeness is equivalent to the requirement that, for every closed element p of A, either p or p' belongs to I. (An element p of A is *closed* if $\exists_I p = p$.) In these terms the celebrated Gödel incompleteness theorem asserts that certain important polyadic logics are either incomplete or inconsistent.

5. *Semantic Completeness.*—Since every simply consistent polyadic logic (A, I) is easily studied in terms of the associated polyadic quotient algebra A/I, there is no essential loss of generality in restricting attention to algebras instead of logics. It is, of course, necessary to keep in mind that the reduction to algebras reduces some non-trivial logical concepts to trivialities; thus, for instance, the only "refutable" element of A/I is the zero element.

"Semantic" concepts refer to the representation theory of polyadic algebras. (The intrinsic, or structural, concepts of the preceding section are usually called "syntactic.") Representation theory proceeds, as always, by selecting a class of particularly simple and "concrete" polyadic algebras and asking to what extent every algebra is representable in terms of algebras of that class. For intuitively obvious reasons the logically important "concrete" algebras are the two-valued functional algebras, i.e., the functional algebras whose elements are functions from a Cartesian power X^I into the two-element Boolean algebra. Such algebras are called *models.*

An *interpretation* of an algebra A in a model B is a polyadic homomorphism from A onto B. An element p of an algebra A is *universally invalid* if it is "false" in every interpretation, i.e., in algebraic language, if $\varphi p = 0$ whenever φ is a homomorphism from A on a model. (Once again, the present discussion is the dual of the usual one.) The algebra A is *semantically complete* if the only universally invalid element is 0. In logical terms, A is semantically complete if every universally invalid element is refutable, or, dually, if every universally valid element is provable.

6. *Constants.*—An important concept in the representation theory of polyadic algebras is that of a constant. Intuitively, a constant is something that can be substituted for a variable; in the functional case a typical example of a constant is an element of the domain. The algebraic essence of the concept is captured by examining (either in terms of intuitive logic or in a functional example) the effect of replacing some of the variables (say the ones corresponding to a finite subset J of I) by an intuitively conceived constant. Abstraction from these special cases suggests the following definition. A *constant* (of a polyadic algebra A with variables I) is a mapping $J \to c(J)$ from finite subsets of I to Boolean endomorphisms of A such that:

(C_1) *If J is empty, then $c(J)p = p$ for all p,*

(C_2) $c(J)c(K) = c(J \cup K)$,

(C_3) $c(J)\,\exists_K = \exists_K\, c(J - K)$,

(C_4) $\exists_J\, c(K) = c(K)\,\exists_{J-K}$,

(C_5) $c(J)\,\tau = \tau c(\tau^{-1}J)$ *for every finite transformation τ on I.*

There are simple examples of polyadic algebras that do not possess any constants. A polyadic algebra A with variables I is *rich* if, whenever $p \in A$ and $i \in I$, there exists a constant c such that $\exists_i p = c(i)p$. (In highly informal language: there is a witness to every existential proposition.) The most important fact about rich algebras is that there are enough of them.

THEOREM 1. *Every polyadic algebra is isomorphic to a subalgebra of a rich algebra.*

The idea of the proof of Theorem 1 is most easily illustrated by sketching the proof of a special case. Suppose that $\boldsymbol{A}$ is a polyadic algebra with variables I, and suppose that $p_0 \in \boldsymbol{A}$ and $i_0 \in I$. Let I^+ be a set containing one more element than I, and consider a free polyadic algebra, with variables I^+, generated by the elements of $\boldsymbol{A}$. If that free algebra is reduced modulo all the relations that hold among the elements of $\boldsymbol{A}$, the result is a polyadic algebra $\boldsymbol{A}^+$ with variables I^+ (and hence, *a fortiori*, with variables I) that includes $\boldsymbol{A}$. If, for every finite subset J of I, $c(J)$ is defined to be the Boolean endomorphism of $\boldsymbol{A}^+$ induced by the transformation on I^+ that replaces the elements of J by the new element of I^+ and is otherwise equal to the identity, then c is a constant of the algebra $\boldsymbol{A}^+$ regarded as having the variables of I only. One more reduction, modulo the relation $\exists_{i_0} p_0 = c(i_0)p_0$, yields a polyadic extension of $\boldsymbol{A}$ that is rich enough, at least, to supply a witness for $\exists_{i_0} p_0$.

7. *Semisimplicity.*—The simplest representation theorem for polyadic algebras follows easily from an algebraic adaptation of the method of Rasiowa and Sikorski;[2] an analogue of this result for cylindric algebras was announced by Tarski.[3]

THEOREM 2. *Every polyadic algebra with an infinite set of variables is isomorphic to a functional algebra.*

Unfortunately Theorem 2 does not yield enough information about the structure of a general polyadic algebra. To state the desired sharper result, it is convenient to introduce the functional counterpart of the concept of richness. A functional algebra $\boldsymbol{A}$ with variables I and domain X is *functionally rich* if, whenever $p \in \boldsymbol{A}$, $i \in I$, and $x \in X^I$, there exists a point y of X^I such that all the co-ordinates of y, with the possible exception of the i co-ordinate, are the same as the corresponding co-ordinates of x and such that $\exists_i p(x) = p(y)$.

THEOREM 3. *Every polyadic algebra $\boldsymbol{A}$ with variables I is isomorphic to a subalgebra of a rich functional algebra the cardinal number of whose domain is less than or equal to the sum of the cardinal numbers of $\boldsymbol{A}$ and of I.*

The proof of Theorem 3 is based on Theorem 1; the method is a modification of the one introduced by Henkin.[4] To construct the promised functional algebra, it is necessary to construct a domain X and a value algebra $\boldsymbol{B}$. The role of X is played by a sufficiently large set of constants for a rich polyadic extension of the given algebra $\boldsymbol{A}$, and the role of $\boldsymbol{B}$ is played by $\boldsymbol{A}$ itself.

The algebraic analogues of the Skolem-Löwenheim theorem and the Gödel completeness theorem are relatively easy consequences of Theorem 3. To discuss the latter, for instance, consider a functionally rich $\boldsymbol{B}$-valued functional algebra, and let φ_0 be an arbitrary Boolean homomorphism from $\boldsymbol{B}$ to the two-element Boolean algebra. The assumption of richness implies that the application of φ_0 to the values of the elements of the functional algebra induces a polyadic homomorphism φ into a two-valued functional algebra; more precisely, φ is defined by $\varphi p(x) = \varphi_0[p(x)]$. This technique, combined with Theorem 3, yields the following basic result.

THEOREM 4. *Every simple polyadic algebra is isomorphic to a model.*

The method of Rasiowa and Sikorski also serves to prove a theorem of this type, and it is in some respects simpler than Henkin's; its disadvantage is that it works under severe infinity and countability restrictions only.

The converse of Theorem 4 is true, easy, and, incidentally, independent of the preceding relatively high-powered representation theorems.

THEOREM 5. *Every model is simple.*

Since, on universal algebraic grounds, the kernel of a polyadic homomorphism is a maximal ideal if and only if its range is simple, semantic completeness is equivalent to the requirement that the intersection of all maximal ideals is $\{0\}$. Since, in analogy with other parts of algebra, it is natural to call a polyadic algebra satisfying this condition *semisimple*, it follows that a polyadic algebra is semantically complete if and only if it is semisimple. It is now easy to state the algebraic version of the Gödel completeness theorem.

THEOREM 6. *Every polyadic algebra is semisimple.*

The assertion of Theorem 6 for a polyadic algebra with the empty set of variables (i.e., for an ordinary Boolean algebra) is the most important part of Stone's theorem on the representation of Boolean algebras. Theorem 6 is, in fact, a relatively easy consequence of Stone's theorem; it is much nearer to the surface than, for instance, Theorem 3. In other words, the hard thing in proving Gödel's completeness theorem is not to prove semisimplicity but to prove the appropriate representation theorem (Theorem 4) that guarantees that semisimplicity is the right thing to prove.

[1] A. Tarski, *J. Symbolic Logic*, **6,** 73–89 (1941); C. J. Everett and S. Ulam, *Amer. J. Math.*, **68,** 77–88 (1946); F. I. Mautner, *Amer. J. Math.*, **68,** 345–384 (1946); A. Tarski and F. B. Thompson, *Bull. Amer. Math. Soc.*, **58,** 65 (1952).

[2] H. Rasiowa and R. Sikorski, *Fund. Math.*, **37,** 193–200 (1950).

[3] A. Tarski, *Bull. Amer. Math. Soc.*, **58,** 65 (1952).

[4] L. Henkin, *J. Symbolic Logic*, **14,** 159–166 (1949)

IX

TERMS AND EQUALITY

PREDICATES, TERMS, OPERATIONS, AND EQUALITY IN POLYADIC BOOLEAN ALGEBRAS

The theory of Boolean algebras is an algebraic counterpart of the logical theory of the propositional calculus; similarly, the theory of polyadic (Boolean) algebras[1] is an algebraic way of studying the first-order functional calculus. The Gödel completeness theorem, for instance, can be formulated in algebraic language as a representation theorem for a large class of simple polyadic algebras, together with the statement that every polyadic algebra is semisimple.[2] The next desideratum is an algebraic study of the celebrated Gödel incompleteness theorem. Before that can be achieved, it is necessary to investigate the algebraic counterparts of some fundamental logical concepts (such as the ones mentioned in the title above). The purpose of this rather technical note is to report (without proofs) the results of such an investigation; the details are being published elsewhere.[3]

1. *Basic Concepts.*—For convenience of reference, this section contains the definitions of the (already known) basic concepts of algebraic logic; all further work is based on these concepts and on their elementary properties.

The following notation will be used for every Boolean algebra $\boldsymbol{A}$: the supremum of two elements p and q of $\boldsymbol{A}$ is $p \vee q$, the infimum of p and q is $p \wedge q$, the complement of p is p', the zero element of $\boldsymbol{A}$ is 0, and the unit element of $\boldsymbol{A}$ is 1. Sometimes it is convenient to write $p \rightarrow q$ instead of $p' \vee q$. The natural order relation is denoted by $\leqq$, so that $p \leqq q$ means that $p \vee q = q$ (or, equivalently, that $p \wedge q = p$ or that $p \rightarrow q = 1$). The simple Boolean algebra $\{0, 1\}$ is denoted by $\boldsymbol{O}$.

A *quantifier* (more precisely, an *existential quantifier*) on a Boolean algebra $\boldsymbol{A}$ is a mapping $\exists$ of $\boldsymbol{A}$ into itself such that (i) $\exists 0 = 0$, (ii) $p \leqq \exists p$, and (iii) $\exists(p \wedge \exists q) = \exists p \wedge \exists q$, for all p and q in $\boldsymbol{A}$. Suppose that $\boldsymbol{A}$ is a Boolean algebra, I is a set, S is a mapping that associates a Boolean endomorphism $\mathsf{S}(\tau)$ of $\boldsymbol{A}$ with every transformation τ from I into I, and $\exists$ is a mapping that associates a quantifier $\exists(J)$ on $\boldsymbol{A}$ with every subset J of I. The quadruple $(\boldsymbol{A}, I, \mathsf{S}, \exists)$ is a *polyadic algebra* if (i) $\exists(\emptyset)$ is the identity mapping on $\boldsymbol{A}$ (here $\emptyset$ is the empty set); (ii) $\exists(J \cup K) = \exists(J)\, \exists(K)$ whenever J and K are subsets of I; (iii) $\mathsf{S}(\delta)$ is the identity mapping on $\boldsymbol{A}$ (here δ is the identity transformation on I); (iv) $\mathsf{S}(\sigma\tau) = \mathsf{S}(\sigma)\, \mathsf{S}(\tau)$ whenever σ and τ are transformations on I; (v) $\mathsf{S}(\sigma)\, \exists(J) = \mathsf{S}(\tau)\, \exists(J)$ whenever $\sigma i = \tau i$ for all i in $I - J$; and (vi) $\exists(J)\, \mathsf{S}(\tau) = \mathsf{S}(\tau)\, \exists(\tau^{-1}J)$ whenever τ is a transformation that never maps two distinct elements of I onto the same element of the set J.

If I is a set, X is a non-empty set, and $\boldsymbol{B}$ is a Boolean algebra, the set of all functions from the Cartesian product X^I into $\boldsymbol{B}$ is a Boolean algebra with respect to the obvious pointwise operations. If τ is a transformation on I, and if x and y are in

X^I, write $\tau_* x = y$ whenever $y_i = x_{\tau i}$ for all i in I. (The value of a function x from I into X, i.e., of an element x of X^I, at an element i of I will always be denoted by x_i.) If p is a function from X^I into $\boldsymbol{B}$, write $\mathsf{S}(\tau)p$ for the function defined by $(\mathsf{S}(\tau)p)\ (x) = p(\tau_* x)$. If J is a subset of I and if x and y are in X^I, write $x\ J_*\ y$ whenever $x_i = y_i$ for all i in $I - J$. If p is a function from X^I into $\boldsymbol{B}$, and if the set $\{p(y)\colon\ x\ J_*\ y\}$ has a supremum in $\boldsymbol{B}$ for each x in X^I, write $\exists(J)p$ for the function whose value at x is that supremum. A *functional polyadic algebra* is a Boolean subalgebra $\boldsymbol{A}$ of the algebra of all functions from X^I into $\boldsymbol{B}$, such that $\mathsf{S}(\tau)p \in \boldsymbol{A}$ whenever $p \in \boldsymbol{A}$ and τ is a transformation on I, and such that $\exists(J)p$ exists and belongs to $\boldsymbol{A}$ whenever $p \in \boldsymbol{A}$ and J is a subset of I. The set X is called the *domain* of this functional polyadic algebra.

It is convenient to be slightly elliptical and, instead of saying that $(\boldsymbol{A}, I, \mathsf{S}, \exists)$ is a polyadic algebra, to say that $\boldsymbol{A}$ is a polyadic algebra, or, alternatively, to say that $\boldsymbol{A}$ is an I-algebra. An element of I is called a *variable* of the algebra $\boldsymbol{A}$. The *degree* of $\boldsymbol{A}$ is the cardinal number of the set of its variables. An element p of $\boldsymbol{A}$ is *independent* of a subset J of I if $\exists(J)p = p$; the set J is a *support* of p if p is independent of $I - J$. The algebra $\boldsymbol{A}$ is *locally finite* if each of its elements has a finite support. Some of the results that follow are true for arbitrary polyadic algebras, but most are not. To simplify the statements, it will be assumed throughout the sequel that $(\boldsymbol{A}, I, \mathsf{S}, \exists)$ is a fixed, locally finite polyadic algebra of infinite degree.

Associated with every subset J of I there is a polyadic algebra $(\boldsymbol{A}^-, I^-, \mathsf{S}^-, \exists^-)$ that is said to be obtained from $\boldsymbol{A}$ by *fixing* the variables of J. The set $\boldsymbol{A}^-$ is the same as the set $\boldsymbol{A}$, and $I^- = I - J$. If τ^- is a transformation on I^-, let τ be its canonical extension to I (i.e., τ is the extension of τ^- to I such that $\tau i = i$ whenever $i \in J$), and write $\mathsf{S}^-(\tau^-)p = \mathsf{S}(\tau)p$ for every p in $\boldsymbol{A}$. If J^- is a subset of I^-, then J^- is also a subset of I; write $\exists^-(J^-)p = \exists(J^-)p$ for every p in $\boldsymbol{A}$.

Suppose that c is a mapping that associates a Boolean endomorphism of $\boldsymbol{A}$ with every subset K of I; denote the value of c at K by $\mathsf{S}(K/c)$. The mapping c is a *constant* of $\boldsymbol{A}$ if (*i*) $\mathsf{S}(\emptyset/c)$ is the identity mapping on $\boldsymbol{A}$; (ii) $\mathsf{S}(H \cup K/c) = \mathsf{S}(H/c)\ \mathsf{S}(K/c)$, (iii) $\mathsf{S}(H/c)\ \exists(K) = \exists(K)\ \mathsf{S}(H - K/c)$, and (iv) $\exists(H)\ \mathsf{S}(K/c) = \mathsf{S}(K/c)\ \exists(H - K)$, whenever H and K are subsets of I; and (v) $\mathsf{S}(K/c)\ \mathsf{S}(\tau) = \mathsf{S}(\tau)\ \mathsf{S}(\tau^{-1}K/c)$ whenever K is a subset of I and τ is a transformation on I. (The notation here is different from the one used before for constants; the innovation has a beneficial unifying effect.)

2. *Predicates.*—An n-place *predicate* of $\boldsymbol{A}$ $(n = 1, 2, 3, \ldots)$ is a function P from I^n into $\boldsymbol{A}$ such that if $(i_1, \ldots, i_n) \in I^n$ and if τ is a transformation on I, then

$$\mathsf{S}(\tau)\ P(i_1, \ldots, i_n) = P(\tau i_1, \ldots, \tau i_n).$$

THEOREM 1. *If P is an n-place predicate of $\boldsymbol{A}$ and if $(i_1, \ldots, i_n) \in I^n$, then $\{i_1, \ldots, i_n\}$ supports $P(i_1, \ldots, i_n)$.*

THEOREM 2. *If $p \in \boldsymbol{A}$ and if $j_1, \ldots, j_n$ are distinct variables such that $\{j_1, \ldots, j_n\}$ supports p, then there exists a unique n-place predicate P of $\boldsymbol{A}$ such that $P(j_1, \ldots, j_n) = p$.*

Corollary. *Every element of* $\boldsymbol{A}$ *is in the range of some predicate of* $\boldsymbol{A}$.

Whereas, by Theorem 1, a value of a predicate depends on no more variables than the visible ones, it may happen that it depends on fewer. If, for example, Q is a 1-place predicate, and if $P(i, j) = Q(i)$ for all i and j, then P is a 2-place predicate such that $\{i\}$ supports $P(i, j)$. It is interesting to ask what happens if the roles of P and Q in this discussion are interchanged. Suppose, in other words, that Q is a 2-place predicate and that j is a particular element of I, and write $P(i) = Q(i, j)$ for all i. The function P is not quite a predicate; it is a $\{j\}$-predicate in the sense of the following definition. If J is a finite subset of I, and if $\boldsymbol{A}^-$ is the algebra obtained from $\boldsymbol{A}$ by fixing the variables of J, then an n-place J-*predicate* of $\boldsymbol{A}$ is an n-place predicate of $\boldsymbol{A}^-$. What was called a predicate before is a $\emptyset$-predicate in the present terminology.

Theorem 3. *If* $j_1, \ldots, j_m$ *are distinct elements of* I, *if* $J = \{j_1, \ldots, j_m\}$, *and if* P *is an* n*-place* J*-predicate of* $\boldsymbol{A}$, *then there exists a unique* $(n + m)$*-place predicate* Q *of* $\boldsymbol{A}$ *such that*

$$P(i_1, \ldots, i_n) = Q(i_1, \ldots, i_n, j_1, \ldots, j_m)$$

whenever $(i_1, \ldots, i_n) \in (I - J)^n$.

It is easy to see that the construction described in Theorem 3 always does yield a J-predicate.

3. *Terms.*—Suppose that M and N are finite subsets of I. A transformation σ on I will be said to be of *type* (M, N) if

(i) σ is one-to-one on M,

(ii) $\sigma M \subset I - (M \cup N)$,

and (iii) $\sigma = \delta$ outside M.

If J is a finite subset of I, and if $\boldsymbol{A}^-$ is the algebra obtained from $\boldsymbol{A}$ by fixing the variables of J, then (by definition) a J-*constant* of $\boldsymbol{A}$ is a constant of $\boldsymbol{A}^-$. If c is a J-constant, then $\mathbf{S}(K/c)$ is not defined for every subset K of I; this is the main point in which the concept of a J-term (to be defined next) differs from that of a J-constant. If J is a finite subset of I, a J-*term* is a mapping t that associates with every subset K of I a Boolean endomorphism of $\boldsymbol{A}$, denoted by $\mathbf{S}(K/t)$, so that the restriction of t to the subsets of $I - J$ is a J-constant of $\boldsymbol{A}$, and so that if p is an element of $\boldsymbol{A}$ with finite support L, if K is a finite subset of I, and if σ is a transformation of type $(K, L \cup J)$, then $\mathbf{S}(K/t)p = \mathbf{S}(\sigma K/t)\,\mathbf{S}(\sigma)p$. Every constant is a $\emptyset$-term (and hence a J-term for every finite set J). If $j \in I$ and if t is defined by writing $\mathbf{S}(K/t) = \mathbf{S}(K/j)$ for every subset K of I, then t is a $\{j\}$-term. (The transformation (K/j) is the one that maps every element of K onto j and every element of $I - K$ onto itself.) In this sense, the concept of a term is a simultaneous generalization of the concept of a constant and the concept of a variable.

Theorem 4. *If* c *is a* J*-constant, then there exists a unique* J*-term* t *such that* $\mathbf{S}(K/t) = \mathbf{S}(K/c)$ *whenever* $K \subset I - J$.

4. *Properties of Terms.*—The algebraic behavior of terms is similar to, but necessarily somewhat more complicated than, the algebraic behavior of constants. The chief complication arises in the relation between terms and transformations.

To get a concise statement of that relation, it is convenient to introduce a new item of notation. If τ is a transformation on I and if J is a subset of I, then let τ_J be the transformation such that $\tau_J = \tau$ on $\tau^{-1}J$ and $\tau_J = \delta$ outside $\tau^{-1}J$.

THEOREM 5. *If t is a J-term of A, if H and K are subsets of I, and if τ is a transformation such that $\tau = \delta$ on J, then*

(T 1) $\mathsf{S}(\emptyset/t)$ *is the identity mapping on* A,
(T 2) $\mathsf{S}(H \cup K/t) = \mathsf{S}(H - J/t)\,\mathsf{S}(K - J/t)\,\mathsf{S}((H \cup K) \cap J/t)$,
(T 3) $\mathsf{S}(H/t)\,\exists(K) = \exists(K - J)\,\mathsf{S}(H - K/t)\,\exists(K \cap J)$,
(T 4) $\exists(H)\,\mathsf{S}(K/t) = \exists(H \cap J)\,\mathsf{S}(K/t)\,\exists(H - (J \cap K))$,
(T 5) $\mathsf{S}(K/t)\,\mathsf{S}(\tau) = \mathsf{S}(\tau_{I-J})\,\mathsf{S}((\tau_{I-J})^{-1}K/t)\,\mathsf{S}(\tau_J)$.

THEOREM 6. *If t is a J-term, if K is a subset of I, and if p is an element of A with support L, then $\mathsf{S}(K/t)p = \mathsf{S}(K \cap L/t)p$ and $\mathsf{S}(K/t)p$ has support $J \cup (L - K)$.*

The following result concerning terms is the deepest one; it is used repeatedly in the construction of terms satisfying various conditions.

THEOREM 7. *If J and K are finite subsets of I, if g is a Boolean homomorphism from the range of $\exists(K)$ into A, and if i is an element of $I - (J \cup K)$ such that* (i) $\exists(i)\,g\exists(K) = g\exists(K)$, (ii) $g\exists(K)\,\exists(i) = \exists(K)\,\exists(i)$, (iii) $g\exists(K)\,\exists(j) = \exists(j)\,g\exists(K)$ *whenever $j \neq i$ and $j \in I - J$, and* (iv) $g\exists(K)\,\mathsf{S}(\tau) = \mathsf{S}(\tau)\,g\exists(K)$ *whenever τ is a transformation that maps some finite subset of $I - (\{i\} \cup J \cup K)$ into itself and is equal to δ otherwise, then there exists a J-term t of A such that $\mathsf{S}(i/t)\,\exists(K) = g\exists(K)$.*

The next result shows that the term t of Theorem 7 is unique.

THEOREM 8. *If J and K are finite subsets of I, if s and t are J-terms, and if i is an element of $I - (J \cup K)$ such that $\mathsf{S}(i/s)p = \mathsf{S}(i/t)p$ whenever p is independent of K, then $s = t$.*

A transformation on I acts in a natural way not only on the variables but also on the terms of A. If τ is a transformation and if t is a J-term, let τ_0 be the transformation such that $\tau_0 = \tau$ on J and $\tau_0 = \delta$ outside J, and let i be an element of $I - J$; an application of Theorems 7 and 8 shows that there exists a unique τJ-term, to be denoted by τt, such that $\mathsf{S}(i/\tau t)p = \mathsf{S}(\tau_0)\,\mathsf{S}(i/t)p$ whenever p is independent of J. This definition of τt is unambiguous (i.e., it does not depend on the particular choice of i).

THEOREM 9. *If t is a J-term, then $\delta t = t$; if σ and τ are transformations, then $(\sigma\tau)t = \sigma(\tau t)$; if $\sigma = \tau$ on J, then $\sigma t = \tau t$.*

If P is an n-place predicate and if $t_1, \ldots, t_n$ are terms, then there is a natural way of defining $P(t_1, \ldots, t_n)$. To do so, suppose that t_1 is a J_1-term, $\ldots$, t_n is a J_n-term, let $i_1, \ldots, i_n$ be distinct elements of $I - (J_1 \cup \ldots \cup J_n)$, and write

$$P(t_1, \ldots, t_n) = \mathsf{S}(i_1/t_1) \ldots \mathsf{S}(i_n/t_n)\,P(i_1, \ldots, i_n).$$

This definition of $P(t_1, \ldots, t_n)$ is unambiguous (i.e., it does not depend on the particular choice of $i_1, \ldots, i_n$).

THEOREM 10. *If P is an n-place predicate, if $t_1, \ldots, t_n$ are terms, and if τ is a transformation, then*

$$\mathsf{S}(\tau)\,P(t_1, \ldots, t_n) = P(\tau t_1, \ldots, \tau t_n).$$

5. *Operations.*—An n-place *operation* ($n = 1, 2, 3, \ldots$) of $\boldsymbol{A}$ is a function T on I^n whose values are terms of $\boldsymbol{A}$, such that if $(i_1, \ldots, i_n) \in I^n$ and if τ is a transformation on I, then

$$\tau T(i_1, \ldots, i_n) = T(\tau i_1, \ldots, \tau i_n).$$

The theory of operations is quite similar to the theory of predicates. All the results of Section 2 have their operation analogues; the following result, for example, is the analogue of Theorem 1.

THEOREM 11. *If T is an n-place operation of $\boldsymbol{A}$ and if $(i_1, \ldots, i_n) \in I^n$, then $T(i_1, \ldots, i_n)$ is an $\{i_1, \ldots, i_n\}$-term.*

The useful part of the theory of operations is based on the fact that there is a natural sense in which the variables of a term can be replaced by terms. (In this rough, descriptive language, the theory of transforms of terms furnishes a natural sense in which the variables of a term can be replaced by other variables.) If s is an H-term, t is a J-term, and K is a subset of I, let i be an element of $I - (H \cup J)$; an application of Theorems 7 and 8 shows that there is a unique $(H \cup (J - K))$-term, to be denoted by $(K/s)t$, such that $\mathbf{S}(i/(K/s)t)p = \mathbf{S}(J \cap K/s)\ \mathbf{S}(i/t)p$ whenever p is independent of J. This definition of $(K/s)t$ is unambiguous.

If T is an n-place operation and if $t_1, \ldots, t_n$ are terms, then there is a natural way of defining $T(t_1, \ldots, t_n)$. To do so, suppose that t_1 is a J_1-term, $\ldots$, t_n is a J_n-term, let $i_1, \ldots, i_n$ be distinct elements of $I - (J_1 \cup \ldots \cup J_n)$, and write

$$T(t_1, \ldots, t_n) = (i_1/t_1) \ldots (i_n/t_n)\ T(i_1, \ldots, i_n).$$

This definition of $T(t_1, \ldots, t_n)$ is unambiguous.

THEOREM 12. *If T is an n-place operation, if $t_1, \ldots, t_n$ are terms, and if τ is a transformation, then*

$$\tau T(t_1, \ldots, t_n) = T(\tau t_1, \ldots, \tau t_n).$$

6. *Equality.*—An *equality* for $\boldsymbol{A}$ is a 2-place predicate E that is *reflexive* (i.e., $E(i, i) = 1$ for all i) and *substitutive* (i.e., $p \wedge E(i, j) = \mathbf{S}(i/j)p \wedge E(i, j)$ for all i and j and for all p in $\boldsymbol{A}$). (The transformation (i/j) is the one that maps i onto j and everything else onto itself.) An equality E is necessarily *symmetric* (i.e., $E(i, j) = E(j, i)$ for all i and j) and *transitive* (i.e., $E(i, j) \wedge E(j, k) \leqq E(i, k)$ for all i, j, and k). If E is an equality for $\boldsymbol{A}$, then

$$\exists(i)\ (p \wedge E(i, j)) \wedge \exists(i)\ (p' \wedge E(i, j)) = 0$$

whenever $p \in \boldsymbol{A}$ and $i \neq j$, and

$$\exists(i)\ (E(i, j) \wedge E(i, k)) = E(j, k)$$

whenever $i \neq j$ and $i \neq k$. These two equations, together with the other elementary properties of an equality, can be summed up by saying that a polyadic algebra with an equality is a cylindric algebra in the sense of Tarski.

THEOREM 13. *If E is an equality for $\boldsymbol{A}$ and if $\boldsymbol{D}(i, j) = \{p \in \boldsymbol{A}\colon\ \mathbf{S}(i/j)p = 1\}$, then the set $\boldsymbol{D}(i, j)$ has an infimum in $\boldsymbol{A}$ for each i and j, and that infimum is equal to $E(i, j)$.*

It is a consequence of Theorem 13 that a polyadic algebra can have at most one equality. The theorem does not say that $\boldsymbol{D}(i, j)$ always has an infimum in $\boldsymbol{A}$ (it does not) or that, if it does, then $\boldsymbol{A}$ must have an equality (it need not). The parenthetical negative assertions are established by the construction of explicit counterexamples.

THEOREM 14. *Every polyadic algebra is a polyadic subalgebra of an algebra with equality.*

It follows from this result that every polyadic algebra can be imbedded into a cylindric algebra. (Here the given algebra may have finite degree, or else it need not even be locally finite; if, however, it is locally finite, then the imbedding algebra can be made to be locally finite also.) The logical counterpart of Theorem 14 asserts the consistency of equality. One way to put it is this: a consistent first-order functional calculus remains consistent when a sign of equality is adjoined to its symbols, and the usual requirements on an equality are adjoined to its axioms.

7. *Algebras with Equality.*—Suppose that I is a set, X is a non-empty set, and $\boldsymbol{B}$ is a Boolean algebra. If i and j are in I, let $E_0(i, j)$ be the function from X^I into $\boldsymbol{B}$ such that $E_0(i, j)\ (x) = 1$ or 0 according as $x_i = x_j$ or $x_i \neq x_j$. The function-valued function E_0 is called the *functional equality* associated with I, X, and $\boldsymbol{B}$. The terminology is justified by the fact that whenever all the values of E_0 (i.e., all the functions $E_0(i, j)$) belong to a $\boldsymbol{B}$-valued functional I-algebra with domain X, then E_0 is an equality for that algebra.

If E is an equality for a $\boldsymbol{B}$-valued functional I-algebra $\boldsymbol{A}$ with domain X, it is not necessarily true that E coincides with the associated functional equality E_0. It is true, however, at least in the important $\mathbf{O}$-valued case, that a suitable modification converts E into E_0. The process (an algebraic modification of known logical methods) involves the introduction of an equivalence relation in X induced by E and the reduction of X modulo that relation. The end-product of the process is conveniently described in terms of a new definition: the equality E will be said to be *reduced* if (for all i and j in I and for all x in X^I) $E(i, j)\ (x) = 1$ implies that $x_i = x_j$. The result is that $\boldsymbol{A}$ is isomorphic to a $\boldsymbol{B}$-valued functional I-algebra with a reduced equality. The most useful part of this result is the following special case.

THEOREM 15. *Every $\mathbf{O}$-valued functional polyadic algebra with equality is isomorphic to an $\mathbf{O}$-valued functional algebra $\mathbf{A}_0$ such that the associated functional equality E_0 is an equality for $\mathbf{A}_0$.*

These results (together with the known representation theorems for polyadic algebras) imply the algebraic version of the completeness theorem for the first-order functional calculus with equality.

8. *Unique Existence.*—An existential assertion ("there is at least one i such that q") corresponds in algebraic terms to an element of the form $\exists(i)q$. The algebraic correspondent of an assertion of uniqueness ("there is at most one i such that q"), to be denoted by $!(i)q$, is defined (in an algebra with equality E) by

$$!(i)q = \forall(i)\ \forall(j)\ (q \wedge \mathbf{S}(i/j)q \rightarrow E(i, j)).$$

(If J is a subset of I, the operator $\forall(J)$ is defined by $\forall(J)p = (\exists(J)p')'$.) The auxiliary variable j here is to be distinct from i and such that q is independent of j; the definition is easily shown to be independent of the particular choice of j. The algebraic correspondent of an assertion of unique existence ("there is exactly one i such that q"), to be denoted by $\exists!(i)q$, is defined by $\exists!(i)q = \exists(i)q \wedge !(i)q$.

THEOREM 16. *If J is a finite set not containing i, if $J \cup \{i\}$ supports q, and if $\exists!(i)q = 1$, then there exists a unique J-term t of $\mathbf{A}$ such that $\mathbf{S}(i/t)q = 1$.*

The term t constructed in Theorem 16 ("the i such that q") will be denoted by $\perp(i)q$. (The traditional symbol is an inverted iota instead of an inverted $\mathbf{T}$.)

The similarity between the theory of predicates and the theory of terms is clearest in the presence of an equality. There is, in fact, a natural one-to-one correspondence between all operations and some predicates. The relevant predicates are the single-valued ones. An $(n + 1)$-place predicate P is *single-valued* (with respect to its first n arguments) if $\exists!(j)P(i_1, \ldots, i_n, j) = 1$ whenever $(i_1, \ldots, i_n) \in I^n$ and $j \in I - \{i_1, \ldots, i_n\}$.

THEOREM 17. *If T is an n-place operation and if*

$$P(i_1, \ldots, i_n, j) = E(T(i_1, \ldots, i_n), j)$$

whenever $(i_1, \ldots, i_n, j) \in I^{n+1}$, then P is a single-valued predicate (with respect to its first n arguments). If, conversely, P is an $(n + 1)$-place predicate that is single-valued with respect to its first n arguments, then there exists a unique n-place operation T satisfying the preceding equation; the operation T is such that

$$T(i_1, \ldots, i_n) = \perp(j)\, P(i_1, \ldots, i_n, j)$$

whenever $(i_1, \ldots, i_n) \in I^n$ and $j \in I - \{i_1, \ldots, i_n\}$.

[1] "Polyadic Boolean Algebras," these PROCEEDINGS, **40**, 296–301, 1954.

[2] "Algebraic Logic. I. Monadic Boolean Algebras," to appear in *Compositio Math.*; "Algebraic Logic. II. Homogeneous Locally Finite Polyadic Boolean Algebras of Infinite Degree," to appear in *Fundamenta Math.*

[3] They will be Parts III and IV of a sequence of papers appearing under the main title "Algebraic Logic."

X

BRIEF SUMMARY

POLYADIC BOOLEAN ALGEBRAS

The purpose of this work is to define and study a class of algebraic systems whose relation to the first order functional calculus is the same as that of Boolean algebras to the propositional calculus. (These systems are related to, but not identical with, the cylindric algebras introduced by Tarski and Thompson, Bull. A. M. S. 1952, p. 65.)

A *quantifier* (properly: existential quantifier) on a Boolean algebra **A** is a mapping $\exists$ of **A** into itself such that (i) $\exists 0 = 0$, (ii) $p \leqq \exists p$ for all p in **A**, (iii) $\exists\exists p = \exists p$ for all p in **A**, and (iv) if p and q are in **A** and if $p = \exists p$, then $\exists(p \wedge q) = p \wedge \exists q$.

Let I be a set and let ε be the identity mapping of I onto itself. Let T be the semigroup of all those transformations τ of I into itself which agree with ε outside some finite set.

A *polyadic Boolean algebra* is a Boolean algebra **A** and a set I such that to each finite subset J of I there corresponds a quantifier $\exists_J$ on **A** and to each τ in T there corresponds an endomorphism $\bar{\tau}$ of **A**, subject to the following conditions. (1) If J is empty, then $\exists_J p = p$ for all p in **A**, (2) $\exists_J \exists_K = \exists_{J \cup K}$, (3) if p is in **A**, then there exists a finite subset J of I such that $\exists_K p = p$ whenever $K \cap J$ is empty, (4) $\bar{\varepsilon} p = p$ for all p in **A**, (5) if σ and τ are in T and if $\pi = \sigma\tau$, then $\bar{\pi} = \bar{\sigma}\bar{\tau}$, (6) if τ is one-to-one on $\tau^{-1}J$, then $\exists_J \bar{\tau} = \bar{\tau}\exists_{\tau^{-1}J}$, (7) if $\sigma = \tau$ outside J, then $\bar{\sigma}\exists_J = \bar{\tau}\exists_J$.

If X and I are arbitrary sets, and if **B** is a (lattice-theoretically) complete Boolean algebra, then the set **A** of all functions from X^I into **B**, and also suitable subsets of **A**, form in a natural way polyadic Boolean algebras. A relatively deep theorem asserts that every simple polyadic Boolean algebra can be obtained in this way, with $\mathbf{B} = \{0, 1\}$. (The method of proof is modelled after related work by Henkin, J. Symb. Logic 1941, pp. 159—166, and Rasiowa and Sikorski, Fund. Math. 1950, pp. 193—200.) In view of this result, the algebraic analogue of Gödel's completeness theorem becomes the assertion that every polyadic algebra is semisimple (i.e., that the intersection of all maximal ideals consists of the zero element only).

Via polyadic algebras many other results of modern logic (e.g., Gödel's incompleteness theorem, the consistency of the continuum hypothesis, and Shepherdson's work on inner models) become susceptible of a purely algebraic formulation.

UNIVERSITY OF CHICAGO

ADDITIONAL BIBLIOGRAPHY

ADDITIONAL BIBLIOGRAPHY

The following list contains a few recent additions to the literature of polyadic algebras that appeared about the same time as or slightly later than the papers that constitute this volume.

COPELAND, A. H., SR., *Note on cylindric algebras and polyadic algebras,* Mich. Math. J. 3 (1956) 155-157.

DAIGNEAULT, A., *Products of polyadic algebras and of their representations,* Thesis, Princeton University (1959).

GALLER, B. A., *Cylindric and polyadic algebras,* Proc. A.M.S. 8 (1957) 176-183.

HIZ, H., *A warning about translating axioms,* Amer. Math. Monthly, 65 (1958) 613-614.

LEBLANC, L., *Non-homogeneous and higher order polyadic algebras,* Thesis, University of Chicago (1960).

LEBLANC, L., *Dualité pour les égalités booléennes,* C. R. Acad. Sci. Paris 250 (1960) 3552-3553.

LEBLANC, L., *Les algèbres booléennes topologiques bornées,* C. R. Acad. Sci. Paris 250 (1960) 3766-3768.

LEBLANC, L., *Les algèbres de transformation,* C. R. Acad. Sci. Paris 250 (1960) 3928-3930.

LEBLANC, L., *Représentations des algèbres polyadiques pour anneau,* C. R. Acad. Sci. Paris 250 (1960) 4092-4094.

STONE, M. H., *Some algebraic aspects of logic,* OSR Technical Report (1956) (mimeographed).

WRIGHT, F. B., *Ideals in a polyadic algebra,* Proc. A.M.S. 8 (1957) 544-546.

WRIGHT, F. B., *Some remarks on Boolean duality,* Portugaliae Math. 16 (1957) 109-117.

WRIGHT, F. B., *Polarity and duality,* Pac. J. Math. 10 (1960) 723-730.

INDEX

INDEX